# HIV/AIDS: Immunochemistry, Reductionism and Vaccine Design

Marc H V Van Regenmortel

# HIV/AIDS: Immunochemistry, Reductionism and Vaccine Design

## A Review of 20 Years of Research

 Springer

Marc H V Van Regenmortel
School of Biotechnology
University of Strasbourg
Illkirch, France

ISBN 978-3-030-32458-2      ISBN 978-3-030-32459-9   (eBook)
https://doi.org/10.1007/978-3-030-32459-9

This Springer imprint is published by the registered company Springer Nature Switzerland AG.
The registered company address is: Gewerbestrasse 11, 6330 Cham, Switzerland

# Introduction

According to UNAIDS, during the year 2016, 36.7 million people on our planet were living with HIV, while 1.8 million people were newly infected by the virus. Since the start of the epidemic in the mid-80s, 35 million people have died of AIDS-related illnesses. Although considerable progress has been made in the treatment of HIV patients with antiretroviral drugs, it is now increasingly accepted that only an effective HIV vaccine will be able to end the AIDS pandemic. A few large-scale HIV vaccine trials have been undertaken, but none, so far, has elicited an adequate level of protection against infection.

It may seem paradoxical that although HIV has become the virus we know most about, after three decades of unparalleled high-quality research funded by billions of dollars, we are not yet sure whether a preventive HIV vaccine will ever be developed. This eBook brings together 24 scientific papers published by the author in the last 20 years which propose explanations for why it has been so difficult to develop a vaccine against what is probably the worst pandemic that has confronted humanity in recent times.

The book is divided into three parts: Immunochemistry (4 chapters), Reductionism (7 chapters), and the Search for a preventive HIV vaccine (12 chapters).

Part 1 consists of five chapters that introduce basic concepts in the immunochemistry of viral proteins which are important for understanding the nature of the reactions between viral antigens and antibodies. These are the key players involved in the immunological mechanisms that protect vertebrates against infections caused by viruses, although they are not the sole actors in immune systems that help fighting off virus infections. Another important player, known as the cellular arm of the immune system, consists of different types of immune cells such as T cells (i.e., thymus-derived), T regulatory cells, and dendritic cells which together with other immunoregulatory mechanisms are important for protecting organisms against viral infections. These other players will not be described in detail because it would have required writing an introduction to the whole of immunology which seemed not to be indispensable for understanding how vaccinologists have tried to harness neutralizing antibodies for protecting humans against HIV infection. The only immune cells that will be described in detail are the so-called B cells (i.e., bone marrow-derived) which are cells that in their membranes possess B-cell receptors (BCRs) that correspond to antibodies possessing an additional tail that anchors them in the B-cell membrane. When a host is immunized with an antigen, certain BCRs in the host immune system interact with complementary antigens and the corresponding B cells become activated by interleukin-4 cytokines released by helper T cells. The B cells are then transformed into plasma cells that subsequently release antibodies which correspond to BCRs devoid of their anchoring tail.

The following two books are recommended to readers interested in a more comprehensive introduction to immunology: *Immunity* (Paul 2015) and *Immunity: The Making of a Modern Science* (Gallagher et al. 1995).

**Part 1. Immunochemistry**

Chapter 1 describes the chemical structure of small regions in proteins, called epitopes, which are recognized by the binding sites of antibodies, called paratopes. So-called discontinuous epitopes in a protein are made up of 10–20 amino acid residues located on 3–5 separate segments of the peptide chain that are brought together at the protein surface by the folding of the chain. Continuous epitopes are short peptide fragments of 5–8 residues which are often part of a larger discontinuous epitope and which can bind to antibodies raised against the protein antigen. Classifying epitopes as either continuous or discontinuous may give the erroneous impression that the smallest units of immunological recognition are amino acid residues, although it is at the level of individual atoms that interactions take place. Paratopes consist of 10–25 residues assembled from short stretches of residues located on the six complementarity determining regions of antibodies which together form a discontinuous binding site. Short peptide regions of 3–5 residues present on these six hypervariable antibody loops are often able on their own to bind to continuous epitopes, and these short loop sequences could then be viewed as continuous paratopes. Such binding is usually caused by the type of hydropathic complementarity between hydrophilic and hydrophobic residues that is commonly observed between the sense and antisense peptides encoded by complementary sense and antisense messenger RNA strands. This remarkable pattern arises because RNA codons (sense) and anticodons (antisense), without any exception, always code for amino acids of opposite hydropathy, i.e., either hydrophilic or hydrophobic residues that have a strong affinity for each other. This phenomenon gave rise to the hypothesis that the genetic code initially evolved to favor the simultaneous emergence from complementary RNA strands of sense and antisense peptides that could assume the role of receptors with their complementary ligands (Biro 2005).

Other protein epitopes of 5–10 residues, known as mimotopes, are able to react with short paratope sequences from antibodies raised against the protein even if they show little or no sequence similarity with the protein used to induce the antibody. This type of binding arises when the two peptides possess opposite hydropathic patterns of hydrophilic and hydrophobic residues in their respective sequences.

Chapter 2 analyzes the common assumption that our current knowledge of the chemical structure of epitopes and paratopes should make it possible to apply molecular design strategies for developing synthetic vaccines. The major difficulty that has been encountered in such attempts is that the capacity of the immune system to induce a protective neutralizing antibody response depends more on numerous biological properties and regulatory mechanisms of the immune system of the host than on the elucidated chemical complementarity of single epitope–paratope pairs. Since a large number of necessary preconditions in the immune system have to be met before neutralizing antibodies can be obtained by vaccination, there are plenty of reasons why the molecular design of an HIV vaccine is not at present a realistic scientific enterprise.

Chapter 3 summarizes the numerous attempts that have been made to develop synthetic peptide vaccines by making continuous epitopes adopt the structures observed when viruses are bound to neutralizing antibodies. Although peptide vaccines would have the advantage of increased safety and stability as well as lower cost, it is remarkable that hundreds of studies using synthetic peptides had not led in 2006 to a single synthetic peptide vaccine marketed for human use (Hans et al. 2006).

A major reason for all these repeated failures is the fact that investigators relied mainly on continuous epitopes that are mostly short linear peptide fragments of more complex discontinuous epitopes which are the large majority of neutralization epitopes found in viral proteins. It must also be stressed that most peptides are immunogenic in the sense that they readily elicit antibodies that react with the peptide immunogen, although they do not cross-react with the parent protein and are not able to neutralize viral infectivity.

Chapter 4 analyzes the binding specificity of antibodies that arises from a limited number of residues in the six antigen-binding regions (ABRs) that together form the numerous paratopes present in every antibody molecule. Each ABR differs significantly in its amino acid composition and tends to bind different types of amino acids present at the surface of proteins. In spite of these differences, the combined collective preference of the six ABRs does not allow epitopes to be chemically distinguished from the remainder of the antigen surface. When the amino acid composition of the six ABRs in 200 different antibody-antigen complexes was examined, it was found that there was no noticeable difference between the amino acid composition of epitopes and that of the entire protein surface. No amino acid was significantly overrepresented in epitope regions compared to the rest of the protein surface, and since the entire surface is an antigenic continuum, most of the surface could actually be part of epitopes. These findings explain why the many algorithms that were developed over the last 20 years for predicting the location of epitopes at the surface of proteins were not very successful (Van Regenmortel 1999a, pp. 53–68).

Antibody polyspecificity refers to the fact that every antibody molecule always harbors numerous paratopes that allows it to bind to a large variety of epitopes present in different antigens. Antibody heterospecificity arises when an antibody reacts better with another antigen than with the one used to raise the antibody. Such a phenomenon is not rare and can lead to the unexpected finding that the antibody appears to have been elicited by an antigen with which it is unable to react, although this may only reflect the fact that the binding reactivity was below the detection level of the immunoassay that was used. Antibody heterospecificity explains the common observation that early antisera obtained soon after immunization with any antigen frequently contain levels of total induced immunoglobulins that far exceed the level of antibodies able to react with the immunizing antigen. The heterospecificity of antibodies demonstrates that immunogenicity and antigenic reactivity are not always present simultaneously in the same region of a protein which is one of the reasons why the strategy of structure-based reverse vaccinology has not been successful for developing an effective HIV vaccine.

## Part 2. Reductionism

Chapter 5 analyzes the appeal of reductionist thinking for trying to understand the relationship between the structure and function of antibody molecules. The structure of a protein arises from the selective attention to the visual experience of that object at a specific time. Such a static definition excludes the dimension of time and conceals the fact that pictures of biomolecules are visual time-slides of dynamic systems. The structure of a binding site, furthermore, cannot be identified independently of its interaction with a ligand since these two relational entities are defined by their mutual steric complementarity.

Epitopes and paratopes are thus not intrinsic constituents of proteins and antibodies that exist independently of this relational nexus. Only a small fraction of the residues that are assigned by functional assays to a binding site contribute to the binding free energy and residues that are not in contact at the interface can also affect binding affinity. The reductionist approach of analyzing the structure of epitope–paratope complexes by crystallography has therefore been of limited use for designing functional entities that will exhibit a neutralizing activity when placed in the context of a particular cellular environment in the infected host.

Chapter 6 analyzes the reductionist expectation that assumes that the biological phenomenon of protection against infection achieved with neutralizing antibodies can be reduced to the level of chemistry. It is not possible to predict the neutralizing activity of an antibody simply from its chemical structure because protective immunity only occurs in the context of whole organisms since tissues and organs cannot be vaccinated. The structures of epitopes and paratopes present in a crystallographic complex correspond to a terminal state after a dynamic process of induced fit and conformational selection, and they give little information on the epitope conformation present in the immunogen-bound BCR that triggered the appearance of a neutralizing antibody activity. This activity is not selected by the BCRs during immunization since the neutralizing capacity of the antibody will be revealed only subsequently in the context of a complex cellular environment.

Chapter 7 summarizes the debates that took place during a conference on Reduction and Emergence held in Paris in 2003. William Wimsatt explained that a reductionist starts by choosing and designating a system for analysis which partitions the world of study into that system and its environment. Assumptions will be made about which variables must be controlled or randomized and the description of the environment will be greatly simplified compared to the description of the entities internal to the system under study. Fitness, for instance, will be described as if it were a property of phenotypes or genes and the fact that it is a relation between organism and environment will tend to be ignored. When environmental variables are not monitored, the investigator will be unaware of the context dependence of biological phenomena.

Causal explanations are reductive because one factor is singled out for attention and is given undue explanatory weight on its own. However, in biological systems any observed effect always results from a complex network of interactions and an explanation in terms of a single cause is never satisfactory. Although molecular biology has clarified the intricate mechanisms that allow genes to be translated into proteins, it has provided little insight on the innumerable causal chains that link genes to phenotypic traits. A common fallacy is to assume that the function of a gene is to produce whatever the system fails to do when that gene is absent. As a result, genetic determinism has not been able to unravel the complex series of events that link phenotypes to genes.

Chapter 8 discusses why it is so difficult to analyze the complexity of biological systems in a satisfactory manner. One reason is that complex systems always possess emergent properties that are absent when the parts are studied separately and which cannot be predicted or deduced from the properties of the parts. Interactions between the parts as well as inputs from the environment give rise to novel features such as network behavior that lead to the characteristic homeostasis of biological systems. Homeostasis allows biological systems to maintain a dynamic stability as well as an ability to self-repair which is largely independent of specific control mechanisms.

Biological systems cannot be described adequately without referring to the functional roles that their constituents play in keeping them alive. Darwinian evolution explains the existence of a present-day function by saying that it exists because it contributed to the reproductive fitness of the organism in the past. Biologists thus favor evolutionary explanations for presently observed structures and do not look for structural explanations for a presently observed function.

Design is defined as the deliberate and intelligent conceiving of an artificial thing or process. Since the human mind is the most efficient goal-directed system that exists, human intentionality is perceived as the cause of all purposive human actions. This leads to the anthropomorphic fallacy that the behavior and activities of all biological systems are explainable in terms of goals and purposes. Since any functional part that contributes to the working of the system can be interpreted as achieving a certain goal, this leads to the view that the system was designed by a thinking mind to function in a preordained way. This encourages the use of metaphors of end-directed features for describing biological processes in terms of design, purposes, and functions. It also encourages the expectation that the human mind has the capacity of intentionally designing an integrated immunological system that will protect against HIV infection. Since natural selection has not yet succeeded in achieving such a feat, there is considerable skepticism that a preventive HIV vaccine can be rationally designed by reverse engineering (Green 2013).

Chapter 9 discusses the negative impact that reductionist thinking has had on HIV vaccine development. Although methodological reductionism which dissects biological systems into their constituent parts was initially very successful in molecular biology, it was later found to possess many limitations for studying complex biological systems. For instance in the search for an effective HIV vaccine, the use of Mabs for dissecting HIV immune responses made investigators focus on individual epitopes as elicitors of neutralizing antibodies. This made them neglect the synergistic effects that are observed in polyclonal immune responses directed against individual epitopes of a major HIV antigenic site when a variety of neutralizing antibodies elicited by different epitopes are acting together.

Reductionist thinking also blurs the distinction between the chemical nature of antigenicity and the biological nature of immunogenicity. This leads to the assumption that when an HIV epitope is reconstructed by rational design to better fit a neutralizing antibody, it produces an immunogen able to elicit polyclonal antibodies with the same neutralizing capacity as the neutralizing Mab. Reductionism also fosters the belief that protection against HIV infection can be fully analyzed at the level of molecular interactions between an epitope and a paratope which in most cases is the only parameter that vaccinologists take into account when they design a vaccine. Numerous features of the host immune system are crucial for the induction of neutralizing antibodies, and because they ignored these contributory factors, investigators became engaged in hundreds of abortive attempts to transform HIV epitopes of known structure into effective vaccines. Using a convergence argument, it has been argued that so many negative outcomes justify the conclusion that the reverse vaccinology approach that was used is not capable of developing a preventive HIV vaccine.

Chapter 10 responds to a commentary (King 2016) which rejected the convergence argument that reverse vaccinology is actually unable to develop an effective HIV vaccine. It was argued that such a convergence argument is based on inductive reasoning from a limited number of cases that can never lead to a logically certain conclusion. It is indeed true that no amount of experimental evidence can ever lead to the absolute certainty achievable by deductive reasoning and the convergence argument is therefore only a backup argument. The fundamental reason why reverse vaccinology failed in the case of HIV is that it did not have a sound theoretical basis derived from our current knowledge of immunological specificity and of anti-HIV immune responses (see also Chaps. 4 and 21). King (2016), however, was right to point out that it may be considered unethical to pursue attempts to develop an HIV vaccine by a consistently unsuccessful approach if scarce resources could be used more effectively to combat the AIDS epidemic by other approaches.

Chapter 11 discusses the nature and consequences of biological reductionism in the immunological study of infectious diseases. The major message of this review is that useful information does not depend only on data input but requires procedures that include both reductionist and non-reductionist steps in order to detect distinct patterns that subsequently need to be examined experimentally.

**Part 3. The Search for a Preventive HIV Vaccine**

Chapter 12 discusses the limitations of structure-based design of HIV vaccine immunogens which arise from (1) misconceptions regarding the nature of epitopes,

(2) disregarding the analytical bias that arises when Mabs are used to dissect the structure of antigens, and (3) assuming that effective HIV vaccine immunogens can be identified by analyzing the structure of epitope–paratope complexes by X-ray crystallography. Epitopes and paratopes are relational entities defined by their mutual complementarity, and they depend on each other for acquiring a recognizable identity. Once a neutralizing Mab has been isolated from the serum of an infected individual, it usually becomes conceptually associated with a single discrete epitope. However, since an antibody always contains numerous paratopes, there is no reason why a vaccine containing that one particular epitope should also be able to induce a polyclonal, neutralizing antibody response. A large number of paratopes binding to the major antigenic sites of the HIV envelope (Env) have been thoroughly investigated, but this has not helped the design of an HIV vaccine.

Chapter 13 discusses the two different meanings of the term reverse vaccinology used for developing bacterial and viral vaccines. Rino Rappuoli (2001) developed a strategy, based on bioinformatic analyses of entire bacterial genomes, for identifying all the antigens that a bacterial pathogen is able to express. He called it reverse vaccinology because he started from the genome instead of from the organism for establishing which surface-exposed proteins should be investigated as potential vaccine immunogens. Whereas classical approaches investigate only the small number of purified antigens that can be obtained by fractionating bacterial extracts, reverse vaccinology made it possible to evaluate hundreds of expressed bacterial proteins and to develop successful bacterial vaccines.

In virology, the reverse vaccinology approach of Burton (2002) refers to a completely different strategy. Instead of the usual approach of generating antiviral antibodies by vaccination, it uses a so-called reverse approach for trying to generate a vaccine from the known 3D crystallographic structure of one paratope identified in neutralizing anti-HIV antibodies. From the structure of epitope–paratope complexes, attempts are made to elicit neutralizing antibodies by reverse engineering epitopes in order to endow them with the capacity to induce neutralizing antibodies.

Calling these two completely different strategies "reverse vaccinology" makes little sense, and it would have been preferable to call them genome-based and structure-based reverse vaccinology, respectively. Calling them reverse vaccinology 1 and reverse vaccinology 2 is also not particularly instructive because it obscures the fact that the genome-based approach has been successful, whereas the structure-based approach has systematically failed for more than 20 years (see Chap. 22).

Chapter 14 describes the structure of many continuous and discontinuous HIV epitopes identified in Env spikes and of their complementary paratopes as well as the many unsuccessful attempts that were made to modify the epitope structures in order to improve their vaccine potential.

Chapter 15 summarizes the discussions that took place during a workshop held in Baltimore in 2013. The participants proposed several new research paradigms that better fit our increasing knowledge of HIV immunopathology and which could possibly be more helpful for guiding future HIV vaccine research than past unsuccessful approaches. It was stressed that the development of a successful HIV vaccine required new prescriptive knowledge in the form of a practical invention rather than increased propositional and fundamental knowledge about the inner workings of the immune system.

Chapter 16 is the Introductory Editorial to a Research Topic entitled "Paradigm changes are required in HIV vaccine research" published in 2014. In his influential book *The Structure of Scientific Revolutions*, Kuhn (1962) argued that researchers in every field of scientific enquiry are always guided by theoretical assumptions, presuppositions, and hypotheses that constitute the prevailing scientific paradigm under which they operate at a given time. This paradigm commits them to using certain strategies and experimental approaches considered to be essential for trying to solve the problem at hand. The implicit presuppositions that gave rise to a particular paradigm are rarely stated, and when investigators obtain results that are not compatible with the hypothesis underlying the paradigm, they may fail to appreciate that the paradigm has been refuted and should be abandoned or revised. Scientists, however, tend not to abandon their guiding paradigm when they obtain contradictory results because their main goal is not to try to confirm the validity of the assumptions underlying the paradigm they follow. Instead, they will invent new *ad hoc* hypotheses for resolving contradictions between theory and experimental observations which allow them to pursue their investigations within the framework of their chosen paradigm, even at the risk of pursuing unfruitful lines of investigation. The following five paradigms had a detrimental effect on HIV vaccine research because they were based on erroneous assumptions: (1) Vaccine immunogenicity can be predicted from viral antigenicity, (2) there is a primary and intrinsic epitope for each BCR and its corresponding antibody, (3) unraveling large numbers of antibody maturation pathways will necessarily allow the identification of numerous HIV-1 immunogens suitable for repeatedly vaccinating large populations of genetically variable individuals, (4) rational design of HIV-1 immunogens is more effective than classical empirical screening of immunogens, and (5) reactions of viral antigens with protective Mabs are more specific than the combined reactivity of antibodies in a polyclonal antiserum.

Chapter 17 describes in detail one of the main reasons for the failure of structure-based reverse vaccinology (SBRV) to develop a preventive HIV vaccine. Adepts of SBRV tend to focus their attention mainly on residues in HIV epitopes and paratopes that make contact with each other, and they ignore the fact that the binding activity of these sites is often influenced by structural features distant from the sites themselves. In addition, the structure of the epitopes and paratopes observed in the complex is the result of the mutual adaptation and induced fit that occur when the partners interact during the immunization process and they are unlikely to correspond to the structure of the free sites. Another common assumption is that all antibodies and Mabs are monospecific for a single epitope, although they always contain numerous paratopes able to bind a variety of epitopes as well as autoantigens with various degrees of fit (Mouquet and Nussenzweig 2012). Once it is accepted that the epitope structure observed in an nMab-HIV1 Env complex is only one of the many epitopes that can be accommodated by a polyspecific Mab, there is no reason to assume that the epitope of known structure is the one that elicited the nMab. The central assumption of SBRV is therefore invalid since the one epitope whose structure was established is not necessarily able to elicit antibodies that neutralize HIV-1.

Chapter 18 is a commentary on the paper published by Andrieu et al. (2014) in the Research Topic mentioned in Chap. 16. These authors reported results obtained in Chinese macaque experiments that explored a new vaccine concept aimed at inducing tolerance to the simian immunodeficiency virus (SIV). They administered inactivated SIV intragastrically together with living bacterial adjuvants such as *Lactobacillus* bacteria with the goal of inducing tolerance to SIV antigens. Although this approach did not elicit SIV-specific antibodies nor cytotoxic T-lymphocytes, it protected 23 out of 24 experimental animals from mucosal and parenteral challenges. A previously unrecognized population of non-cytolytic MHV Ib/E-restricted CD8+ regulatory T cells was identified which suppressed the activation of SIV-positive CD4+ lymphocytes (Lu et al 2016). In view of the dearth of novel approaches in HIV vaccine research, such intriguing and unexpected results clearly need to be confirmed or refuted by further investigations with Chinese macaques (Carnathan et al. 2018).

Chapter 19 analyzes the common assumption that Mabs derived from HIV-infected individuals are the most effective reagents for dissecting antibody responses to HIV infection. It is often not appreciated that the apparent specificity of a Mab very much depends on the selection process that was used to obtain it. For instance, when a Mab is selected for its ability to bind a linear peptide in a peptide library that mimics a viral epitope, it is not astonishing that it will bind more strongly to the selecting peptide than to the virions that initially induced the Mab production in the infected host. Since most Env epitopes are discontinuous, their binding activity cannot be assessed by extracting all the constituent epitope residues from the protein to show that they are able to bind on their own when they are not embedded in the native protein structure. A short linear peptide that is part of such a discontinuous epitope may of course retain some binding activity, and this may lead to the peptide being called a continuous epitope.

It is often assumed that Mabs are better reagents for studying immune responses to HIV infection than polyclonal antibodies in spite of the fact that most effective antibody responses are polyclonal and directed to the multiple neutralizing epitopes that constitute the antigenic sites of HIV Env. Although the paratope structures in many nMabs have been elucidated, this knowledge has not helped the design of effective vaccine immunogens. Although it is possible by rational design to improve the ability of one epitope to bind to one paratope in an nMab (i.e., improving the antigenicity of a protein), the claim that what is being designed is an HIV vaccine immunogen able to trigger a protective immune response is the basic flaw of the SBRV approach because it assumes that when an HIV epitope reacts with an nMab, it should also be able to induce similar neutralizing antibodies in an immunized host.

Chapter 20 summarizes the discussions that took place during a workshop on new strategies in HIV vaccinology held in Rome in 2016. Eleven participants were asked to respond to the following five questions: (1) What new approaches should be followed in HIV-1 vaccinology? (2) Should HIV-1 inactivation be reconsidered? (3) Can therapeutic vaccines help the development of a preventive HIV vaccine?

(4) Can therapeutic vaccines lead to a functional cure? (5) If you had the authority to do it, what vaccine concepts would you support for testing? The variety of answers that these questions elicited clearly indicates that novel research strategies for developing a possible HV vaccine are urgently needed.

Chapter 21 explains the many reasons why the rational design of a preventive HIV-1 vaccine by SBRV failed to deliver an effective vaccine. The review describes in considerable detail the structure and dynamics of epitopes and paratopes as well as the considerable improvement of our understanding of antibody specificity in recent years. It has for instance been clearly demonstrated that the amino acid compositions of epitopes and of the entire surface of proteins are very much the same, which finally explained why it had not been possible for more than 25 years to predict accurately the location of epitopes in proteins on the basis of amino acid propensities such as hydrophilicity or accessibility (see Chap. 4).

Although it is frequently stated that rational design is the best strategy for developing new vaccines, this does not seem to be the case for HIV vaccine development. Rational design applies mainly to drug design when candidate molecules are designed to fit the 3D structure of a biological target in order to bind to it with high affinity and inhibit its biological function. Such a computer-assisted approach based on molecular docking succeeds because the complementarity between a drug and its target molecules is fairly unique. In contrast, molecular recognition processes between an antigen and the many antibodies that are able to bind to it are much less specific than drug–receptor interactions mainly because of the conformational flexibility of epitopes and paratopes, the degeneracy of the immune system, and the polyspecificity of antibodies. It also seems inappropriate to use design terminology for describing the SBRV attempts made by investigators when they try to have an antigen produce protective antibodies since they only improve the antigenic binding capacity of viral epitopes and not their immunogenic capacity to trigger the immune system to elicit protective antibodies. SBRV disregards accepted immunological theory pertaining to the degeneracy of the immune system and the polyspecificity of antibodies and also underestimates the extreme plasticity of the HIV Env protein as well as the extensive conformational flexibility of epitope and paratope binding sites.

Chapter 22 points out that when HIV epitopes are called vaccine immunogens, this could be interpreted to mean that they are able to generate immune responses in the immunized host, although it should be evident that they only trigger in the host a series of reactions with BCRs that eventually may lead to the immune system producing antibodies, some of which may be neutralizing. The type of antibody that is produced depends on numerous properties of the immune system such as its antibody gene repertoire, the presence of helper and suppressor T cells, the secretion of cytokines, self-tolerance, and a variety of other immunoregulatory mechanisms. Vaccinologists who claim to be designing a vaccine immunogen by SBRV tend to ignore these contributions from the immune system because they focus on the recognition processes between single epitope–paratope pairs instead of investigating by trial and error which components of the immune system must be controlled to ensure that neutralizing antibodies are elicited. Only when it was found that HIV Env epitopes recognized by affinity-matured antibodies obtained from HIV chronically infected individuals did not bind the germline predecessors of these antibodies was it realized that potential vaccine immunogens may only be discovered if one took into account the slow but extensive antibody maturation that is needed to obtain neutralizing antibodies. Additional immune correlates of protection are needed to guide HIV vaccine development, but these can be inferred only retrospectively when an efficacious vaccine has been developed empirically. Since we know very little about which features of the immune system regulate the production of protective antibodies, it seems unavoidable that empirical vaccination trials will continue to be required when trying to develop an effective HIV vaccine.

Chapter 23 describes the nature of the inverse problems that vaccinologists need to solve when trying to develop an HIV vaccine. Inverse problems differ from the usual direct problems in science that are solved by determining experimentally what are the effects that follow certain causes. Biological systems are made up of successive levels of increasing complexity from genes to RNAs, proteins, organelles, cells, tissues, and organs. At all these different levels, innumerable interactions occur involving genetic, epigenetic, biochemical, and physiological factors which nowadays are mainly analyzed using system biology approaches. Numerous bottom-up and top-down causal links occur across the different constituents and levels of the immune system, and as a result, our theoretical understanding of the mechanisms responsible for the appearance of neutralizing antibodies against HIV is practically nonexisting.

We are unable to imagine what are the multiple causes that could lead to protection against HIV infection because we lack an integrated theoretical model of the entire immune system. Such a model is actually required for conceiving plausible solutions to inverse problems which then need to be verified experimentally.

Solving an inverse problem may for instance require developing a model that could explain the multiple causes that produce a desired effect such as the absence of a deleterious HIV infection in elite controllers, and then subsequently demonstrating that by adjusting certain parameters in the immune system, the desired outcome can be obtained in a genetically heterogeneous population.

Whereas many direct problems can be solved because of our knowledge of the biological mechanisms that bring about certain effects, this is not feasible in the case of inverse problems that cannot be solved in this manner. Because of the complexity of the immune system, vaccinologists mostly do not have access to all the information that would be required in order to make entirely rational decisions based on a complete knowledge of all the relevant parameters. Since several inverse problems may first need to be solved at different levels of the immune system, it may in fact turn out not to be possible to rationally develop an HIV vaccine in this manner. In the last 20 years, investigators have mainly studied the rather ineffective antibody responses that occur several years after the initial acute phase of HIV infection, although these antibodies are usually unable to control the infection in the individuals from whom they had been isolated.

Since the initial acute phase of HIV infection is caused by so-called transmitter/ founder (T/F) viruses that are not predominant in the individual donors, it could be argued that the immune response against T/F viruses may be most relevant for developing a preventive vaccine. Since new and more reliable inactivation methods have been developed that could be applied to T/F viruses, immunization studies with inactivated T/F virions should be undertaken in the near future. Chemically inactivated immunogens have been used successfully for developing vaccines against several animal retroviruses (Beczkowski et al. 2015), but this approach is still considered to be too risky in the case of HIV. It is nevertheless astonishing that such a time-honored strategy has never been tested as a possible way to stop the AIDS epidemic (Sheppard and Dorman 2015; Rios et al. 2016).

Chapter 24 briefly discusses the complexity of biological systems and its compatibility with determinism. Although determinism assumes that every natural phenomenon is entirely determined by preexisting causes, in biological systems no single event can ever be said to be the necessary and sufficient cause for a biological effect to occur. Numerous contributory causes always exist, and it is not possible to single out one factor and give it undue explanatory weight on its own (Van Regenmortel 2007).

Biological systems consist of hierarchical compartments of increasing complexity which all communicate with each other via translocating molecules. This organization leads to upward as well as downward causality, for instance when the constituents at the lower levels are controlled by new emergent properties that arise at the higher levels (Van Regenmortel 2012). One of the aims of systems vaccinology (Pulendran 2014) is to understand how emergent immunological phenomena arise from the multiple interactions that occur in complex immune systems. At each successive level of organization from genes to proteins and to cellular tissues, innumerable interactions occur between genetic, epigenetic, biochemical, and environmental factors which need to be analyzed separately before they could be integrated in a functional complex system. Numerous individual, fully deterministic causal processes follow separate trajectories, but when they impede each other, following a contingency pattern, unpredictable effects are obtained. Our understanding of the entire integrated system remains severely limited because of our inability to measure accurately the on-and-off-kinetic rates of individual interactions. The reason for this is that the kinetic constants of an interaction are highly dependent on the active concentration of the two partners in their cellular environment, a parameter that is extremely difficult to measure accurately (Richalet-Secordel et al. 1997; Zeder-Lutz et al. 2009). The active concentration is always considerably lower than the nominal concentration (that can be measured for instance by spectrophotometry), because of the inevitable presence of an unknown amount of incorrectly folded molecules. As a result, kinetic measurements of the numerous individual interactions are extremely inaccurate which prevents us from understanding the dynamics of fully integrated complex immune systems.

Physicists used to believe that the universe was ruled by mathematical laws that would make it possible to predict the future behavior of a system if one had an intimate knowledge of its initial conditions. That assumption became untenable in the twentieth century with the development of chaos theory which established that extremely small differences in the initial conditions of a dynamic complex system had a huge impact on the subsequent state of the system (Du Sautoy 2016). Making very precise quantitative measurements of the innumerable initial conditions of a complex biological system is of course even more unattainable than in the case of physical systems, and we may have to accept that making accurate predictions about the future states of a dynamic biological system remains completely intractable (Green 2015).

# Contents

# About the Author

**Marc H V Van Regenmortel** was a Professor of Virology at various universities in South Africa (Stellenbosch and Cape Town) and France (Strasbourg) and was head of the Immunochemistry Laboratory at the CNRS Molecular Biology Institute in Strasbourg for 22 years. He served as President of the International Committee on Taxonomy of Viruses and published a dozen books and 420 papers and reviews on the immunochemistry of peptides and viruses, viral taxonomy, and biosensor technology. He was an advisor to the South African AIDS Vaccine Initiative and is currently an associate editor of *Journal of Molecular Recognition*, *Archives of Virology*, *Advances in Virus Research*, *Journal of Immunological Methods*, *Analytical Biochemistry*, and *Frontiers in Immunology*.

He is an Adjunct Professor of the Medical University of Vienna.

# Part I
# Immunochemistry

# What Is a B Cell Epitope

## Introduction

Since most biologically important antigens are proteins, I will discuss only the antigenicity of proteins and will not consider carbohydrate and nucleic acid antigens. The antigenic specificity of a protein resides in restricted areas of the molecule, known as antigenic determinants or epitopes, which are recognized by the combining sites or paratopes of certain immunoglobulin molecules. Once an immunoglobulin has been shown to bind to an antigen, it becomes known as an antibody specific for that antigen. Since epitopes are able to bind antibody molecules both in their free form and as a membrane-bound B-cell receptors, they are often called B-cell epitopes to distinguish them from the T-cell epitopes, which are proteolytically cleaved peptides of the antigen that interact with the receptors of T cells. This chapter deals only with B-cell epitopes, henceforth abbreviated to epitopes.

## Types of Epitopes

Epitopes of proteins are usually classified as either continuous or discontinuous depending on whether the amino acids included in the epitope are contiguous in the peptide chain or not. The distinction between these two types of epitopes is not clear-cut since discontinuous epitopes often contain short segments of a few contiguous residues that are able to bind to antibodies raised against the protein and which could be given the status of continuous epitopes.

---

In: Schutkowski M., Reineke U. (eds) Epitope Mapping Protocols. Methods in Molecular Biology (Methods and Protocols), vol 524. Humana Press, 2009,
Marc H V Van Regenmortel
Copyright © 2009 Springer Nature

© Springer Nature Switzerland AG 2019
Marc H V Van Regenmortel, *HIV/AIDS: Immunochemistry, Reductionism and Vaccine Design*, https://doi.org/10.1007/978-3-030-32459-9_1

## Continuous Epitopes

Any linear peptide fragment of a protein that is able to bind to antiprotein antibodies is called a continuous epitope. Since the criterion for identifying such an epitope is its binding activity, continuous epitopes are actually defined in a functional manner and no evidence is presented to show that each residue in the peptide makes contact with residues of the paratope and is recognized as such by the antibody. The contribution of individual residues to the epitope can be assessed by measuring the binding capacity of peptide analogs presenting single residue replacements. When this is done, it is found that most continuous epitopes contain a number of indifferent residues that seem not to be implicated in the binding interaction and can be replaced by any other amino acid without impairing antigenic activity (Getzoff et al. 1988). Such continuous epitopes can therefore be said to be structurally discontinuous, although it cannot be excluded that replaceable residues play a scaffolding role or are recognized through their backbone atoms. Residues that cannot be replaced in an epitope without causing a major loss in binding activity tend to be regarded as being part of a smaller entity called "functional epitope" in which each residue is assumed to contribute to the free energy of interaction (Cunningham and Wells 1993). However, measurements of perturbations are not the same as energy determinations (DeLano 2002) and substitutions of residues that do not themselves interact with the antibody may induce structural perturbations that propagate beyond the mutated region and affect the activity of a nearby epitope.

There is no reason to assume that all the residues of a continuous peptide epitope correspond to residues present in the epitope of the intact protein since only a limited degree of similarity between the two structures is sufficient to allow the peptide to bind to antiprotein antibodies. Much of our knowledge of protein antigenicity, which is derived from the study of short linear peptides, gives information on which structures are needed for a peptide to possess antigenic cross-reactivity but it does not clarify the exact structure of the actual epitopes in the intact protein. Peptide fragments of a protein are not faithful copies of antigenic regions in protein molecules, one reason being that they do not retain the conformation present in the folded protein.

Many investigators take the view that the majority of continuous epitopes of proteins described in the literature are likely to correspond to unfolded regions of denatured protein molecules and are not genuine epitopes of native proteins (Laver et al. 1990; Van Regenmortel 2006).

They argue that it is very difficult to know whether the immunoassays used to identify continuous epitopes actually measure antibodies specific for the native state of the cognate protein or whether they measure antibodies directed to the denatured protein. Antiprotein antisera frequently contain both types of antibodies because some of the molecules used for immunization are denatured before or after being injected in the animal. It is, of course, not possible to know the exact conformation of the protein molecule when it is interacting with a B-cell receptor during the immunization process.

In the reciprocal situation where antibodies raised to peptides are allowed to react with the cognate protein, it is possible that the antibodies recognize the protein because some of the protein molecules used in the immunoassay were denatured for instance when they were adsorbed to a solid-phase in the assay (Van Regenmortel 1999c). It is now accepted that earlier erroneous claims that immunization with peptides always elicits high levels of antibodies that crossreact with the native cognate protein (Lerner 1984) arose because it was not realized that the protein used in solid-phase immunoassays had become denatured by adsorption to plastic (Darst et al. 1988).

## Discontinuous Epitopes

The second type of epitope known as discontinuous epitope corresponds to the vast majority of epitopes found in proteins. They consist of atoms from surface residues of the protein that are brought together by the folding of the polypeptide chain, and their antigenic reactivity depends on the native conformation of the protein. The atomic groups that form a discontinuous epitope are not held together by internal chemical bonds and they possess a collective identity recognized by antibodies only because the entire peptide chain acts as a scaffold. If the scaffold is perturbed, the epitope ceases to exist. Although such an epitope lies in a molecule and acts like a molecule, it is not actually a molecule (see p. 273 in Rosen 1991). A discontinuous epitope, therefore, cannot be isolated as an entity independent from the rest of the molecule in which it is embedded and it cannot be shown experimentally to possess binding activity on its own, outside of the protein context.

This places severe limits on the functional characterization of discontinuous epitopes since they can only be defined in structural terms. Their structure is established by analyzing antigen-antibody complexes using X-ray crystallography or NMRspectroscopy and identifying the set of atoms of the antigen that make contact with residues of the antibody. Usually a contact between two residues is said to occur if the interatomic distance between their atoms is less than 4 Å, and this criterion leads to the conclusion that discontinuous epitopes consist of 10–22 residues. These residues originate from between two and five separate segments of the polypeptide chain that are often surface loops and are brought together by the folding of the chain. When the protein is fragmented into peptides, residues from distant parts of the sequence are scattered and the individual constituents of discontinuous epitopes are usually no longer recognized by antiprotein antibodies, although short segments of a few residues may sometimes bind to antiprotein antibodies.

Discontinuous epitopes are often called conformational epitopes because their structure depends on the intact conformation of the native protein. This terminology may lead to the erroneous conclusion that continuous epitopes, in contrast, are conformation-independent. This is, of course, not the case since linear peptides constituting continuous epitopes necessarily also have one or, more likely, a number of different conformations.

Peptide fragments of a protein very rarely retain the conformation initially present in the corresponding residues of the correctly folded protein. On the other hand, it is equally unlikely that if a discontinuous epitope could somehow be excised from a protein, it would retain its original conformation.

Studies with monoclonal antibodies (Mabs) have shown that only about 10% of the Mabs that react with a native protein are able to bind short peptide fragments of the protein. Since the range of specificities observed with a panel of Mabs is very similar to that found in a polyclonal antiserum raised against the same antigen (Quesniaux et al. 1990), it is reasonable to assume that also about 10% of the antibodies present in an antiprotein antiserum are able to recognize peptide fragments of the protein (Barlow et al. 1986; Van Regenmortel 1996). These peptides are the ones that will be considered continuous epitopes of the antigen. Although much less common than discontinuous epitopes, continuous epitopes have been studied extensively because they have many applications, for instance as diagnostic reagents to replace infectious agents in immunoassays (Leinikki et al. 1993), as immunogens to obtain antibody reagents useful for isolating gene products (Walter 1986; Muller 1999b), or as potential synthetic vaccines (Van Regenmortel 1999b; Hans et al. 2006).

## Additional Epitope Types

The quaternary structure of polymerized proteins introduces an additional level of antigenic complexity in macromolecular assemblies such as viruses. This has led to the recognition of two other types of epitopes, cryptotopes and neotopes, which are nowadays easily identifiable by using Mabs (Van Regenmortel 1992a). Cryptotopes are epitopes hidden in polymerized proteins or in virus particles because they are present on the surface of the protein subunits that become buried when the subunits aggregate. Cryptotopes of viruses are antigenically active only after dissociation of virus particles. Cryptotopes are fairly conserved in a group of related viruses because the intersubunit surfaces which control virus assembly tend to vary less than the outer surface of virions. As a result, antibodies to cryptotopes will emphasize antigenic similarities between the members of a virus family and they are therefore useful diagnostic reagents for detecting a wide range of related viruses instead of a single member of a virus genus or family (Shepard et al. 1974; Dougherty et al. 1985).

The term neotope was coined (Van Regenmortel 1966) to refer to epitopes that are specific for the quaternary structure of virus particles and are absent in dissociated viral subunits. Neotopes may arise from the juxtaposition of residues from neighboring subunits that are recognized by the antibody as a single epitope. For instance, a neotope of the serotype 1 of poliovirus consists of residues 221–226 of protein VPl, together with residues 164–172 and 270 of protein VP2 (Boeyé and

Rombaut 1992). Neotopes can also arise through the conformational changes in protein subunits that result from intersubunit interactions. Because the quaternary structure of virus particles is not static but can undergo major rearrangements following small changes in the environment (Bothner et al. 1998), neotopes are often transitional epitopes (Zolla-Pazner 2004) that can assume different conformations. This is one of the reasons why it is difficult to mimic them by chemical synthesis, for instance when attempts are made to develop synthetic vaccines (Van Regenmortel 1999b).

## Mimotopes

The term mimotope coined by Geysen (1986) was originally defined as a peptide able to bind to a particular antibody, although unrelated in sequence to the protein antigen used to induce the antibody, often because the antibody is directed to a discontinuous epitope. Currently, the term mimotope is applied to any epitope mimic irrespective of whether the protein epitope being mimicked is continuous or discontinuous. Mimotopes are usually identified by testing combinatorial peptide libraries obtained by chemical synthesis or phage display and selecting peptides able to bind antiprotein Mabs. When mimotopes are selected for use as reagents in the diagnosis of virus infections, it is possible to screen phage libraries with sera collected from individuals who recovered from a viral infection and had mounted an immune response against the infectious agent, even in the absence of any knowledge of which antigens are involved (Folgori et al. 1994).

Mimotopes often show a limited amount of sequence similarity with the protein immunogen that gave rise to the antibodies used in the mimotope screening. It may also happen that the mimotope shows no sequence similarity whatsoever with the protein immunogen, although it is able to induce antibodies that cross-react with the protein (Delmastro et al. 1997). Such findings underline the fact that the recognition between epitope and paratope does not take place at the level of whole residues but at the level of individual atoms. It is always a minority of the atoms of a given residue that participate in the interaction, a situation that is obscured by the division of epitopes into continuous and discontinuous classes which may give the impression that the elementary units of recognition are amino acid residues.

The presence of hydropathic complementarity between short segments of residues in an epitope and in individual CDRs of a paratope may be sufficient to give rise to an antigen-antibody interaction. It has been demonstrated, for instance, that peptides corresponding to short sequences present in one CDR loop of an antilysozyme antibody were able to bind the antigen with similar specificity as the whole antibody, albeit with lower affinity (Laune et al. 1997). In analogy with the binding observed by so-called continuous epitopes that are actually part of more complex discontinuous epitopes, these short segments in paratopes have been called continuous paratopes (Van Regenmortel 1998). Hydropathic complementarity arises from an inverted hydropathic pattern in two short peptide sequences and is caused by the attraction between hydrophilic and hydrophobic groups (Blalok 1990; Tropsha et al. 1992). Peptide analogs that retain the original hydropathic profile present in a continuous epitope but possess no longer any sequence similarity with it may still bind the same antibody, a phenomenon that could be responsible for the binding activity of certain mimotopes (Boquet et al. 1995; Hanin et al. 1997).

To qualify as a mimotope, a peptide should not only be able to bind to a particular antibody but it should also be capable of eliciting antibodies that recognize the epitope being mimicked. This requirement stems from the fact that a single immunoglobulin molecule always harbors a number of partly overlapping or nonoverlapping paratopes, each one capable of binding to related or unrelated epitopes. The potential binding pocket of an immunoglobulin comprises as many as 50–70 hypervariable residues distributed over the six CDRs, although each individual paratope consists of only 10–20 CDR residues. This means that about two thirds of the CDR residues could bind to other epitopes that bear little or no resemblance to the first epitope, a situation that explains the considerable multispecificity of antibodies. Furthermore, different paratopes may partly overlap, in which case binding to one epitope may prevent a second unrelated epitope from being accommodated at a nearby location. Therefore, when a peptide is labeled a mimotope of epitope A because of its capacity either to bind to an anti-A antibody or to inhibit the binding of epitope A to this antibody, it can not be excluded that the so-called mimotope actually binds to a different paratope from the one that interacts with epitope A. This is why it is necessary to show that a peptide is also able to induce antibodies that crossreact with epitope A, to demonstrate that it really is a mimotope of epitope A (Van Regenmortel 2001a).

Although the study of mimotopes has been of little value for investigating the structure of protein epitopes, it has led to the identification of many antigenically active peptides that are useful as immunochemical reagents and are sometimes viewed as potential synthetic vaccines (Meloen et al. 2000; Mullen et al. 2006; Larralde et al. 2007).

## Epitopes Are Relational Entities and Not Intrinstic Features of Proteins

It is important to realize that the epitope nature of a set of amino acids can only be established if an immunoglobulin able to bind to it has been found. In the same way, the antibody nature of an immunoglobulin defined by one of its paratopes becomes apparent only when a complementary epitope has been identified. The immunoglobulin is then called an antibody specific for the antigen that harbors the epitope. This sometimes leads to confusion since most antigens possess many different epitopes and an antibody cannot be specific for the multiepitopic antigen as a whole but only for one of its epitopes (Van Regenmortel 1998).

Epitopes and paratopes are relational entities defined by their mutual complementarity and they depend on each other to acquire a recognizable identity. This means that an epitope is not an intrinsic structural feature of a protein that could be identified in the absence of a particular interaction with a paratope.

Since epitopes acquire an identity by virtue of a relational nexus with complementary paratopes, the number of epitopes in a protein can be equated with the number of different Mabs that can be raised against it. In the case of the small insulin molecule, this number was estimated to be around 100 (Schroer et al. 1983). It is now accepted that the entire accessible surface of a protein harbors many overlapping epitopes, which can be recognized only if a sufficiently large panel of Mabs is available (Benjamin et al. 1984; Berzofsky 1985). Because of this relational dependence, analyzing the antigenic diversity of a protein corresponds to analyzing the size of the immunological repertoire of a host immunized with that protein.

The same residues at the surface of a protein can be part of different overlapping epitopes recognized by different paratopes and it is therefore not possible to draw clear boundaries between individual epitopes. There is no clear-cut minimum difference in atomic positions at epitope-paratope interfaces or in the binding affinity of interactions, that could be used as an absolute yardstick for deciding that two epitopes are the same or not.

Epitopes have been called "fuzzy" binding sites (Van Regenmortel 1998), and they share this fuzziness with all protein binding sites. As pointed by Moodie et al. (1996), the shape and electrostatic complementarity between two binding sites can be achieved by not just a single arrangement of amino acids but by a large number of alternative arrangements. Continuous epitopes have fuzzy boundaries because there are many ways to interpret the effect that removing or adding residues has on the antigenic activity of peptides.

For instance, a longer peptide may be more active because the added residues are part of the epitope or because they induce a more active conformation in a nearby epitope. However, longer peptides are not necessarily more active than shorter peptides (Muller et al. 1986). Sometimes the shortest peptide that retains significant binding activity in an immunoassay is called the epitope, but this is also unsatisfactory since different immunoassays have very different sensitivity thresholds and can induce different conformations in the peptide (Muller et al. 1986).

## Antigenic Cross-Reactivity and Antibody Specificity

Antigenic cross-reactivity is a common phenomenon caused by the ability of an antibody to recognize not only the epitope against which it was elicited but also a variety of related epitopes that possess some structural similarity with it. This type of cross-reactivity has been called "true cross-reactivity" (Berzofsky and Schechter 1981), to distinguish it from the less common "shared cross-reactivity," which occurs when an antibody recognizes the same epitope in two different multiepitopic antigens. Usually, a paratope reacts with higher affinity with the homologous epitope used for raising the antibody than with cross-reacting epitopes, although a paratope can also bind more strongly to heterologous epitopes, a phenomenon known as heterospecificity or heteroclitic binding (Mäkelä 1965). Heterospecificity is potentially widespread but is only observed when it is looked for, for instance if an antibody is tested against a series of analogs related to the epirope used for immunization (Al Moudallal et al. 1982; Underwood 1985; Harper et al. 1987). Heterospecificity can be of practical use when attempts are made to obtain from a single hybridoma fusion experiment a number of Mabs specific for different members of a family of related proteins (Frison and Stace-Smith 1992). Heterospecificity is due to the fact that the clonal selection of a B cell, which eventually leads to antibody secretion, can be triggered by an immunogen endowed with only moderate affinity for the B-cell receptor. Because a high degree of fit between epitope and paratope is not required for initiating B-cell differentiation and because of antibody multispecificity, there is nothing strange in the finding that low-affinity antibodies may react better with related epitopes endowed with a superior degree of complementarity with the paratope. It can even happen that the antibody has such a low affinity for the immunogen that it does not react with it at all and only binds to a related antigen. This was commonly observed with antibodies raised against tobacco mosaic virus (TMV), which reacted with a mutant harboring a single proline residue substitution in the viral coat protein but not with TMV itself (Van Regenmortel 1982, p. 198; Loor 1971; Von Sengbusch and Wittman 1965).

As discussed earlier in section "Types of Epitopes", the description of protein antigenicity in terms of continuous epitopes is based on the ability of short peptides to cross-react with antiprotein antibodies. These cross-reactions occur even when only a few of the peptide residues correspond to interacting residues in an epitope of the protein immunogen and in spite of major differences in conformation between the peptide and the corresponding region in the intact protein. The structural basis of antigenic cross-reactivity has been studied extensively (Roberts et al. 1993), and it is known that the flexibility of CDR loops greatly facilitates the ability of antibodies to adapt to a variety of epitopes (James et al. 2003).

Biological specificity has been defined as the exact complementary relationship between an agent and something acted on (see p. 199 in Medawar and Medawar 1978), and such a definition is valid for the specificity of enzyme-substrate, receptor-ligand, and antigen-antibody interactions. The term specificity is derived from the word species and describes what is characteristic of a species. Biologists believed for centuries that biological species were separated by clear-cut discontinuities, and bacteriologists later turned to serology in the hope that it would allow them to distinguish between different species of bacteria by using specific antisera raised against different bacteria. A belief in the absolute separation between bacterial species led Paul Ehrlich to accept the absolute immunological specificity of antibacterial antibodies (Van Regenmortel 1998; Mazumdar 1995). This view was questioned by Landsteiner (1962), who demonstrated experimentally that serological cross-reactions between different cell types were caused by antibodies that reacted to different degrees with a wide range of cells. He showed that there was no one-to-one relationship between an antigen and its antibody and that antigens were able to elicit a wide spectrum of antibodies capable of crossreacting with many related antigens.

The widespread occurrence of cross-reactions between epitopes and paratopes may seem to contradict the accepted view that antigen-antibody interactions are very specific. However, a perfect fit between an epitope and its paratope is not a meaningful concept, since it would imply that heterospecific binding or additional affinity maturation of the antibody during prolonged immunization could not occur. Antibody specificity is often believed to be correlated with high affinity, since it is expected that highly specific antibodies will possess a better stereochemical complementarity with their antigens than will antibodies of lower affinity. However, there is no necessary link between affinity and specificity, and antibodies of low affinity may in fact discriminate better between two antigens than do antibodies of high affinity. The reason is that low-affinity antibodies may detect fewer cross-reactions than do antibodies of high affinity since weaker cross-reactions will tend more quickly to be below the level of detection in the case of low-affinity antibodies (Van Regenmortel 1998).

It is generally more meaningful to speak of the discrimination potential of antibodies rather than of their specificity. Since proteins harbor many different epitopes, different degrees of crossreactivity will be found depending on the epitope that is singled out by a particular Mab. It is, in fact, the wish of the investigator to differentiate between two proteins that provide the criterion for deciding whether a particular antibody is specific or not, usually because it recognizes an epitope present in only one of the two proteins. If the antibody recognizes an epitope present in both proteins, it would be called nonspecific. Antibody reagents are thus considered specific if they achieve the level of discrimination that is required in any particular case and the same antibody will be considered specific or nonspecific depending on what the investigator is trying to achieve (Van Regenmortel 1998). Antibodies, of course, are only specific for individual epitopes and not for antigens. Although epitope-paratope recognition phenomena possess a fair degree of specificity, this is not due to the existence of specific "immunological" bonds different from the bonds observed in other protein interactions. Extensive shape complementarity at the surface of the two partners in an antigen-antibody complex leads to the formation of physicochemical bonds consisting of electrostatic forces and polar forces, such as van der Waals and hydrogen bonds that are present in all protein interactions.

The equilibrium affinity constants of antibodies lie in the range of $10^6$–$10^{10}$ L/mol (Foote and Eisen 1995). The interpretation of antigen antibody binding energies is complicated by the role played by entropy and hysteresis in the interaction. It was initially assumed that all water molecules are extruded from antigen-antibody interfaces in a complex, but it was later found that many interstitial water molecules remained at the interface because of imperfect steric complementarity. The reaction is in fact often driven by enthalpy and not by entropy (Braden and Poljak 1995).

Hysteresis, the phenomenon whereby more energy is needed to dissociate most antigen-antibody bonds than is required to prevent their formation, is caused by the formation of additional secondary bonds subsequent to the initial primary bonds. For an extensive discussion of antigen-antibody bonds, see Van Oss (1995).

It is important to appreciate that the discrimination potential of antibodies is optimal only in a limited range of experimental conditions. When they are used at high concentrations, antibodies tend to react nonspecifically with many molecules, especially when the antigen is also present at a high local concentration, for instance on a solid-phase or in an immunoblot assay (Ghosh and Cambell 1986; Zimmermann and Van Regenmortel 1989).

One method for ascertaining the specificity of an interaction is to measure its stoichiometry for instance by biosensor assays, since nonspecific interactions do not possess a unique stoichiometric binding ratio when tested at different concentrations of the two reactants. However, it is always good practice to include various controls in any immunoassay to establish that the observed reaction is specific for the molecule of interest.

## Antigens Versus Immunogens

The antigenicity of proteins is a chemical property describing interactions between epitopes and paratopes in terms of chemical and structural complementarity. Such a description takes the existence of antibodies for granted and does not consider the biological origin and synthesis of antibodies by the immune system. In contrast, immunogenicity, which is the ability of a protein to give rise to an immune response in a competent, vertebrate host, is a biological property definable only in the biological context of an immune system (Van Regenmortel 2001a). Immunogenicity always depends on extrinsic factors such as the host immunoglobulin repertoire and self-tolerance, the production of chemokines and cytokines as well as numerous cellular and regulatory mechanisms of the immune system.

The difference between antigens and immunogens is a crucial one in immunology but is often not sufficiently appreciated. For instance, when a peptide fragment of a protein is found to crossreact antigenically with antibodies raised against the protein, this does not in any way guarantee that it will be able to elicit antibodies that cross-react with the protein (Van Regenmortel 2006).

Most peptides are immunogenic in the sense that they readily elicit antibodies that react with the peptide immunogen. However, this type of immunogenicity is mostly irrelevant since the purpose of peptide immunization is usually to obtain antibodies that crossreact with the cognate, native protein. What is needed, therefore, is so-called cross-reactive immunogenicity, i.e., the ability to induce antibodies that cross-react with the cognate protein. In addition, if the peptide is to have vaccine potential, it must elicit antibodies that neutralize the infectivity of the pathogen harboring the cognate protein; i.e., it must also possess so-called cross-protective immunogenicity (Van Regenmortel 2001a, 2006). This type of immunogenic capacity is not necessarily present when the peptide is antigenically active and able to bind to a neutralizing antibody.

An antiprotein antibody used in an immunoassay may be able to select one of the conformations present in a peptide or may induce such a cross-reactive conformation by an induced fit mechanism, whereas a B-cell receptor that possesses no prior specificity for the cognate protein will not be able during the immunization process to carry out such a selection or induction. The peptide will bind to various B-cell receptors that recognize some of the peptide conformations, but it will not preferentially bind to those rare receptors which, in addition to recognizing the peptide, also cross-react with the epitope present in the native protein. There is thus no reason why most of the elicited anti peptide antibodies should also react with the cognate protein.

Since B-cell epitopes are usually defined as regions of the antigen that bind both free and membrane-bound antibodies, little attention is given to the fact that the structural context in which an epitope recognizes a free antibody molecule or a B-cell receptor embedded in a membrane is not the same. This difference is one of the reasons why knowledge of the structure of a viral epitope bound to an antibody does not necessarily provide relevant information on its immunogenic potential since this depends on a reaction occurring in the membrane environment of a B-cell receptor. It has been pointed out by Zwick (2005), for instance, that the type of lipid that surrounds the gp41 membrane-proximal external region (MPER) antigen of HIV, when it is presented to membrane-bound B-cell receptors, is likely to affect its immunogenicity in ways that are not predictable from the structure of MPER antigen-antibody complexes studied by X-ray crystallography outside such a lipid environment.

Furthermore, the structure of epitopes and paratopes seen in a complex may be different from the structure of the respective binding sites in the free antigen and antibody molecules, before they have been altered by the mutual adaptation that occurs during the binding interaction (Wilson and Stanfield 1994; Halperin et al. 2002). As a result the structure of an epitope after complexation with a neutralizing Mab may be an unreliable guide for identifying the exact epitope structure that was recognized by B-cell receptors during the immunization process and which should be present in a vaccine immunogen intended to elicit the same type of neutralizing antibodies.

# Epitope Prediction

The main purpose of predicting epitopes is to replace the epitope regions of intact antigen molecules by linear synthetic peptides that could be used as reagents for detecting antiprotein antibodies in an immunoassay (Leinikki et al. 1993) or as immunogens for raising antipeptide antibodies able to cross-react with the protein (Muller 1999b). A further purpose is to develop synthetic peptide vaccines in which case the predicted epitopes must also be able to elicit antibodies that neutralize the infectivity of the pathogen harboring the protein antigen (Van Regenmortel 2006). Unfortunately, neutralization epitopes that elicit antibodies that protect against infection do not have special physicochemical properties that allow them to be recognized and predicted separately from other epitopes.

Most attempts at predicting epitopes have been restricted to continuous epitopes since any predicted structure could then easily be synthesized chemically or inserted into a recombinant protein. Since the epitopes of native proteins are located on the surface of the molecules, initial prediction attempts analyzed protein sequences using certain amino acid propensity scales which identified segments of the protein that protruded at the surface, were hydrophilic or possessed a high mobility (Hopp and Woods 1981; Westhof et al. 1984; Thornton et al. 1986; Novotny et al. 1987).

Many different scale-based prediction methods were developed and compared (Pellequer et al. 1991; Ponomarenko and Van Regenmortel 2009), but none of them, even when used in combination, gave high rates of successful prediction (Odorico and Pellequer 2003; Blythe and Flower 2005; Greenbaum et al. 2007). The apparent success rate of a prediction depends very much on the method used to measure its effectiveness (Van Regenmortel and Pellequer 1994), and at present, a method measuring the values of the area under the receiver operating characteristic curve ($A_{ROC}$) is the one most commonly used (Greenbaum et al. 2007; Van Regenmortel and Pellequer 1994; Haste-Andersen et al. 2006).

As an increasing number of 3D structures of proteins is becoming available, new prediction methods are being developed which incorporate information from docking algorithms and 3D structures (Ponomarenko and Van Regenmortel 2009). Unfortunately, many investigators claim that they are able to predict discontinuous epitopes when in fact they only predict that certain surface residues are likely to be part of a discontinuous epitope. They do not predict which minimum set of neighbouring residues must be assembled in a defined configuration to achieve a structure with antigenic or immunogenic activity. Such an alleged prediction of discontinuous epitopes is actually a misnomer since it does not entail predicting that a particular group of atoms or residues possesses the characteristic immunological activity of an epitope.

Since the usual purpose of discontinuous epitope prediction is to be able to replace the protein epitope by a synthetic construct that possesses the same immunological activity, many investigators have attempted to design linear peptides that mimic the binding activity of a surface patch of the protein. One such approach is the Mapitope strategy (Bublil et al. 2007), which consists in assembling clusters of connected amino acid pairs that lie within the footprint of an epitope. Residue pairs originating from distant regions of the protein sequence are affinity-selected from a random peptide library and retained if they are present at the protein surface.

Other approaches have also been used to reconstitute discontinuous epitopes by aligning putative epitope residues along a synthetic peptide (Timmerman et al. 2005). However, it may be insufficient to include only the solvent-exposed surface of proteins when mapping epitopes, since certain buried residues can be involved in the paratope interaction following conformational rearrangements (Alexander et al. 1992).

Since it is well-known that synthetic peptides are less rapidly degraded and more immunogenic if they are constrained into stable secondary structures by cyclization or other chemical procedures, there have been many attempts to constrain peptides into helix, turn, or cyclic conformations (Hans et al. 2006; Shepherd et al. 2004;

Sundaram et al. 2004). However, constraining peptides will not necessarily make them adopt a conformation that closely mimics the immunogenic structure in the cognate protein.

In some cases unconstrained peptides possessing intrinsic disorder can be superior immunogens (Uversky et al. 2005). For the same reason, terminal segments of proteins are often correctly predicted to be continuous epitopes because these regions tend to be surface-oriented (Thornton and Sibanda 1983) and are more hydrophilic and mobile than internal regions (Pellequer et al. 1994).

As discussed elsewhere, there are many reasons for the low success rate of epitope predictions (Ponomarenko and Van Regenmortel 2009; Greenbaum et al. 2007; Van Regenmortel and Pellequer 1994). Unsuccessful epitope prediction may partly explain why, after several decades of intensive research efforts, no synthetic peptide vaccine has yet been developed. More than a thousand synthetic peptides have been examined as potential vaccines in numerous preclinical studies. About 125 peptides have progressed to phase I clinical trials and about 30 to phase II trials but not a single peptide vaccine passed phase III trials and is presently marketed for human use (Hans et al. 2006). This striking lack of success leads to the inescapable conclusion that some of our underlying assumptions regarding what constitutes an effective "protective" B-cell epitope must have been incorrect (Van Regenmortel 1999b, 2007). One can only hope that the new bioinformatics tools being developed at present (Ponomarenko and Van Regenmortel 2009) will improve our understanding of the nature of epitopes and allow us in future to develop better immunological intervention strategies.

# Molecular Design Versus Empirical Discovery in Peptide-Based Vaccines: Coming to Terms with Fuzzy Recognition Sites and Ill-Defined Structure-Function Relationships in Immunology

## Introduction

X-ray crystallographic studies of antigen-antibody complexes and peptide/MHC/ TCR interactions have given us considerable insight into the structural basis of immunological recognition (Davies and Cohen 1996; Garcia and Teyton 1998). This has led in some quarters to the belief that it should be possible to apply molecular design strategies to the development of synthetic vaccines. I will argue that this is unlikely and that an empirical approach to vaccine development will continue to be the most profitable strategy.

In recent years it has become fashionable to give the label 'rational' to any biological research that makes extensive use of molecular data. For instance, in research on peptide-based viral vaccines, the common sense decision to focus on known antigenic regions of a viral protein as primary targets for investigation would be hailed as a rational approach (Obeid et al. 1995). In a similar vein, the decision to introduce random amino acid substitutions in a peptide epitope in an attempt to have it mimic better the functional activity of the parent viral antigen would be described as rational design. What the term rational is meant to convey in such contexts is not clear, since other respectable approaches that address the same scientific problems are not derided as nonrational or irrational. In many instances, it seems that the term rational is meant to emphasize that the approach that is used is based on recently acquired scientific knowledge rather than on empirical, trial and error investigations. In a recent review on structure-based drug design, Amzel (1998) epitomizes this point of view when he states: "One of the goals of research in biotechnology is to transform the process of developing a drug from a trial and error empirical operation into a rational, structure-based process." One unfortunate consequence of such

Vaccine, 2000, 18, 216–221
Marc H V Van Regenmortel

© Springer Nature Switzerland AG 2019
Marc H V Van Regenmortel, *HIV/AIDS: Immunochemistry, Reductionism and Vaccine Design*, https://doi.org/10.1007/978-3-030-32459-9_2

statements is that it tends to accredit the view that empiricism and rationality are two separate approaches to problem solving and that the rational approach is to be preferred. That such a view is mistaken is clearly demonstrated by the crucial role which empirical investigations played in the development of science over the last three centuries.

## Rational Design Versus Empirical Discovery

A scientific procedure is said to be rational if it is based on reason, logic and scientific theory. A rational design implies that it is possible to predict the outcome of an experiment on the basis of logic and theoretical principles. In contrast, an empirical approach implies that the outcome cannot be predicted from pre-existing knowledge and must be derived from the experimental observations themselves. Opposing rational and empirical approaches in the practice of experimental science is actually misleading since all scientific knowledge is derived from empirical observations gathered in the course of experiments that are necessarily designed and analyzed in a rational manner. Any attempt to glorify rationality and denigrate empiricism in science contradicts the historical fact that modern science blossomed after the seventeenth century because of the primacy of empirical observations over purely rational analysis. Although logic and rationality are needed to set up a research program, it is not deductive thinking but the unpredictable outcome of controlled experimental observations that leads to novel findings and discoveries. Furthermore, as stated by Root-Bernstein (1991, p. 111): "Discovery is the recognition that things don't fit some pattern you have in your head so that you must invent a new pattern." Empirical discovery is thus also a conceptual process involving the invention of new interpretations of unexpected experimental results.

Rational design consists of modifying a molecule in order to improve its biological activity in a predictable manner and it is usually based on the analysis and engineering of a single reactant pair. Although considerable success has been reported with the structure-based design of drugs, optimizing the structural complementarity of a single peptide-monoclonal antibody couple will not necessarily lead to a peptide vaccine of adequate immunogenicity. A successful rational design strategy implies that it is possible to predict how a biologically active molecule can be obtained by chemical synthesis. However, this is feasible only if the structure-activity correlates of the molecule under study are fully known, which is rarely the case in immunology. In fact, it is our inability to modify at will the immunogenic activity of a peptide that plays havoc with our attempts to rationally design synthetic vaccines. To make this clear, it will be helpful to analyze what is meant by structure-function relationships.

# The Fuzzy Nature of Structure-Function Relationships

Most molecular biologists adhere to some form of reductionism according to which the sequence of a protein causally determines its three-dimensional structure and this in turn causes and explains its activity and biological function. This leads to the widely held view that uncovering a structure-function relationship consists of describing and explaining what a molecule does solely in terms of its known structure, i.e. without considering the molecule's integrated interactions with other components of a more complex system. It is a reductionist fallacy to believe that there is a simple causal link between the structure and function of a biomolecule and that it may be possible to find a unique type of connection between them. On the contrary, it is well-established that a single chemical structure or protein fold can have a multiplicity of functions and conversely, that a single function can be generated by a variety of structures or protein folds (Martin et al. 1998).

The mistaken belief that structure and function are causally linked arises from a misunderstanding of the notion of causality itself. Causality is a relation existing between successive events and not between material objects. A cause is an event that is a necessary and sufficient condition for the occurrence of a later event, called the effect. Such a simple pattern of linear causality is of little relevance in biology since biological effects always result from a complex network of interactions. Many variable factors are always found to simultaneously influence the features of a biological system (Sattler 1986). Furthermore, a biological event such as a binding reaction cannot be the inevitable consequence of something that is not an event, i.e. the structure of a molecule. In the case of protein-protein interactions, a binding process is a dynamic event that cannot be described satisfactorily outside of the relational nexus of two partners and without dealing with various contributions from the environment, for instance interstitial water (Van Oss 1995).

The idea that the activity of a protein is simply caused by its structure is contradicted by the fact that no unique amino acid sequence is required to obtain a protein with a specific activity. There is also no unique relation between a drug and its receptor. As stated by Kauvar and Villar (1998): "A drug is not a unique key that can open only one lock. Rather, all drugs show varying degrees of interaction with a plethora of proteins. A successful drug is a compound for which proteins at the top of its preference list contribute to the desired therapeutic effect while those that are detrimental are at the bottom of the list".

In the case of protein-protein interactions, Moodie et al. (1996) introduced the notion of fuzzy recognition motifs to account for the finding that the shape and electrostatic complementarity between two partners can be achieved by not just a single arrangement of specific amino acids but by a large number of alternative arrangements. The idea of fuzziness is useful because it admits of the inherent imprecision of terms and concepts used in both ordinary language and in science (McNeill and Freiberger 1993). The binding sites that give rise to immunological specificity consist of a number of epitopes which have also been called fuzzy (Van Regenmortel 1998) because there is no clear-cut minimum difference in binding affinity or in atomic positions at the antigen-antibody interface that can serve as an absolute yardstick for deciding that two epitopes are the same or not. The advent of biosensor technology has made it possible to detect very small differences in binding affinity and kinetics (Karlsson and Roos 1997) and the task of relating these differences to very small changes in atomic positions at the antigen-antibody interface has turned out to be very difficult (Janin 1997; Rauffer-Bruyère et al. 1997). In the absence of a direct causal link between structure and binding activity, uncovering mere correlations between structure and activity is not easy to achieve (Kelley 1996).

A further ambiguity in the definition of structure-function relationships arises from the term function itself. According to Kitcher (1998) the function of an entity is what it is designed to do. In modern discussions of the subject, only two sources of design are considered: (1) The intention of a cognitive agent, for instance, when a machine or artifact is being designed and (2) the action of natural selection when a useful function is retained during evolution. In the latter case, a functional explanation is given for the observed structure rather than a structural explanation for a conserved function. Evolution is seen to operate on the protein sequence through feedback from its effects (Rosenberg 1985).

It is commonly accepted that a function must confer some advantage to the biological system. The term function then conveys the meaning of a purpose or a role and implies that it has some value for the system as a whole. However, any relationship between function and structure is then even more tenuous and ill-defined.

Another possibility is to equate the function of an entity with what it does, but in this case the terms 'functioning' and 'activity' would seem more appropriate than the term 'function'. It is also possible to view structure and function in an integrated manner and to consider them as a form-function complex (Bock and Von Wahlert 1965). It has been argued that a complete account of antigenic specificity demands the integration of both structural and binding activity data and that this can be achieved only through a spatiotemporal four-dimensional analysis (Van Regenmortel 1996). In order to avoid the complications that arise when teleological considerations are introduced in the analysis (Rosenberg 1985; Nagel 1979), it often seems preferable to speak of structure-activity relationships rather than of structure-function relationships.

# Empirical Discovery Rather than Molecular Design Will Bring Forth Synthetic Vaccines

There are many reasons why the molecular design of synthetic vaccines is not a realistic scientific enterprise:

1. The shape and chemical complementarity of antigen-antibody interfaces is poorer than that of other protein ligand interfaces (Braden and Poljak 1995; Stanfield and Wilson 1994). This is due to the conserved architecture of the antibody-combining site and to the absence of optimization of individual epitope-paratope pairs through evolution.

2. Immune recognition is mediated by multispecific binding sites. A single immunoglobulin molecule harbours many different combining sites or paratopes. each one consisting of only about one third of the CDR residues of the molecule. This leads to many potential antigenic cross-reactivities between related as well as unrelated antigenic determinants or epitopes. As a result, a peptide may react with an antibody directed to the cognate protein without actually binding to exactly the same paratope as the protein. Such a peptide may appear to mimic an antigenic site of the protein but could be unable to elicit antibodies that recognize the protein (Van Regenmortel 1998).

3. Whereas the lock-and-key model has been useful for describing the molecular docking of drugs (Gschwend et al. 1996), it fails in the case of antigen-antibody docking (Carneiro and Rethinking 1994), because the induced fit (Rini et al. 1992), and solvent effects (Bhat et al. 1994) that occur during epitope-paratope interactions cannot be modeled satisfactorily (Doniach 1997).

4. It is not possible to translate the atomic coordinates of an antigen-antibody complex into interaction binding energy (Janin 1997). Although higher equilibrium affinity constants or slower dissociation rate constants of antibodies could be useful for increasing the capacity of antibodies to neutralize infectivity (Van Cott et al. 1994), such constants cannot be engineered at will.

5. Structural and functional approaches to the study of protein antigenicity lead to different perceptions of the nature of epitopes (Lambert and Hughes 1988; Van Regenmortel 1989a). Structural epitopes defined in terms of contact residues usually comprise about 10–15 amino acid residues whereas energetic epitopes defined in terms of their binding energy are usually limited to no more than four residues (Novotny et al. 1989; Jin and Wells 1996). Information on the binding characteristics of peptide epitopes can only be obtained experimentally from affinity measurements and cannot be derived from their structure (Kauvar et al. 1995; Van Regenmortel 1995).

6. Epitopes and paratopes are relational entities defined by their partners and not by intrinsic features that are identifiable independently of this relationship (Van Regenmortel 1995). A single atomic substitution in an epitope necessarily leads to an altered complementary paratope by virtue of its relational definition. Molecular design may succeed in optimizing one epitope recognized by a single monoclonal antibody but cannot be applied to all the relevant epitopes involved in the immune response to a pathogen.

7. Antigenic sites and antibody combining sites are fuzzy recognition sites, each one consisting of several individual epitopes and paratopes respectively. Fuzzy sites are not amenable to synthesis by a molecular design strategy which, in principle, is applicable only to a single pair of interacting molecules.

8. Structural analysis of an immune complex involves a single epitope-paratope pair. However, neutralization antigenic sites usually consist of several overlapping or non-overlapping epitopes that are functionally active in a collective manner. Cooperativity effects in infectivity neutralization by antibodies (Dimmock 1993) cannot be dissected at the molecular level in terms of individual structures.

9. Infectivity neutralization by antibody involves a ternary interaction between the pathogen, the antibody and the host cell. Mechanisms of neutralization and of escape from neutralization are poorly understood (Dimmock 1993). None of the existing successful vaccines has had its mechanism of action fully elucidated.

10. Modifying binding sites by mutagenesis does not amount to a molecular design strategy. The effects of site-directed mutagenesis on binding activity are not predictable, partly because of nonadditivity (cooperativity) between individual mutations (Rauffer-Bruyère et al. 1997).

    Residue substitutions tend to produce small structural shifts that cannot be detected at the current resolution of structural data although they affect binding activity (Sturtevant 1994).

11. The recognition between epitopes and paratopes does not take place at the level of whole residues but at the level of individual atoms. Peptides showing little sequence similarity may cross-react with the same antibody because only a minority of the atoms of any residue actually participate in the interaction. Sequence information derived from the pathogen is also of little use for the design of peptides that mimic discontinuous epitopes (Kaumaya et al. 1994). Mimotopes possessing the required antigenic and immunogenic activity cannot be designed and must be identified by screening peptide libraries (Delmastro et al. 1997; Steward et al. 1995).

12. There are no general rules for designing peptide immunogens that elicit neutralizing rather than nonneutralizing antibody. For instance, it is customary to block the unnatural charges present at the termini of peptides which are absent in the cognate protein, but this does not guarantee that such blocked peptides will be able to elicit neutralizing antibodies. In the case of retro-all-D peptide analogues of an immunodominant epitope of foot-and-mouth disease virus, it was found that only some of the methods used for removing the charges of the peptides were successful and led to peptides able to induce neutralizing antibodies (Briand et al. 1997). Clearly, the optimal solution to this end-group problem has to be determined for each peptide by trial and error (Van Regenmortel and Muller 1999). When immunodominant epitopes of influenza virus and of poliovirus were cyclized in order to improve the level of conformational mimicry between peptide and intact protein, it was found that conjugation of the peptides to a carrier affected the induction of neutralizing antibodies in an unpredictable manner (Friede et al. 1994; Muller et al. 1990; Van der Werf et al. 1994). It is a sobering fact that the best results obtained with synthetic peptide vaccines so far concern epitopes located in disordered regions of viral proteins which cannot be designed to have the putative correct conformation (Brown 1994; Langeveld et al. 1994).

13. The optimal size of a peptide and the need to incorporate helper T cell and cytotoxic T cell epitopes in the construction of a synthetic peptide vaccine can only be established empirically (Partidos and Steward 1992).

14. MHC polymorphism dictates the use of several peptide epitopes of the pathogen for effective memory response (Berzofsky et al. 1991). Clinical trials are needed to assess the efficacy of each particular combination of peptides (Ahlers et al. 1996).

15. The best adjuvant and delivery route for each vaccine cannot be predicted and must be established by trial and error (Berzofsky and Berkover 1995; Obeid et al. 1996).

## Immunogenicity Is Not Amenable to Molecular Design

The ability of antigens to react specifically with complementary antibodies is known as antigenic reactivity or antigenicity, and it is at least plausible that a molecular design approach could improve the antigenicity of an individual peptide by optimizing the degree of steric complementarity between the peptide and a single monoclonal antibody. The situation is quite different with immunogenicity, which is the ability of an antigen to induce an immune response in a competent host. In addition to being dependent on intrinsic properties of the peptide, the immunogenic capacity of a peptide depends also on the potentialities of the host being immunized, i.e. on extrinsic factors such as the immunoglobulin gene repertoire, self-tolerance and a variety of cellular and regulatory mechanisms that have no meaning outside the host context (Berzofsky 1985). A molecular design approach is simply not adapted to the task of optimizing these various biological and cellular interactions in order to achieve a strong protective immune response. It has been shown, for instance, that the sequential order of T and B cell epitopes in a synthetic peptide construct can significantly alter the immunogenicity of the peptide without altering its antigenicity (Denton et al. 1994).

It must also be stressed that although immunization with peptides readily induces a strong antipeptide response (Friede et al. 1993), this is of little relevance for vaccine design (Van Regenmortel 1989a). What is required is the induction of antibodies that recognize the cognate protein and neutralize the infectious agent in vivo. Furthermore, the induction of neutralizing antibodies on its own is also not a sufficient criterion to gauge the effectiveness of a synthetic vaccine since titres of neutralizing antibody in the serum of vaccinated subjects often do not correlate with protection against infection (Obeid and Steward 1994).

## Conclusion

The preceding discussion amply demonstrates that peptide-based vaccines cannot be developed using a molecular design strategy. The capacity of a peptide to induce a protective immune response depends on many extrinsic factors and regulatory mechanisms in the recipient host which cannot be controlled following molecular design principles. Since the immunogenic effectiveness of peptides cannot be predicted, the inescapable conclusion is that the development of peptide-based vaccines will continue to be driven by empirical discovery rather than by so-called rational design.

Since all scientific knowledge and applications are ultimately derived from empirical observations, such a conclusion should not detract from the renewed interest in the potential of synthetic vaccines (Briand et al. 1997; Meloen et al. 1995).

# Synthetic Peptide Vaccines and the Search for Neutralization B Cell Epitopes

## Introduction

The B cell epitopes of proteins are the regions that are recognized by the binding sites or paratopes of antibody molecules, when these are present either in their free form in serum or as membrane-bound B cell receptors. They are called B cell epitopes to distinguish them from the T cell epitopes of proteins which are proteolytically cleaved peptides of the antigen that interact with the receptors of T cells. The present review will discuss only B cell epitopes and they will be referred to simply as epitopes.

In the context of vaccines, an important category of epitopes are the so-called neutralization B cell epitopes (sometimes mistakenly called neutralizing epitopes) that are able to elicit the formation of antibodies that neutralize the infectivity of pathogens. Such antibodies are known as neutralizing antibodies and the epitopes they bind to are mostly identified using neutralizing monoclonal antibodies (Mabs).

Attempts to develop synthetic peptide vaccines usually assume that it is possible to synthesize effective vaccine immunogens by making short linear peptides adopt the structures observed when epitopes of pathogens are bound to neutralizing antibodies. In theory, peptide-based vaccines could have many advantages compared to conventional vaccines such as increased safety and stability and lower cost (Van Regenmortel and Muller 1999). More than a thousand synthetic peptides have been examined as potential prophylactic vaccines against viral, bacterial and parasitic infections (Arnon 1987; Nicholson 1994; Arnon and Ben Yedidia 2003) and as

The Open Vaccine Journal, 2009, 2, 33–44.
Marc H V Van Regenmortel

© Springer Nature Switzerland AG 2019
Marc H V Van Regenmortel, *HIV/AIDS: Immunochemistry, Reductionism and Vaccine Design*, https://doi.org/10.1007/978-3-030-32459-9_3

therapeutic vaccines for chronic infections and non-infectious diseases as well as cancer (Sundaram et al. 2002). It was reported by Hans et al. (2006) that although 125 peptides had progressed to phase I clinical trials and 30 peptides had undergone phase II trials, not one peptide vaccine had passed phase III trials and is currently marketed for human use. This striking lack of success in developing synthetic peptide-based vaccines suggests that some of the assumptions underlying these efforts were misguided.

In the present review a number of misconceptions prevalent in this research area will be analysed. These include 1) an excessive reliance on so-called continuous epitopes as vaccine candidates, 2) an exaggerated confidence in the specificity of antibodies, 3) the failure to recognize that an operational bias is introduced when monoclonal antibodies are used to characterize epitopes and 4) a tendency to underestimate the difference between antigenicity and immunogenicity.

In view of the recent failure of the Merck HIV-1 vaccine trial aiming at cell-mediated immunity (Watkins et al. 2008) it seems timely to review prospects for peptide vaccines that attempt to induce humoral immune responses.

## The Nature of Protein Epitopes

Epitopes of proteins are usually classified as continuous or discontinuous depending on whether the amino acids that constitute the epitope are contiguous in the peptide chain or not (Atassi and Smith 1978). The vast majority of protein epitopes are discontinuous epitopes made up of atoms from residues located on two to five separate segments of the peptide chain brought together by the folding of the chain. These epitopes arise because the chain acts as a scaffold to bring distant residues together and if the scaffold is perturbed, the epitope ceases to exist (Van Regenmortel 1999a).

Discontinuous epitopes can only be identified by crystallography on the basis of atomic contacts with a paratope and not by showing experimentally that a set of non-contiguous residues in a protein, when positioned correctly in space, possess binding activity.

The situation is completely different in the case of continuous epitopes of proteins since these epitopes are identified by showing experimentally that short linear peptide fragments of the protein are able to bind to antibodies raised against the protein. A continuous epitope is identified only by the binding activity of a peptide and not by showing that all the residues in this peptide interact with antiprotein antibodies.

It is unfortunate that when one talks of the continuous epitopes of a protein, the impression is created that these epitopes exist as such in the native protein. In reality the label "continuous epitope" is given to any linear peptide that is able to react, usually only weakly, with antibodies directed to more complex discontinuous epitopes. Such peptides are not faithful copies of epitopes present in native protein molecules, one reason being that they do not retain the conformation present in the corresponding part of the folded protein; in most cases they possess only limited structural similarity with portions of the protein surface. It is not unusual for a short linear peptide such as a tripeptide or a pentapeptide to be called a continuous epitope, even if its binding activity is increased considerably when flanking residues are added to it. In the absence of structural information about which residues of the peptide are in contact with the antibody, is not clear if the longer peptides are more active because they possess a more appropriate conformation or because the added residues actually interact with the antibody (Van Regenmortel 1999a). As a result, continuous epitopes always have ill-defined boundaries since they are defined functionally rather than structurally. Functional epitopes are usually delineated by establishing which residue replacements in a protein or peptide affect its capacity to bind to a Mab. Such an approach not only identifies epitope residues that are in contact with the paratope but also residues that affect the epitope activity indirectly by altering the conformation of the peptide chain. Structural and functional approaches to epitope characterization therefore lead to different perceptions of the nature of epitopes (Van Regenmortel 1989a).

Although large numbers of poorly characterized continuous epitopes are listed in many databases such as the Immune Epitope Database (Zhang et al. 2008), the vast majority of them are very poor mimics of the actual epitopes present in native proteins. They become known as epitopes because the relationship between an epitope and its paratope is never of an exclusive nature and because antibodies are always able to cross-react with a wide variety of epitopes that may possess only a limited degree of structural similarity (Roberts et al. 1993). Another reason for the large number of reported continuous epitopes is that many of them correspond to unfolded regions of denatured protein molecules that are not antigenically active in native proteins (Laver et al. 1990). Antiprotein sera used for detecting continuous epitopes often contain antibodies specific for denatured proteins because some of the protein molecules used for immunization became denatured before of after being injected in the animal (Van Regenmortel 2006). Such antibodies usually do not react with the native protein but are able to bind to various linear fragments of the protein.

Since the vast majority of continuous epitopes reported in the literature do not correspond to the actual epitopes present in native proteins, it is not astonishing that continuous epitopes have not been successful as candidate peptide vaccines (Hans et al. 2006; Van Regenmortel 1999b).

The inability of linear peptides to effectively mimic the discontinuous epitopes of proteins should come as no surprise since this has been the finding in hundreds of immunochemical analyses of protein antigens over the last two decades (Van Regenmortel 1995; Van Regenmortel and Muller 1999). It is unfortunate that so many investigators still rely on continuous epitopes for developing synthetic vaccines against pathogens since there is little experimental evidence that such a strategy has ever been effective. In recent years, crystallographic evidence has also been obtained showing that short linear peptides cannot be effective structural mimics of discontinuous epitopes (Saphire et al. 2007).

## Neotopes and Mimotopes

Two additional types of epitopes are important in the context of vaccines. Neotopes are epitopes specific for the quaternary structure of virus particles that arise only after the assembly of coat protein subunits into capsids. The term neotope was coined in 1966 (Van Regenmortel 1966) to describe epitopes that result either from the conformational changes in protein subunits induced by intersubunit interactions or from the juxtaposition of residues of neighbouring subunits that are recognized by an antibody as a single epitope (Van Regenmortel 1992a). Neotopes have been shown to be present in the capsids and membrane proteins of many viruses (Neurath and Rubin 1971) and since the quaternary structure of virions can undergo major rearrangements following small changes in the environment (Bothner et al. 1998), neotopes are often transient epitopes (Zolla-Pazner 2004) that can assume several conformations and are present for only short periods of time. The trimeric form of the envelope proteins of HIV-1 possesses transient neotopes, absent in the monomeric form, that are able to induce neutralizing antibodies (Zwick and Burton 2007).

The term mimotope was coined by Geysen et al. (1986) to refer to a peptide that is able to bind to a particular antibody but shows little or no sequence similarity with the protein antigen used to induce the antibody, usually because the antibody is directed to a discontinuous epitope. Mimotopes are mostly identified by testing combinatorial peptide libraries obtained by chemical synthesis or phage display and selecting peptides that bind to antiprotein antibodies. Even if the mimotope shows no sequence similarity whatsoever with the protein immunogen, it may sometimes be able to induce antibodies that cross-react with that protein (Delmastro et al. 1997). The capacity of mimotopes to cross-react with antibodies to continuous and discontinuous epitopes of proteins demonstrates that epitope-paratope recognition does not occur at the level of whole amino acid residues but is mediated through individual atomic interactions that may occur through main chain atoms rather than side chain atoms of residues. The phenomenon of hydropathic complementarity also explains why peptide sequences that appear to have little in common are able to bind to the same antibody. Hydropathic complementarity arises from an inverted hydropathic pattern in two peptide sequences and is due to the attraction between hydrophilic and hydrophobic groups (Tropsha et al. 1992; Van Oss 1995; Boquet et al. 1995). Peptide analogs that show little or no sequence similarity but retain the original hydropathic profile of the original peptide may be able to react with the same antibody.

## Antibody Multispecificity

The existence of mimotopes illustrates the ability of antibodies to react with a wide range of antigenic structures possessing little or no sequence similarity. The potential binding pocket of an immunoglobulin molecule comprises 50–70 hypervariable residues distributed over the six complementarity determining regions (CDR). However, an individual paratope consists mostly of only 10–20 CDR residues, which means that about two thirds of the CDR residues may be able to bind additional epitopes that bear little or no resemblance to a first bound epitope. A single immunoglobulin will thus always harbour a number of partly overlapping or nonoverlapping paratopes although, after binding to one epitope, it may not be able, because of steric hindrance, to accommodate a second epitope at a nearby location. This means that when an epitope is labelled a mimotope of epitope A because of its capacity either to bind to an anti-A antibody or to inhibit the binding of epitope A to the antibody, it cannot be excluded that the putative mimotope actually binds to a different paratope from the one that interacts with epitope A. This is the reason why the mimotope nature of a peptide can only be established by showing that it is also able to elicit antibodies that cross-react with epitope A, i.e. with the epitope being mimicked (Van Regenmortel 2001a; Meloen et al. 2000; Larralde et al. 2007).

The ability of antibodies to react with a large number of different epitopes that share only a limited degree of sequence or structural similarity may seem to contradict the accepted view that antigen-antibody interactions are very specific. In reality, antibody molecules are always able to react with many related antigens as evidenced by the large numbers of reported continuous epitopes and mimotopes.

Antibodies are often said to be specific for the particular antigen that was used for eliciting them in the immunized host. However, such a formulation can lead to considerable confusion. Proteins always harbour many different epitopes and each of them is able to elicit its own set of specific Mabs. When two viral strains are compared using a panel of Mabs, the strains will be indistinguishable if a Mab is used that recognizes an identical epitope that exists in both strains.

On the other hand, if a Mab is used that recognizes an epitope present in only one of the strains, the two viruses will appear to be unrelated (Van Regenmortel 1998). If an investigator wishes to differentiate between the two strains, he will call the first type of Mab non-specific since it reacts with both viruses while the second Mab will be called specific because it discriminates between the two strains. Instead of speaking of specificity, it is therefore preferable to speak of the discrimination potential of antibodies since this underlines the fact that it is the wish of the investigator to distinguish between two antigens that determines which antibody will be considered specific (Van Regenmortel 1998). In fact, the paratopes of antibodies can only be said to be specific for their complementary epitopes and it makes little sense to say that an antibody is specific for a multi-epitopic antigen.

Another factor that contributes to the extensive crossreactive potential of antibodies is the ability of the CDRs to adopt various conformations in their free states and when bound to different antigens. This flexibility of the CDR loops in fact increases the effective size of the antibody repertoire (James et al. 2003). Furthermore, since all protein antigens exist as dynamic distributions of different conformers, this also increases the range of antibodies they can recognize and therefore also increases the number of cross-reactions that can occur (Demchenko 2001; Ma et al. 2002).

When the antigenicity of a protein is described in terms of the many peptides that are able to cross-react with antibodies raised against the protein, this does not imply that all the residues in these peptides correspond to residues of the protein immunogen that are able to react with antibodies. Peptide fragments and the corresponding regions in the intact protein usually differ considerably in conformation and the many analogs of continuous epitopes that are recognized by a single Mab further demonstrates the considerable multispecificity of antibodies (Getzoff et al. 1988).

Antibody specificity is often believed to be correlated with high affinity since it is expected that highly specific antibodies will possess a better stereochemical complementarity with their antigens than antibodies of low affinity. However, when the discrimination potential of antibodies is considered more relevant than their specificity, one often finds that antibodies of low affinity are able to discriminate better between two antigens than antibodies of high affinity.

This is due to the fact that low affinity antibodies usually will detect fewer cross-reactions than antibodies of high affinity since weaker cross-reactions will tend more quickly to be below the level where they can be detected with low affinity antibodies (Van Regenmortel 1998).

## The Operational Bias of Monoclonal Antibodies

It is now widely accepted that the entire accessible surface of a protein antigen contains a very large number of overlapping epitopes (Benjamin et al. 1984). The same residues at the protein surface can be part of neighbouring epitopes recognized by different antibodies and no sharp boundaries exist between these different epitopes which together form an antigenic continuum. The situation is somewhat analogous to the fuzzy boundaries between the colours in a rainbow which do not prevent us from distinguishing colours conceptually (Fig. 1), in spite of the continuous nature of the spectrum of electromagnetic waves (Van Regenmortel 1998).

It is only because antigenic sites of proteins are defined with Mabs that antigenicity appears to be located in discrete regions rather than in an antigenic continuum. Since in most cases only one or a very small number of Mabs have been used to characterize the epitopes of a protein by crystallography, it is difficult to assess if epitope regions differ from the remaining antigen surface that harbours additional uncharacterized epitopes (Rubinstein et al. 2008). In the case of lysozyme, five epitopes have been located by means of their respective Mabs and they were found to cover about two thirds of the lysozyme surface. When compared to the epitope areas, the nonepitopic surface was somewhat less accessible to solvent and contained fewer charged, polar and aromatic residues (Rubinstein et al. 2008). When the immune response to viral antigens is dissected with neutralizing Mabs, artificial boundaries are also created in what may be a functional continuum of several neighbouring neutralizing epitopes. The use of Mabs leads investigators to focus on single epitopes as elicitors of neutralizing antibodies instead of analyzing the protective immune response to a pathogen in terms of the collective neutralizing activities of

**Fig. 1** A sculpture called "Rainbow Crash" by artist Federica Marangoni in the Chianti sculpture park, near Siena in Tuscany. Broken fragments of the rainbow appear as coloured pebbles which give the impression that there are clear boundaries between portions of the rainbow. In an analogous way, epitopes delineated with Mabs may hide the fact that the protein surface is an antigenic continuum. Epitope dissection with Mabs leads investigators to study immune responses elicited by single epitopes instead of analyzing the neutralizing activity observed in polyclonal immune responses

antibodies directed to several epitopes. It is well-known that a normal polyclonal protective immune response is particularly effective because different neutralizing antibodies act in synergy and similar synergistic effects are observed when mixtures of neutralizing Mabs are used (Tilley et al. 1992; Laal et al. 1994; Mascola et al. 1997; Zeder-Lutz et al. 2001).

Different mechanisms have been suggested to explain the observed synergy between different neutralizing antibodies (Zwick et al. 2001b), one of them being that the binding of a first antibody to a viral protein induces conformational changes that expose new epitopes leading to enhanced binding by other antibodies. As a result, combinations of different neutralizing antibodies achieve a higher degree of neutralization than expected from the additive effect of each antibody taken individually.

Dissecting protective immune responses with Mabs leads investigators to concentrate on a single epitope that induces neutralizing antibodies instead of investigating which combinations of epitopes and immunogens would be most effective for eliciting a synergistic protective effect. In this respect, it is unfortunate that few investigators have analyzed in detail the immune response induced by whole HIV particles inactivated by heat treatment or various chemical treatments (Poon et al. 2005a, b). Studies using inactivated viruses or virus-like particles (Crooks et al. 2007) endowed with an increased expression of oligomeric envelope proteins in a particular conformation could help to unravel which combinations of immunogenic epitopes are able to mediate a more potent neutralization than is obtained with purified recombinant envelope proteins or single peptide epitopes (Crooks et al. 2007).

The epitope nature of a set of amino acids can only be revealed when an immunoglobulin that binds to it has been found. Similarly, the antibody nature of an immunoglobulin becomes apparent only when a complementary epitope to its paratopes has been identified. Epitopes and paratopes are relational entities defined by their mutual complementarity and they depend on each other to acquire a recognizable identity (Van Regenmortel 1998, 1999a). An epitope is thus not an intrinsic structural feature of a protein that could be recognized in the absence of a particular interaction with a paratope. Epitopes acquire their identity by virtue of a relational nexus with complementary paratopes and this relational dependence means that analyzing the antigenicity of a protein amounts to analyzing the size of the immunological repertoire of the host immunized with that protein. The number of epitopes present in a protein can be equated with the number of different Mabs that can be raised against it. Using that criterion, the insulin molecule was found to possess at least 115 different epitopes [60] while the BLyS molecule possesses as many as a thousand epitopes (Edwards et al. 2003).

Another bias introduced when Mabs are used to characterize protein epitopes is that the specificity of the antibody is determined by the selection process which was used to obtain the Mab. For instance, if a Mab has been selected using a library of linear peptides hypothesized to mimic parts of a continuous epitope of a virus protein, it would be somewhat of a self-fulfilling prophecy to discover subsequently that such an antibody binds more strongly to peptides than to intact virus particles.

It seems that many investigators do not always fully appreciate the bias that is introduced by the selection process used to obtain Mabs, since in many publications describing the binding properties of Mabs, no mention is made of the antigen that was actually used in the selection process. In the case of human Mabs 2F5, 4E10 and Z13 that recognize the gp41 MPER of HIV-1, various peptides were used for selecting the antibodies and some short peptides became known as the core epitopes recognized by these Mabs (Zolla-Pazner 2004). Considerable work was then undertaken to try to convert these peptides into immunogens capable of eliciting antibodies with the same neutralizing capacity as the original Mabs (Montero et al. 2008; Zwick et al. 2005; Brunel et al. 2006; Nelson et al. 2007). As discussed above, these misguided attempts failed because these so-called epitopes were not recognized for what they actually were: short peptides corresponding to small parts of complex discontinuous epitopes.

# The Cross-Protective Immunogenicity of Peptides cannot Be Predicted from their Antigenicity

Many investigators do not fully appreciate the considerable difference between the antigenicity and immunogenicity of peptides and proteins. The antigenicity of peptides is a purely chemical property describing the interactions between epitopes and paratopes in terms of structural and chemical complementarity. In contrast, the immunogenicity of peptides is their ability to give rise to an immune response which is a biological property that has meaning only in the context of a competent host.

When a protein epitope recognizes a free antibody molecule or a B cell receptor embedded in a membrane, the chemical environment is not the same (Zwick 2005) which means that the antigenic epitope bound to a free antibody molecule may not be identical to the immunogenic epitope that interacts with a B cell receptor.

It is commonly found that when a peptide fragment of a protein is able to bind to antibodies raised against the protein, this does not guarantee that the peptide will necessarily be able to elicit antibodies that react with the native protein. When the peptide reacts in an immunoassay with an antibody directed to a native protein, the antibody may be able to select one conformation of the peptide or it may induce a reactive conformation in the peptide by an induced fit or mutual adaptation process, the result in both cases being the occurrence of a cross-reaction between the peptide and the antiprotein antibody. In contrast, during the immunization process, when the same peptide is confronted with a variety of B cell receptors, different conformations of the peptide may be recognized by separate B cell receptors. However, there is no reason why the peptide would bind preferentially to those rare receptors which in addition to recognizing one conformation of the peptide also cross-react with a related epitope present in the native protein. It is thus to be expected that most elicited antipeptide antibodies will not react with the native, cognate protein, unless the peptide used for immunization was constrained and mimicked exactly the conformation of the corresponding region in the native protein immunogen.

Attempts to determine the conformation of epitopes present in immunogens are mostly based on X-ray crystallography or NMR studies of antigen-antibody complexes. However this approach is bedevilled by the fact that the structures visualized in the complexes may be different from the structures of the binding sites in free antigen and antibody molecules before the process of mutual adaptation that occurs when the two partners interact (Rini et al. 1992; Churchill et al. 1994; Wilson and Stanfield 1994; Berger et al. 1999; Bosshard 2001; Goh et al. 2004; Stanfield et al. 2006). In spite of this limitation, many groups continue to use this approach and are not deterred by the fact that structural data obtained from complexes with neutralizing Mabs will not necessarily reveal the epitope structure recognized by B cell receptors during the immunization process and which is therefore likely to be required for eliciting neutralizing antibodies. Since it is known that all antibodies including Mabs are multispecific, it follows that the epitope structure observed in a complex is only one of the many antigenic binding sites that could be accommodated by the antibody. The epitope structure observed in a complex with a neutralizing

Mab is thus not necessarily the one that corresponds to the immunogenic epitope which the investigator is trying to elucidate.

Most peptides are immunogenic in the sense that they readily elicit antibodies that react with the peptide immunogen (Muller 1999a). However, this type of immunogenicity is irrelevant for vaccination purposes since what is required is both cross-reactive and cross-protective immunogenicity, i.e. the induction of antibodies that recognize the parent protein and neutralize the infectivity of the pathogen harbouring the antigen (Van Regenmortel 2006). Cross-reactive immunogenicity can often be obtained by increasing the conformational similarity between peptide and intact protein for instance by cyclization of the peptide but such an approach is rarely successful for achieving crossprotective immunogenicity (Van Regenmortel 1999b). It is sometimes possible to obtain a more suitable epitope conformation by inserting a peptide at certain locations in a recombinant protein (Benito and Van Regenmortel 1998; Chakraborty et al. 2006; Rybicki 2009). However, in the absence of information regarding which precise conformation is required and is actually present in the recombinant construct, such a strategy remains entirely empirical.

The difficulties encountered when one tries to utilize a suitable peptide conformation for eliciting the formation of neutralizing antibodies are illustrated by the V3 peptide of HIV-1. This semiconserved and rather flexible loop of about 35 residues, which determines which co-receptor (CCR5 or CXCR4) is used by the virus to gain entry into cells, is recognized by many HIV-1 neutralizing antibodies present in infected inviduals. It is known that the V3 loop is able to induce anti HIV-1 neutralizing antibodies (Zolla-Pazner 2004; Gorny et al. 1993, 2004) but the precise V3 conformation that should be present in a V3 synthetic peptide vaccine has not been established. NMR studies of linear peptides complexed with Mabs specific for the conformation of the V3 loop present in intact virus and in gp120 protein showed that the conformation of the antibodybound V3 peptides was dictated by a process of induced fit to each Mab (Rosen et al. 2005). Alternative β-hairpin conformations could be induced in the same V3 peptide depending on whether it binds to a Mab neutralizing a broad spectrum of virus isolates or to a Mab able to neutralize only a single type CXCR4 virus. Such findings once again demonstrate that the epitope conformation observed in Mab-peptide complexes is actually induced by the binding process and that structural analysis of complexes does not necessarily indicate which immunogen conformation is required for eliciting neutralizing antibodies.

## Epitope Prediction and the Synthetic Reconstruction of Discontinuous Epitopes

The main objective of epitope prediction is to be able to replace a complete protein antigen by a small fragment of the molecule corresponding to a single epitope. Ideally such a fragment can subsequently be synthesized as a peptide and used either to detect specific antibodies for immunodiagnostic purposes or to elicit antipeptide antibodies for antigen detection or as a potential synthetic vaccine (Van Regenmortel and Muller 1999; Leinikki et al. 1993; Gomara and Haro 2007). Most attempts to predict epitopes have analyzed protein sequences in an effort to identify short surface-exposed regions of 5–10 residues that are particularly hydrophilic and accessible to the solvent and therefore likely to be recognized by antibodies (Pellequer et al. 1991; Ponomarenko and Van Regenmortel 2009). Since sequence-based prediction methods will only identify continuous epitopes that are known to be poor mimics of the epitopes present in native proteins, it is not astonishing that the success rate of these prediction methods rarely exceeds 60% correct predictions (Pellequer et al. 1993; Blythe and Flower 2005; Greenbaum et al. 2007). Furthermore the relevance of predicting such epitopes in the context of vaccine development is doubtful, since relying on continuous epitopes for developing synthetic vaccines has not been an effective strategy.

It has been claimed (Hopp 1993) that epitope prediction methods based on the analysis of protein sequences using amino acid propensity scales also allows the prediction of discontinuous epitopes. However, if the 3D protein structure is unknown, such an approach only provides a list of contiguous and noncontiguous residues that are likely to be accessible at the protein surface, without an indication of how these residues must be positioned in space to make up a discontinuous epitope (Van Regenmortel and Pellequer 1994). If prediction of discontinuous epitopes is limited to predicting that certain surface residues are likely to be part of an epitope, this does not amount to predicting the epitope itself since it is necessary to predict which residues from distant parts of the sequence must be brought together in a precise configuration to form an antigenically active site. Since predicting an epitope means predicting something that has a functional activity, many of the published methods for predicting discontinuous epitopes are actually misnomers.

Several prediction methods that take into account the 3D structures of antigens have been published (Ponomarenko and Van Regenmortel 2009). The CEP (Conformational Epitope Prediction) web server predicts discontinuous epitopes by collapsing predicted continuous epitopes for which the Cα atoms are within a distance of 6 Å (Kulkarni-Kale et al. 2005). The DiscoTope method uses a combination of hydrophilicity parameters, amino acid statistics, numbers of contacts and area of relative solvent accessibility (Haste-Andersen et al. 2006). The MIMOP computational prediction tool (Moreau et al. 2006) identifies key residues from sets of mimotope sequences and matches them with accessible amino acids on the antigen surface. The PEPOP tool (Moreau et al. 2008) uses the 3D coordinates of proteins to predict clusters of surface accessible segments of the peptide chain. The Mapitope algorithm (Bublil et al. 2007) also uses information from sets of mimotopes to identify key residue pairs and then maps these on the known surface of the antigen. The PEPITO predictor uses a combination of amino acid propensity scores and half sphere exposure values at multiple distances and achieves "area under the curve" prediction values of 68.3% (Sweredoski and Baldi 2008). All these methods succeed in predicting a number of residues that are part of known discontinuous epitopes but it remains to be seen whether they can provide information useful for the development of peptide vaccines.

Instead of relying on epitope prediction methods to infer which synthetic peptides are likely to be the best candidates for developing a vaccine, some investigators have used a more empirical approach and synthesized peptides that were constrained in a particular conformation in an effort to make them resemble the conformation of the corresponding region in the parent protein (Hans et al. 2006; Muller 1999a, b). The ability of such peptides to induce antibodies cross-reactive with the cognate protein and endowed with neutralizing activity can then be determined experimentally. Attempts to reconstitute discontinuous epitopes in this manner have so far only been moderately successful (Oomen et al. 2003; Enshell-Seijffers et al. 2003; Villen et al. 2004; Dakappagari et al. 2005; Misumi et al. 2006) although it is undeniable that such experiments provide highly relevant information on the vaccination potential of synthetic peptides that mimic complex discontinuous epitopes. A particularly promising approach is the CLIPS-technology (Chemical Linkage of Peptides onto Scaffolds) which combines the chemical linkage of a linear peptide to a synthetic scaffold with conformational fixation of the peptide (Timmerman et al. 2005, 2007). This method can be used for constraining the conformation of free peptides in solution as well as for obtaining solid phase-attached peptides in microarrays useful for mapping epitopes recognized by neutralizing antibodies. A strategy such as the CLIPS technology is likely to be more effective than approaches that attempt to reconstitute so-called epitope hot spots. Hot spots in binding interfaces are regions that contribute most to the interaction binding energy, since when hot spot residues are mutated, the binding constant tends to be decreased about 100-fold (DeLano 2002; Van Regenmortel 2002a). However, residues identified in this way do not represent the full immunogenic site required for eliciting neutralizing antibodies and reconstructing hot spots by synthesis is unlikely to lead to effective vaccines.

## Conclusion

The search for neutralization epitopes that could be used to develop synthetic peptide vaccines has so far met with little success. Some of the misconceptions discussed in this review together with the dynamic nature of epitopes are probably responsible for this state of affairs although another factor is the common assumption that it should be possible to develop successful peptide vaccines by structure-based rational design (Van Regenmortel 2002a, 2004b, 2007). It is unfortunately the case that we still do not understand why some epitopes are able to elicit neutralizing antibodies while others are not and this prevents us from producing neutralization epitopes by design. This means that only an empirical, trial and error, approach may eventually succeed in identifying which synthetic immunogens are capable of protecting against disease.

Immunizing animals with a variety of well-chosen immunogenic constructs and analyzing the resulting immune responses with respect to neutralizing capacity may well be the strategy that is most likely to succeed for developing an HIV vaccine. It is sometimes claimed that the animal experiments needed for studying immunogenicity are unacceptably expensive.

However, such experiments may in fact be less costly than some of the existing and well funded programs that attempt to elucidate the 3D structures of hundreds of antigen-antibody complexes and it is conceivable that they might provide information that will be highly relevant for vaccine development. In view of our inadequate understanding of structure-function relationships in neutralizing antibody molecules (Van Regenmortel 1995, 2002a, 2004b), the current emphasis on structural analysis of vaccine antigens should be counterbalanced and complemented by more extensive studies of the functional activity of candidate vaccine immunogens and the antibodies they elicit.

# Specificity, Polyspecificity and Heterospecificity of Antibody-Antigen Recognition

## Introduction

Adaptive immune responses involve B cells that recognize native protein antigens and differentiate into antibody-secreting plasma cells and T cells that recognize unfolded peptide fragments of the antigen that have been processed in antigen-presenting cells. These peptide fragments known as T cell epitopes are presented in the groove of surface-exposed major histocompatibility complex class I and II molecules recognized by CD8+ and CD4+ T cells respectively. The pathways of antigen processing in antigen-presenting cells are fairly well understood (Vyas et al. 2008) but fall outside the scope of this review.

The primary antibody repertoire consists mainly of IgM molecules, generated independently of exogenous antigen, which arise in the bone marrow following immunoglobulin (Ig) variable (V) gene rearrangements and imprecise joining at the borders of integrated gene segments.

Following encounter with antigen, B cells are activated and migrate into germinal center follicles where they proliferate (Mc Lennan 1994). Under the influence of antigen selection, a small number of these B cells undergo an Ig heavy chain class-switch to IgG, IgA or IgE molecules and produce the secondary antibody repertoire following a process of somatic hypermutation that introduces point mutations in the V regions at a rate of about $10^{-4}$–$10^{-3}$ per base pair per generation (Rajewsky 1996). Most B cells that fail to be selected by antigen in the germinal centers undergo apoptosis (Tarlinton and Smith 2000).

Somatic hypermutation produces many amino acid substitutions in the complementarity-determining regions (CDRs) of the V genes but fewer ones in the conserved Ig framework (FR) region. This leads to a considerable increase in the

Journal of Molecular Recognition 2014, 27, 627–639.
Marc H V Van Regenmortel

© Springer Nature Switzerland AG 2019
Marc H V Van Regenmortel, *HIV/AIDS: Immunochemistry, Reductionism and Vaccine Design*, https://doi.org/10.1007/978-3-030-32459-9_4

antibody (Ab) affinity for the antigen. The mechanisms by which B cells with higher affinity receptors (BCRs) are selected are not entirely clear. Progeny cells with higher affinity BCRs may be selected because they bind preferentially to the immunogen as its concentration decreases during antigen clearance. Another possibility is that unprocessed antigen molecules presented by follicular dendritic cells, perhaps via antibodies and Fc receptors picked up from the serum, are able to stimulate memory B cells for long periods of time (Wabl et al. 1999).

Following further proliferation and antigen selection, B cells with increased receptor affinity differentiate into either Ab-producing plasma cells or memory B cells, a fate that is controlled by T cell help and the IL-21 cytokine (Zotos and Tarlington 2012).

The binding capacity of an Ab molecule resides in an antigen-binding cleft of 50–70 residues that is located in the N-terminal regions of the Ig heavy (H) and the light (L) chains. Each cleft harbours several overlapping binding subsites of 15–20 amino acid residues that are called paratopes. These subsites possess a particular structural and chemical complementarity to certain patches of residues present at the surface of protein antigens, known as B cell epitopes. If the context makes it clear that one refers to epitopes recognized by antibodies rather than by T cells, one simply calls them epitopes.

Because an Ab always harbours several individual paratopes able to bind different epitopes present in one or other antigen, an Ab is never monospecific for a single binding partner because this would require that the remaining 50 or so residues in the antigen-binding cleft that are not involved in the one interaction are unable to bind any other antigenic structure, which is unlikely. Steric hindrance may prevent two antigen molecules from binding simultaneously to the same Ab-binding cleft, but if the two paratope subsites do not overlap, it is possible for two small antigens to bind simultaneously to the same Ab molecule (Bhattacharjee and Glaudemans 1978).

## Antibody Specificity

The concept of antibody specificity is widely used in immunology although it has rarely been defined satisfactorily. The term specificity is derived from the word species and "specific" properties were initially considered to be properties that allow the members of one species to be distinguished from those of another. This terminology goes back to the days when individual species were believed to be separated by permanent, sharp boundaries and Darwinian evolution and the transformation of species by selection had not yet become universally accepted. As recounted by Mayr (1982), many biologists in the nineteenth century continued to view species as fixed entities separated by clear-cut discontinuities that gave rise to a belief in absolute immunological specificity. Paul Ehrlich, for instance, believed that antisera raised against members of different species of pathogenic bacteria were completely specific, allowing the members of different species to be distinguished serologically with absolute certainty (Mazumdar 1995). When it was later discovered that antisera raised against cells from different animal species cross-reacted serologically with cells from many other species, it became clear that an antigen was able to elicit not only a single antibody of absolute specificity but also a whole spectrum of antibodies that could cross-react with many related antigens (Landsteiner 1962). The specificity of antigen-antibody recognition was then no longer perceived as an all or none phenomenon but was interpreted as a matter of more or less good fit between molecules that possessed different degrees of stereochemical complementarity (Silverstein 1982).

The most reliable method for identifying epitopes and paratopes is by solving the 3D structure of antigen-antibody complexes and determining which amino acids in the two partners make contact with each other (Sundberg and Mariuzza 2002). The surfaces of proteins always harbour many different epitopes and each of them is able to recognize complementary antibodies. When two antigens are compared with a panel of monoclonal antibodies (Mabs) raised against them, they will appear to be identical if a Mab is used that recognizes an identical epitope present in both antigens (Fig. 1). Such a cross-reaction between two antigens is usually referred to as shared reactivity (Berzofsky and Schechter 1981). On the other hand, if a Mab is used that recognizes an epitope present in only one of the antigens, both antigen molecules will appear to be unrelated. If an investigator wants to differentiate between the two antigens, the first type of Mab would be called non-specific whereas the second Mab would be called specific because it discriminates between the two antigens. Instead of speaking of specificity, it may thus be preferable to speak of the discrimination potential of antibodies because it is the wish of the investigator to distinguish between two antigens that determines whether an Ab is considered to be specific.

A third possibility referred to as true cross-reactivity (Berzofsky and Schechter 1981) occurs when an Ab recognizes an epitope that is only structurally related by

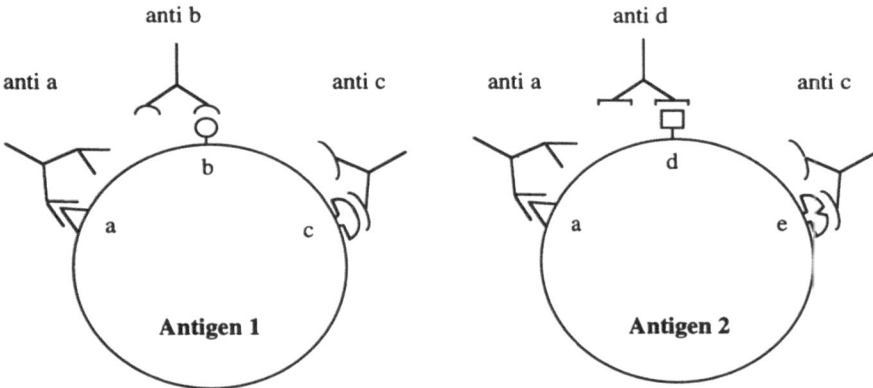

**Fig. 1** Potential cross-reactivity between two related antigens 1 and 2. Epitope a is present on both antigens, and Mab anti-a will react in an identical fashion with antigens 1 and 2. Mabs anti-b and anti-d recognize unrelated epitopes on the two antigens and do not cross-react. They are specific for either Ag 1 or Ag 2. Mab anti-c reacts strongly with the homologous epitope c and cross-reacts weakly with the heterologous epitope (Van Regenmortel 1998)

not identical in both antigens (Fig. 1). In many cases but not always, the Ab will react with higher affinity with the homologous epitope used for raising the Ab than with the cross-reactive heterologous epitope. It is unfortunate that Abs are often given names derived from the antigen they are able to react with, because they cannot be specific for an entire protein antigen that harbours many different epitopes but only for one of its epitopes.

It is often assumed that Abs that possess a high binding affinity are likely to be also more specific because it is believed that they should possess a better stereo-chemical complementarity with their antigens than Abs of lower affinity. However, there is no necessary link between affinity and specificity because Abs of low affinity may be able to discriminate better between two antigens than Abs of high affinity. This is because these Abs may detect fewer cross-reactions between antigens than Abs of high affinity because weaker cross-reactions are more likely to occur below the level where they can still be detected with low affinity Abs (Van Regenmortel 1998).

Whereas Ab affinity is defined by a binary relationship between one epitope and one paratope, Ab specificity is a ternary relational property that has meaning only with respect to a minimum of three partners, for instance one paratope and two epitopes. Specificity is only meaningful with respect to the capacity of an Ab to react differently with two or more antigens and thereby to discriminate between them (Van Regenmortel 1998).

Which selection pressure gave rise during evolution to the huge primary reper-toire of specific paratopes (Marchalonis et al. 2002) has been debated for many years, the most common explanation being that the immune system needed to remove either exogenous pathogens or antibodies to self antigens (Stewart 1992; Cohn 1997; Langman 2000; Efroni and Cohen 2002; Gonzalez et al. 2011). How-ever, this somewhat theoretical debate falls outside the scope of this review.

## The Structural Basis of Paratope-Epitope Recognition

Epitopes and paratopes are relational entities defined by their mutual complemen-tarity and they depend on each other to acquire a recognizable identity. Epitopes and paratopes are thus not intrinsic structural features of an antigen and Ab molecule respectively because they cannot exist in the absence of a relational nexus between the two partners. This means that the number of epitopes in a protein is equal to the number of different Mabs that can be raised against it. In this way, the insulin molecule can be said to have 115 epitopes (Schroer et al. 1983) and the BLyS molecule more than a thousand (Edwards et al. 2003).

This relational dependence means that as soon as an epitope is slightly altered and binding to the paratope is affected, the paratope is also no longer the same. This analysis differs from the classification of binding sites introduced by Cohn (2005) which defines a paratopic clan as a family of paratopes, distinguishable from each other, that are all functionally capable of binding a given single epitope. Cohn also defined a mimotopic array as a set of distinguishable epitopes that are all able to bind to a given single paratope.

Protein epitopes are usually classified as continuous or discontinuous, depending on whether the amino acids that form the epitope are contiguous in the peptide chain or not. The majority of protein epitopes is discontinuous and consists of two to five short stretches of residues that are distant in the protein sequence and are brought together at the surface of the protein by the folding of the peptide chain (Sivalingam and Sheperd 2012).

Discontinuous epitopes are defined structurally by the amino acids that are found to be in contact with paratope residues in a crystallographic complex. However, discontinuous epitopes cannot be extracted from the protein antigen to demonstrate that they possess binding activity outside the context of the native protein and it is also extremely difficult to reconstitute them in active form with sufficient 3D precision by peptide synthesis (Timmerman et al. 2007; Van Regenmortel 2009a).

Continuous epitopes, on the other hand, always have fuzzy boundaries because they are identified functionally by the ability of short peptide fragments of the protein to bind to an antibody and not by establishing that all the residues in the epitope interact with a paratope. In most cases all the residues in a continuous epitope do not interact with a paratope and only some of its residues will be located at the surface of a native protein where they are usually part of a more complex discontinuous epitope (Van Regenmortel 2009a; Chen et al. 2009).

Antibodies recognize antigens through six CDRs that because of their enormous sequence variability are able to form millions of different antigen-binding paratopes. These hypervariable loops are denoted L1, L2 and L3 in the Ig light chain and H1, H2 and H3 in the Ig heavy chain, and they associate non-covalently at the tip of Ig Fab fragments to form the antigen-binding site.

Because most protein epitopes are discontinuous and usually involve numerous residues situated on different accessible loops of the protein, their structure is somewhat analogous to the discontinuous structure of paratopes built up from residues located on the different Ab hypervariable loops. Furthermore, because peptides of 6–12 residues corresponding to the CDRs of an Ab are sometimes able on their own to bind to the cognate intact antigen, they could be viewed as continuous paratopes, reminiscent of the binding capacity of continuous epitopes (Van Regenmortel 1998). Such continuous paratopes that are able to recognize the antigen in a specific manner, albeit with a lower affinity, may contain residues found by crystallographic analysis of the antigen-antibody complex to be located at the paratope-epitope interface (Laune et al. 1997).

The functional binding capacity of a continuous paratope has been illustrated by the construction of a chimeric peptide consisting of one CDR of an antiviral antibody conjugated to a continuous epitope of another virus (Sällberg et al. 1994). This construct made it possible to redirect the specificity of an antiviral Ab to allow it to recognize another virus and illustrates the functional equivalence of continuous paratopes and epitopes because they were able to mimic intact Abs and antigens respectively (Van Regenmortel 1998; Sällberg et al. 1994).

The ability of linear peptides to bind to each other is usually attributed to hydropathic complementarity. This arises when peptides of opposite hydropathy bind to each other because hydrophilic residues in one peptide are oriented towards the aqueous solvent and liberate a space that can accommodate an hydrophobic residue from the other peptide (Markus et al. 1989).

Hydropathic complementarity is regularly observed between the sense and anti-sense peptides encoded by complementarity sense and antisense messenger RNA (Tropsha et al. 1992). There is in fact no exception to the rule that DNA or RNA codons (sense) and anti-codons (anti-sense) always code for amino acids of opposite hydropathicity (i.e. either hydrophilic or hydrophobic residues) (Blalock and Bost 1986). This remarkable pattern suggests that the genetic code may have evolved initially to favour the simultaneous emergence from two complementary nucleic acid strands, of sense and anti-sense peptides that are able to interact with each other as do receptors and ligands (Brentani 1988; Biro 2005).

The specificity of interactions between two peptides is enhanced by the ability of amino acids to form complementary protrusions and cavities and by the presence of amino acids of opposite charge (Biro 2007). In a recent application of the phenomenon of hydropathic complementarity to the study of human immunodeficiency virus type 1 (HIV-1) immune responses, a complementary HIV gp120 anti-sense peptide of inverted hydropathy corresponding to the cA1 T cell epitope peptide was used to induce cellular immunity against HIV-1 (Achour et al. 2007).

In recent years, increasing numbers of antibody structures have been elucidated by crystallographic analysis of antibody-antigen complexes and this has given us a much better understanding of the structural basis of antibody specificity. The CDRs of Abs vary considerably in length whereas the individual paratopes within the antigen-binding site usually consist of 10–20 residues. Residues found to be in contact with the antigen are often referred to as specificity-determining residues (SDRs) and they are more variable than residues that are not in contact with it (Padlan et al. 1995; MacCallum et al. 1996). Recently, SDRs were compared with residues that had undergone somatic hypermutation during affinity maturation. The results based on an analysis of 140 antibody-antigen complexes showed that somatic replacements occurred mainly in residues that were not involved in contacts with the antigen (Raghunathan et al. 2012). This observation is consistent with the fact that the hypermutation process is stochastic and occurs in a non-selective manner irrespective of whether residues are in contact with the antigen or not. However, non-SDRs may contribute to binding activity by helping to maintain the conformation of the binding site (MacCallum et al. 1996).

Initially CDRs were identified by aligning a limited number of antibody sequences and determining the positions of the most variable residues (Wu and Kabat 1970). As increasing numbers of 3D Ab structures became available, the hypervariable loop and constant FR could be located in the Ab structure and the CDRs were found to adopt a restricted set of conformations termed canonical structures (Chothia and Lesk 1987). Different combinations of canonical structures alter the topography of paratopes and determine the size of the antigen surface with which the Ab is able to interact (MacCallum et al. 1996; Almagro 2004).

A third approach to define CDRs was developed using the extended database of variable Ig genome sequences (Lefranc 2003; Lefranc et al. 2005). All these approaches produce slightly different residue numbering systems, mainly because nucleotide insertions are accommodated differently.

More recently, a fourth approach using the Paratome webserver was developed on the basis of a multiple structural alignment of all antibody-antigen complexes available in the Protein Data Base (PDB) (Kunik et al. 2012a, b). This method identified regions of structural consensus called antigen-binding regions (ABRs), roughly corresponding to CDRs, in which the pattern of structural positions that bind the antigen was found to be very similar among all antibodies. The superiority of ABRs compared with previously used CDR identification tools was demonstrated by the fact that the ABRs included as much as 96% of all the residues that actually bind the antigen (Kunik et al. 2012b). It was also found that several residues in the FR region and constant regions of IgGs contributed significantly to antigen binding (Sela-Culang et al. 2013).

Using the Paratome web server; Kunik and Ofran (2013) recently examined the amino acid composition of the 6 ABRs in 200 antibody-antigen complexes, which is the largest number of complexes ever analyzed. The average lengths of the six ABRs and the number of SDRs in each ABR are presented in Fig. 2. ABR H2 was found to have the longest median length of 14 residues, followed by H3 and L2 with a median length of 11 residues. H3 showed the highest length diversity whereas H1, L2 and L3 showed limited length diversity. These lengths vary somewhat from earlier results (Wu and Kabat 1970; Collis et al. 2003; North et al. 2011) which may be due to the different methods that were used to define CDRs. The number of SDRs in each ABR is shown in Fig. 3b. H2 and H3 have the largest median number of SDRs (six residues) but H3 is more diverse because it may contain as many as 14 SDRs. L2 sometimes contains as many as 10 SDRs although its median number of SDRs is only 1. For all ABRs, there are instances where they do not contain a single SDR.

Some authors have concluded that the amino acid composition of protein epitopes does not differ significantly from that of protein surfaces (Kringelum et al. 2013) whereas others suggested that epitopes may be enriched with certain types of amino acids (Rubinstein et al. 2008; Soga et al. 2010). In their study, Kunik and Ofran (2013) determined the frequency of the five most abundant SDRs in each ABR. This is a finer analysis than that carried out in earlier studies which always determined the frequency of SDRs averaged over an entire paratope rather than in individual ABRs (Collis et al. 2003; Kringelum et al. 2013; Soga et al. 2010).

As shown in Fig. 3, Tyr was the most frequent SDR in the ABRs followed by Asp, Asn, Arg and Trp. The five most common SDRs in each ABR covered 63% of L1, 60% of L2, 59% of H3, 58% of H1, 54% of H2 and 48% of L3. Although Tyr, Ser, Asn and Trp have previously been reported to be the most abundant residues in paratopes (Collis et al. 2003; Kringelum et al. 2013; Ofran et al. 2008), exactly the same distribution of residues was not found when the frequencies of each SDR in individual ABRs were considered. Tyr makes the greatest energetic contribution to antigen binding, in line with its ability to mediate different types of contacts (i.e. van der Waals, aromatic interactions and hydrogen bonds). It should be noted that there

**Fig. 2** Lengths of antigen-binding regions (ABRs) and number of antigen-binding residues. (**a**) ABRs length. The black bold lines represent the median length. The second and third quartiles are depicted by light and dark gray boxes, respectively. The range of lengths is indicated by vertical lines. (**b**) Number of antigen-binding residues in each ABR. The black bold lines and the light and dark gray zones are as in (**a**). The maximal and minimal number of antigen-binding residues within each ABR are indicated by vertical lines (Kunik and Ofran 2013)

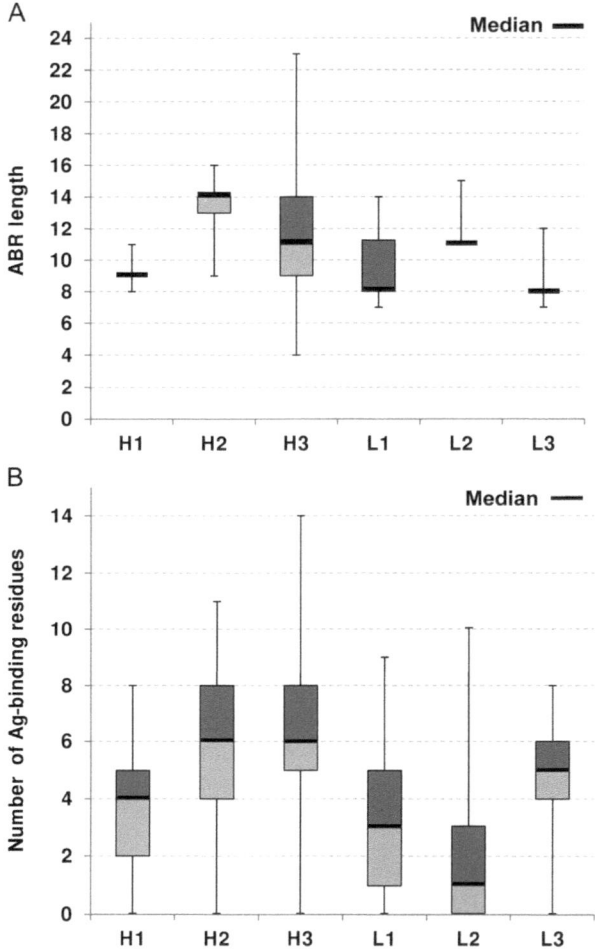

is no link between the abundance of an amino acid in SDRs and its energetic contribution to antigen binding. The most energetically important residues in the ABRs were found to be Tyr, Asp, Asn and Arg. H3 has the highest percentage (29%) of SDRs that are energetically important for binding followed by L1 (24%) and H2 (22%).

The analysis of Kunik and Ofran (2013) for the first time revealed that the six ABRs differ significantly in their amino acid compositions and that each ABR tends to bind different types of amino acids at the surface of proteins. Because the six ABRs, because of their significantly different amino acids compositions, have different contact preferences for certain epitope residues, it seems plausible that epitopes might also possess distinguishable amino acid compositions. However, when the amino acid composition of epitopes was compared with that of entire

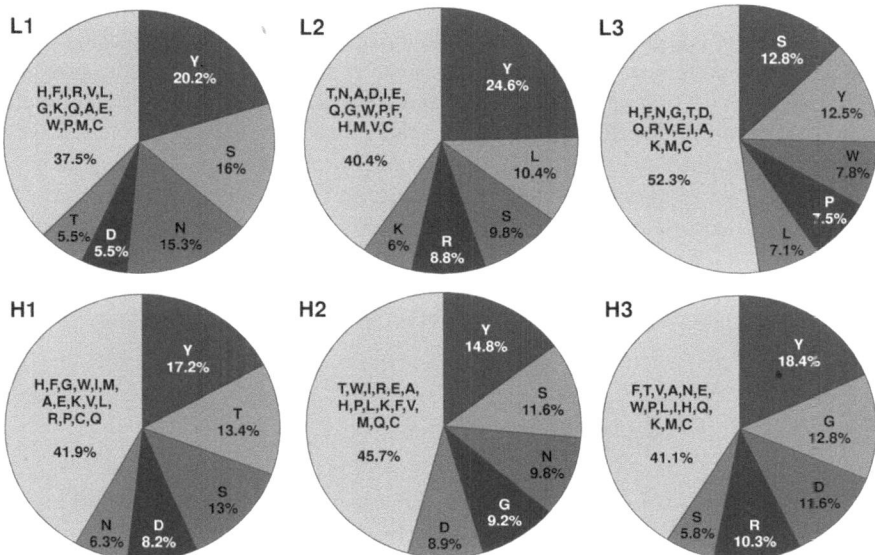

**Fig. 3** The five most frequent antigen-binding amino acids in each antigen-binding region (ABR). For each ABR, the list of residues that contact the antigen is listed, and the frequency of each amino acid is indicated (Kunik and Ofran 2013)

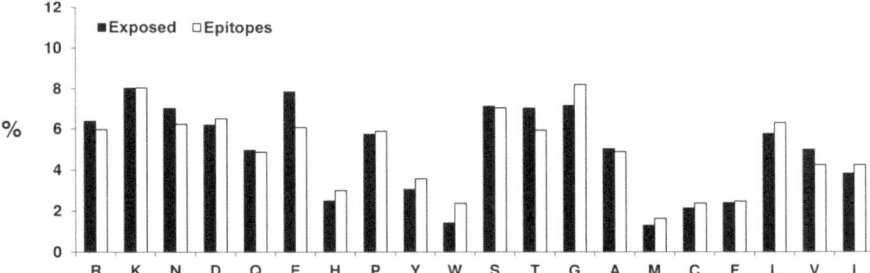

**Fig. 4** Amino acid composition of antigen surface residues and of epitopes. The frequency of each amino acid is calculated for exposed residues and for epitope residues, using a nonredundant set of antibody-antigen complexes (Kunik and Ofran 2013)

protein surfaces, no noticeable differences were observed (Fig. 4). In other words, although each ABR has a unique set of contact preferences favouring certain epitope residues over others, the combination of all these individual ABR preferences yields a collective amino acid composition of epitopes that is very similar to the composition of protein surfaces in general (Kunik and Ofran 2013; Kringelum et al. 2013).

Because the entire accessible surface of a protein is a continuum of potential epitopes (Berzofsky 1985), it could be argued that it would be advantageous for Abs to bind any protein surface patch without requiring specialized sites of increased stickiness. It seems that antibodies are in fact able to achieve this because they have evolved a set of ABRs where each ABR binds different types of amino acids while the combined preference of the entire set is for epitopes that are indistinguishable from the rest of the protein surface (Kunik and Ofran 2013). The situation is different in most protein-protein complexes, for instance virus-host receptor partners, which have optimized their mutual complementarity over long periods of biological evolution. This co-evolution has favoured the selection of small "sticky" areas at the surface of proteins that are enriched in hydrophobic, aromatic and charged residues (Keskin et al. 2008). This is not the case with epitope-paratope partners because paratopes need to optimize their complementarity to epitopes fairly rapidly (i.e. weeks or months) within the context of an individual immune system.

## The Epitope Prediction Conundrum

The recent results of Kunik and Ofran (2013) and Kringelum et al. (2013) have revealed a major difficulty when attempts are made to predict epitopes at the surface of proteins. Because no amino acid was found to be significantly over represented in epitope regions compared with the rest of the surface and because the surface is an antigenic continuum, most of the protein surface can be expected to be part of some epitope potentially recognizable by one or other antibody. Success in predicting that some residues are part of an epitope will thus improve as the number of examined antibodies raised against the antigen increases (Sela-Culang et al. 2014) and not necessarily because some residues are inherently more immunologically active than others. This may be one reason why attempts to predict protein epitopes using various amino acid propensities have been notoriously unsuccessful (Van Regenmortel 2009a; Chen et al. 2009; Hopp and Woods 1981; Thornton et al. 1986; Pellequer et al. 1991; Blythe and Flower 2005; Greenbaum et al. 2007; Liang et al. 2009; El-Manzalawy and Honavar 2010; Zhang et al. 2011; Ponomarenko and Van Regenmortel 2009). However, there may also be other reasons for this lack of success. For many years, investigators concentrated mainly on the prediction of continuous epitopes because these correspond to short peptides that can easily be synthesized and could readily be used to replace pathogenic antigens in diagnostic immunoassays (Hopp 1993; Gomara and Haro 2007) or to act as immunogens for raising anti-peptide antibodies that cross-react with the cognate protein (Muller 1999b). Unfortunately, when one refers to continuous epitopes *of* a protein, the impression is created that these epitopes exist as such in the native protein. In reality, these so-called protein epitopes are mostly short linear peptide fragments of more complex discontinuous epitopes that cross-react only weakly with certain anti-protein Abs and possess only limited structural similarity with small regions of the protein surface that may be as short as dipeptide or tripeptide sequences (Chen et al. 2007, 2009; Yao et al. 2012; Ponomarenko and Van Regenmortel 2009). Predicting continuous epitopes is thus of limited value for analyzing the antigenicity and immunogenicity of native proteins. It is astonishing that in spite of innumerable failed attempts to use continuous epitopes as potential synthetic vaccines (Hans et al. 2006; Van Regenmortel 2006, 2009a) many investigators continue to use short linear peptides, possibly because they are dubbed continuous epitopes of a pathogen, as promising candidates for developing synthetic vaccines.

Because it is increasingly accepted that the vast majority of protein epitopes are discontinuous, many investigators have attempted to develop computational prediction algorithms that they claim allow the prediction of discontinuous epitopes (Van Regenmortel 2009a; Blythe and Flower 2005; Greenbaum et al. 2007; Liang et al. 2009; El-Manzalawy and Honavar 2010; Zhang et al. 2011; Ponomarenko and Van Regenmortel 2009). What these methods do is to predict that a small number of residues located at the protein surface are likely to be part of a discontinuous epitope. What these methods are unable to do is to identify which full set of residues from distant parts of the protein sequence need to be assembled in a precise conformation to form an active site endowed with the antigenic and immunogenic properties of a discontinuous epitope present in the native protein (Chen et al. 2009; Blythe and Flower 2005; Greenbaum et al. 2007; Liang et al. 2009; El-Manzalawy and Honavar 2010; Zhang et al. 2011; Ponomarenko and Van Regenmortel 2009; Hopp 1993; Van Regenmortel and Pellequer 1994). To claim that such methods allow discontinuous epitopes to be "predicted" is injustified because they are only able to "map" or "identify" a limited number of residues present in complex discontinuous epitopes. Discontinuous epitopes can only be defined structurally by determining which residues of an antigen are in contact with paratope residues and not functionally by showing that this complete set of residues, when it is not embedded in the protein, achieves adequate immunological mimicry. Because the main purpose of epitope prediction is to replace a complete protein antigen by a small fragment of the molecule that possesses the antigenic and/or immunogenic properties of one of its epitopes, the feasibility of predicting functional discontinuous epitopes by computation remains at present a rather elusive goal.

Recently, a completely different approach to epitope prediction has been proposed that utilizes the paratope sequences of a series of Mabs raised against an antigen in order to predict which surface regions of the antigen are likely to be recognized by these Mabs (Sela-Culang et al. 2014). Instead of predicting which residues of an antigen possess a superior capacity for binding to any potential Ab, this method attempts to predict patches of about five residues located at the surface of a protein that are likely to be recognized by one or a few available Mabs.

The method is based on residue pairing preferences that have been shown to exist in ABRs and epitopes (Kunik and Ofran 2013) and it predicts potential matches between a given Ab and a given epitope. It also utilizes cross-blocking inhibition assays with different Abs to assess if the epitopes they recognize overlap or not.

Because patches of five residues cannot represent a complete discontinuous epitope, their usefulness to serve as potential diagnostic reagents or immunogens is likely to be rather limited, especially because such applications always involve the participation of a large variety of different Abs. The method also depends on the availability of a large number of Mabs required to sample a representative set of all the epitopes that an antigen may contain.

## Antibody Polyspecificity

Antibody polyspecificity refers to the ability of an Ab molecule to bind a large variety of diverse epitopes in different antigens. As mentioned earlier, the fairly large antibody-binding site of an Ig is able to accommodate a significant number of smaller paratope subsites of 10–20 amino acid residues, some of which may be overlapping. The six CDRs of an antibody never constitute a single paratope specific for one large epitope but comprise several smaller paratopes able to bind individual epitopes in different antigens. The surface of the Ig antibody-binding site is about 2800 nm $^2$ in area whereas the surface of each paratope is 30–110 nm $^2$, depending on whether it binds a small hapten or a protein molecule (Padlan et al. 1995; Raghunathan et al. 2012). Antibodies are thus never specific for a single epitope or antigen although they may of course appear to be monospecific if they are only tested for their capacity to bind the immunizing antigen.

The fact that every Ab always harbours numerous paratopes able to bind many related or unrelated epitopes was recognized as soon as myeloma proteins and Mabs became available and their binding specificity could be analyzed (Eisen and Chakraborty 2010). The polyspecificity of Mabs also became evident when they were tested against peptide libraries and were found to bind many peptides and mimotopes that often possessed only limited sequence similarity with the immunizing antigen used to raise the Mabs (Van Regenmortel 2012a). Instead of referring to their polyspecificity, Abs are also often said to possess plasticity, promiscuity, degeneracy or cross-reactive potential (Cohn 2005; Sperling et al. 1983; Cohen et al. 2004; Parnes 2004; Sercarz and Maverakis 2004). Antibodies have also been called polyreactive when they are able to bind a variety of structurally diverse and unrelated antigens, in opposition to presumed monoreactive Abs that would recognize only a single antigen (Zhou et al. 2007a; Dimitrov et al. 2013). However, referring to the polyspecificity of Abs seems preferable because this term captures two of their essential features: their ability to recognize many different epitopes and their capacity to distinguish between closely related epitopes presenting only small structural differences (Wucherpfennig et al. 2007).

The term cross-reactivity may suggest that there is a primary epitope for each Ab and that the binding of all other epitopes are instances of cross-reactions. In reality, there is no single intrinsic or "specific" epitope for any Ab but only a diverse group of potential epitopes that are able to bind to it with varies degrees of fit (Wucherpfennig et al. 2007). Antigenic cross-reactivity is sometimes said to occur only when antibodies react with other antigens than the immunizing antigen. However, such a restrictive use of the term cross-reactivity implies that an antigen that binds a given Ab must always possess the immunogenic capacity of eliciting that same Ab, which is actually not the case (Van Regenmortel 1989b). It is well-known that when a peptide fragment of a protein is able to bind Abs raised against the protein, there is no guarantee that it will also be able to elicit Abs that react with the native protein.

It is now well-established that the majority of Abs derived from Ab germline genes and expressed by immature B cells before any antigenic simulation (Coutinho et al. 1995) are highly polyreactive and are able to react with a variety of self-antigens such as DNA, cytoskeleton proteins, nuclear antigens and carbohydrates as well as bacterial and viral antigens (Chen et al. 1998; Zhou et al. 2007a). During normal human B cell development, many of these Abs are deleted from the repertoire in the bone narrow before they enter the mature B cell compartment (Wardemann et al. 2003), although as many as 20% of all Abs subsequently produced by mature human B cells may still be polyreactive and self-reactive (Notkins 2004).

The polyreactivity of BCRs in the primary Ab repertoire is in fact what allows a limited repertoire of BCRs possessing considerable structural flexibility to recognize, albeit with low affinity, every antigen the immune system is likely to encounter. A process of somatic hypermutation occurring in the variable regions of Ab genes subsequently produces an increased diversity of mature Abs of increased affinity but decreased cross-reactive potential (Rajewsky 1996; Berek and Milstein 1987; Wedemayer et al. 1997; Manivel et al. 2002; Yin et al. 2003). However, the affinity constant of Abs is rarely higher than $10^{10}$ $M^{-1}$ because this would require that the dissociation half-life of the Ab must be longer than 1 h. which is a property unlikely to be selected for (Foote and Eisen 1995; Batista and Neuberger 1998).

Several mechanisms exist that allow a polyspecific Ab to recognize a multiplicity of epitopes and antigens (Mariuzza 2006). As shown in Fig. 5a, an essentially rigid antigen-binding site may be able, in separate subsites, to make different interactions with structurally distinct antigen surfaces, a mechanism called rigid adaptation which does not involve substantial conformational changes (Mc Farland and Strong 2003). A second mechanism (Fig. 5b) achieves polyspecificity through the considerable conformational flexibility of binding sites which allows distinct binding site topologies that are in dynamic equilibrium to be generated by induced fit (Yin et al. 2003; James et al. 2003; Khan and Salunke 2012). During the subsequent somatic hypermutations that give rise to Ab maturation, the flexibility of the binding site tends to decrease as its affinity increases (Jimenez et al. 2003; Thorpe and Brooks 2007). A third mechanism (Fig. 5c) that also expands the primary Ab repertoire has been called differential ligand positioning and occurs when a single Ab conformer is able to bind diverse antigens at spatially distinct regions of the binding site. An example is the germline Ab 36–65 that is able to bind three separate 12-mer peptides using three different subsites of a single conformational state of the antibody-binding site that differs from the conformation observed in the free Ab (Sethi et al. 2006). This third mechanism (Fig. 5c) differs from the first one (Fig. 5a) because the bound and unbound forms of Ab 36–65 display different site topologies, which emphasizes the role of conformational flexibility for this particular recognition mode. It also differs from the second mechanism (Fig. 5b) where polyspecificity is mediated by conformational diversity, because the three different peptides were able to bind to a single conformer of Ab 36–65 which is absent in the free Ab (Mariuzza 2006: Sethi et al. 2006).

An Ab is never specific for a multiepitopic antigen as a whole but only for one of its epitopes (Fig. 1) and this may give the impression that a Mab directed to an

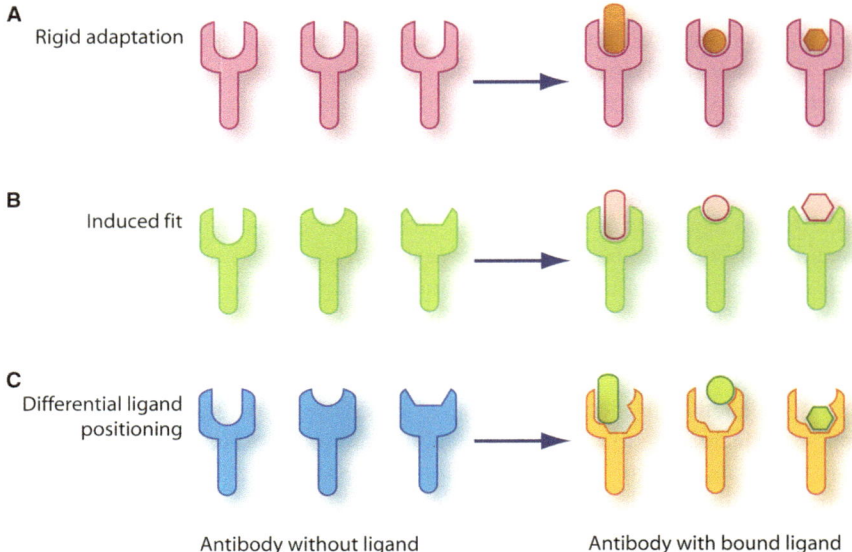

**Fig. 5** Mechanisms for multispecific ligand recognition by antibodies or other receptors. (**a**) Multispecificity through rigid adaptation. An essentially rigid receptor binding site recognizes structurally distinct ligands, without the need for substantial conformational changes in the receptor. (**b**) Multispecificity through conformational diversity. A conformationally flexible receptor binding site exists in dynamic equilibrium between different conformational states. Each conformation generates a distinct binding site topology, allowing the receptor to engage multiple ligands at the same region of the binding site. (**c**) Multispecificity through differential ligand positioning. As in (**b**), the unligated receptor adopts different conformational states. However, an altogether different conformational state mediates recognition of diverse ligands. In this case, multispecificity is conferred not by conformational diversity but by ligand binding to spatially distinct regions of an otherwise topologically identical binding site (Mariuzza 2006)

antigen is a more specific reagent than a polyclonal antiserum elicited against it. An individual Mab may indeed be able to better discriminate between two epitopes of the antigen than a polyclonal antiserum, although the antiserum tends to have a greater collective specificity for a multiepitopic protein. This is because of the additive specificity effect that arises from the presence in the antiserum of Abs directed to several different epitopes of the antigen. The Mab and the different Abs present in the polyclonal antiserum are all polyspecific which means that in addition to binding to one epitope of the antigen, they will all cross-react with a large number of unrelated epitopes although these epitopes are different for each type of antibody in the antiserum. As a result, the collective cross-reactive potential of these different Abs in the antiserum will be diluted out and the individual cross-reactions will therefore be masked (Talmage 1959; Richards et al. 1975). This phenomenon explains the power of the multiepitope-ligand cartography technique developed for mapping the topological location of proteins in multimolecular assemblies present within cells and tissues (Schubert et al. 2006). This technique is based on many automated sequential rounds of protein detection using Mabs conjugated to fluorescent dyes. Each cycle utilizes a Mab specific for a different protein epitope, resulting in a composite set of images that show the distributions of many individual proteins in supramolecular assemblies. This procedure utilizes the additive effect and achieves a remarkable selectivity and specificity of protein detection by diluting out the many cross-reactions that would otherwise be observed with single labelled Mabs (Schubert 2014).

The immune system is able to recognize specifically a very large number of multiepitopic antigens because of the combinatorial effect of many polyspecific Abs that bind separate epitopes on the same antigen and not because of the production of myriads of different Abs, each one recognizing a unique epitope present in only one antigen. Antibody polyspecificity can also be increased when Abs are submitted to low pH or to chemical treatments used to dissociate antibody-antigen complexes (Dimitrov et al. 2010, 2013; McMahon and O'Kennedy 2000; Bouvet et al. 2001).

It must also be pointed out that the polyspecificity of Abs is responsible for the fact that an epitope structure deduced from the crystallographic analysis of an antigen-antibody complex will not necessarily reveal which immunogenic structure was recognized by B cell receptors during the immunization process used to obtain the Ab. The epitope structure observed in the complex is only one of several epitopes that could be accommodated by the polyspecific Ab and any one of them could equally have been assumed to correspond to the immunogen. Because it is impossible to predict which protein epitope identified by crystallography will be able to elicit neutralizing and protective Abs against a pathogen, it has not been so far possible to develop a preventive HIV-1 vaccine by reverse vaccinology (Van Regenmortel 2012b).

# What Are the Implications of Antibodies Being Able to Bind Viral and Microbial Antigens as Well as Self Antigens?

It is now well-established that germline-encoded Abs that react with self antigens in the peripheral B cell repertoire of adult individuals are not necessarily removed by clonal deletion (Dimitrov et al. 2013). Although some germline-encoded Abs may lose their polyreactivity during antigen-driven affinity maturation (Oppezzo et al. 2004), other polyreactive Abs may undergo positive selection during viral and bacterial infections or in the course of autoimmune diseases (Ditzel et al. 1996; Hunziker et al. 2003; Ludewig et al. 2004). It has been found, for instance, that some of the broadly neutralizing Mabs reacting with the HIV-1 Env protein are polyreactive and bind several self antigens. It has been suggested that such neutralizing antibodies are rarely elicited by HIV-1 vaccine candidates because the relevant anti-HIV-1 Ab producing B cells are down-regulated by host B cell tolerance mechanisms involving self antigens such as cardiolipin (Haynes et al. 2005b). It has also been proposed that some anti-HIV-1 neutralizing Mabs such as 2F5 and

4E10 that react with epitopes in HIV-1 gp41 and with virion lipids may be able to bind the virion surface in a monogamous bivalent fashion using so-called heteroligation, whereby one Fab combining site attaches to the gp41 and the other Fab attaches to various self antigens (Mouquet et al. 2010).

Usually such monogamous bivalent binding (Day 1990) is not possible when both Fabs of an anti-HIV-1 Ab must bridge the same epitope in two accessible Env spikes on the virion surface. The small number of Env spikes on the HIV surface (Zhu et al. 2006; Klein and Bjorkman 2010) and the position of the epitopes on one Env trimer prevent both Fabs of a single Ab to bind either the same trimer or two adjacent trimers in a monogamous bivalent fashion (Mouquet and Nussenzweig 2012). The increased avidity arising from monogamous bivalent binding compared with the hundredfold lower affinity that is present when only one Fab of an Ab binds the virion surface (Van Regenmortel and Hardie 1976; Klein et al. 2009), is therefore unlikely to be observed with anti-HIV-1 Abs that are not polyreactive; in such a case it would be easier for the virus to escape antibody-mediated neutralization (Mouquet and Nussenzweig 2012). Whether polyreactive anti-HIV-1 Abs are actually positively selected during anti-HIV-1 immune responses has not yet been clearly established.

There is evidence, however, that polyreactive Abs provide a first line of defense against infections (Zhou et al. 2007a) and may increase the immunogenicity of viruses by Ab-mediated trapping in secondary lymphoid organs (Ochsenbein et al. 1999). The possible roles of polyreactive Abs in viral diseases, autoimmunity and B cell malignancies are currently the subject of intense research but are still poorly understood (Dimitrov et al. 2013; Münz et al. 2009).

## Antibody Heterospecificity

Heterospecific Abs (also called heteroclitic Abs) (Day 1990) are Abs that react more strongly with another antigen than the one used in the immunization process that elicited the Ab. Heterospecificity occurs because B cells can be selected and triggered by immunogens endowed with only minimal affinity for its BCRs. Abs of low or moderate affinity may then be obtained that react better with heterologous epitopes endowed with a superior degree of fit with the paratope.

It may also happen that such Abs have such a low affinity for the immunogen that the reactivity is below the detection level of the immunoassay that is used. It may then appear as if the Ab has been elicited by an antigen with which it is unable to react. This phenomenon is often observed when Abs raised against a virus are tested with viral mutants, serotypes or strains that present substitutions in a particular epitope of the wild type immunogen, because it is then easy to detect the presence of heterospecific Abs that recognize the variants. For instance, anti-T2 phage Abs have been found to react more strongly with the related phage T4 than with T2 itself (Mäkelä 1965) and Mabs against influenza virus subtype H3N2 were found to react better with other influenza virus strains than with the immunogen (Underwood 1985).

A particularly well-studied case is the heterospecificity of Abs found in many rabbits immunized with tobacco mosaic virus (TMV) (Van Regenmortel 1982). All the antisera that were tested contained Abs that did not react with the wild type virus used as immunogen but which reacted with viral mutants that presented a single Pro->Leu substitution at position 156 in TMV protein subunits (Van Regenmortel 1982; Von Sengbusch and Wittman 1965). When all the Abs reacting with TMV were removed from TMV antisera, it was also possible to detect residual heterospecific Abs that reacted with the Dahlemense TMV strain presenting a substitution Thr->Ser at the C-terminal residue (position 158) of the protein (Loor 1971). The C-terminal four residues 155–158 of TMV protein are known to be located in a flexible and disordered region of the protein (Li et al. 2013) and it seems that amino acid substitutions at positions 156 and 158 make epitopes that are not reactive in the wild type virus, more accessible to heterospecific Abs. This phenomenon can be readily demonstrated when intragel absorption is used to completely remove from the antiserum all Abs that react with the homologous TMV and then showing by immunodiffusion that the residual heterospecific Abs still react strongly with various mutants and strains (Van Regenmortel 1967, 1982). Because the specificity of polyclonal antisera arises from a population phenomenon that masks the cross reactivities of each individual Ab in the antiserum, it is to be expected that heterospecific reactivities will be more easily detected with Mabs. This was shown to be the case with heterospecific anti-TMV Mabs that readily reacted with viral mutants presenting substitutions at positions 63, 65 and 66 located close to the C-terminus at the surface of the viral protein (Al Moudallal et al. 1982).

Antibody heterospecificity is a general phenomenon that is responsible for the common observation that early antisera obtained soon after immunization with any antigen very often contains levels of total Igs that far exceed the level of Abs that can be shown to react in immunoassays with the immunizing antigen (Boyd and Bernard 1937; Urbain-Vansanten 1970; De Vos-Cloetens et al. 1971; Eisen 2001). Such Igs or Abs were initially believed to result from a non-specific stimulation of the Ig-secreting cells that greatly increase in number immediately after any immunization (Moticka 1974). It was later accepted that these so-called normal globulins were specifically induced by the immunizing antigen but failed to react with it because their affinity was too low in the immunoassays that were used (Haimovich et al. 1970). However, it is possible to demonstrate the heterospecific reactivity of such globulins when they are tested with other antigens. Because few researchers investigated this phenomenon, the heterospecific reactivity of such low affinity Abs has remained unnoticed for many years.

The failure to look for heterospecific Abs can also lead to much unnecessary experimental work when attempts are made to produce specific diagnostic Ab reagents by hybridoma technology. In order to obtain Mabs intended for viral diagnosis, investigators tend to screen hybridoma cell cultures only for their ability to recognize the virus used for immunization. In so doing, they may unwittingly discard many heterospecific Mabs that could have been useful for detecting other related viruses. Frison and Stace-Smith (1992) showed that when a variety of serologically related viruses were used for screening hybridomas, many additional heterospecific Mabs reacting with other viruses could be obtained from a single fusion experiment. This makes it possible to greatly reduce the number of fusion experiments that would otherwise be required for obtaining panels of Mab reagents suitable for detecting a variety of different viruses or other antigens.

The heterospecificity of Abs also helps to clarify the difference between the antigenicity and the immunogenicity of a protein. It is now well-established that the antigenic reactivity of a viral protein demonstrated by its ability to bind a given Ab is not necessarily accompanied by an immunogenic capacity to induce that same Ab in a particular immune system (Van Regenmortel 2009a). An epitope is defined only by its ability to bind a paratope and not by its capacity to induce certain Abs by immunization. If immunogenicity was also a required property for being considered an epitope, it would make the existence of epitopes in a protein depend on extrinsic factors (BCRs, host Ig repertoire, chemokines etc) and immune regulatory mechanisms of the immunized host (Berzofsky 1985; Van Regenmortel 1989b). However, a protein region that does not induce Abs in a given host may nevertheless be found to be antigenic when tested with other Abs.

Furthermore the immunogenic capacity of an epitope to induce heterospecific Abs that do not react with the protein used for immunization shows that such immunogenicity is not necessarily accompanied by an antigenic reactivity that enables the epitope to bind to the induced Abs. Successful viral vaccines require the use of effective immunogens, that is biologically active entities able to induce a specific protective immune response, rather than antigens that are only chemical entities defined by their ability to bind certain Abs. The failure to distinguish between antigenicity and immunogenicity is one of the reasons why the strategy known as reverse vaccinology has not been successful for developing an effective, preventive HIV-1 vaccine (Van Regenmortel 2012b, 2014b).

# Part II
# Reductionism

# Reductionism and the Search for Structure: Function Relationships in Antibody Molecules

*A lump of iron is a chemical entity, and the word 'iron' stands for a chemical concept. But suppose that the iron has the form of a padlock, then although the iron is still chemically analyzable in the same way as before it cannot still be fully described in terms of chemical concepts. It now has an organization above the chemical level.*

J.H. Woodger (Biological Principles, 1967)

## Introduction

The above quotation from Woodger (1967) refers to the classical antireductionist view that biological systems cannot be fully described in terms of the physico-chemical properties of their constituent parts. Although the image of a padlock is reminiscent of the outdated lock-and-key model of antigen–antibody recognition, a biologically more significant analogy would be the claim that Michelangelo's *David* is composed of nothing but marble (Dupré 1993). Such a statement clearly does not imply that the statue is nothing but a piece of marble.

In view of the self-evident nature of such statements, it is odd that many biologists still adhere to the reductionist view that, since biological systems are solely composed of atoms and molecules, they can be fully described and understood in terms of the physico-chemical properties of their constituent parts. Reductionists tend to disregard the fact that all biological systems possess so-called emergent properties that arise through the multiple interconnections and relations existing between individual components of the system. These emergent, relational properties do not exist in the constituent parts and they cannot be deduced or predicted from the properties of the individual, isolated components (Holland 1994). Examples of emergent properties are the viscosity of water (individual water molecules have no viscosity), the colour of a

Journal of Molecular Recognition 2002, 15, 240–247.
Marc H V Van Regenmortel
Copyright © 2002 John Wiley and Sons
All rights reserved, used with permission

© Springer Nature Switzerland AG 2019
Marc H V Van Regenmortel, *HIV/AIDS: Immunochemistry, Reductionism and Vaccine Design*, https://doi.org/10.1007/978-3-030-32459-9_5

chemical, a melody arising from notes, the saltiness of sodium chloride, the specificity of an antibody and the immunogenicity of an antigen (Van Regenmortel 2002c).

When the immune system is dissected into its constituent parts, the connections that link the various components to each other in a functionally integrated manner are severed. As a result, essential and irreducible aspects of the immune system's behaviour are destroyed and it is no longer possible to understand and explain the workings of the system as a whole.

The simplest biological system that can be said to possess the property of being alive is a single cell. A unicellular organism is alive by virtue of its integrated activities but none of its constituents at the molecular or organelle level can actually be said to be alive. A protein molecule or a ribosome assembly are not alive and if a living cell is reduced to its nonliving constituents, no sense can be made of the fact that a unicellular organism actually becomes alive through the juxtaposition of inanimate components. A living system is an organic whole and not an additive system of independent parts (Nagel 1961, p. 431). What is needed is for the parts to be directively organized in a self-regulated and integrated manner. This gives rise to the autonomy of organisms and to their capacity, in principle, to produce an endless lineage of descendants. The central property of all life is the capacity and necessity to build, maintain and preserve itself, a process known as autopoiesis (Rose 1997, p. 18). The tendency of a regulated system to maintain itself close to some fixed point, like the temperature of a room by a heating system and a thermostat, is called homeostasis. In his book, Lifelines, Steven Rose has argued that, since the equilibrium achieved by the cell is dynamic rather than static, this process of self-regulation should rather be called homeodynamics (Rose 1997, p. 154).

Following the demise of spontaneous generation as a plausible mechanism for the continued creation of living organisms, life has acquired the status of a steady-state, homeodynamic phenomenon which has succeeded in maintaining itself over three and a half billion years of terrestrial evolution. Reducing living organisms to a juxtaposition of non-living constituents is no longer considered a satisfactory analytical framework and contemporary biological analysis relies on the notions of emergence and complexity to explain biological phenomena (Holland 1994; Casti 1994; Weng et al. 1999; Bhalla and Iyengar 1999).

Reductionism is sometimes considered to be only an analytical strategy for describing biological systems in terms of physics and chemistry, rather than an explanatory scheme that can be effective without invoking biological concepts at all. In fact there is nowadays an increasing realization that biology is an autonomous discipline that does require its own explanatory concepts and categories not found in chemistry or physics (Dupré 1993; Rosenberg 1994; Rose 1997).

## Structures

It is a truism that living activities do not occur in the absence of a material structure and that specific structures do set bounds to the kind of biological activities that are possible. Although structure and function are inseparable aspects of biological organization, this does not mean that a structure logically entails a particular function or that a structure can possess causal efficacy in bringing about a certain function. Woodger (1967) defined structure as the result of selective attention to the visual experience of an object at a specific time. Such a static definition which excludes the dimension of time conceals the fact that pictures of biomolecules are visual time-slices of dynamic systems. Since biomolecules possess a variety of structures rather than a single structure, the term structure is best declined in the plural rather than the singular and should be understood as a collective notion.

Thinking in terms of static instead of dynamic structures reinforces the appeal of lock-and-key models of the process of molecular recognition and makes it more difficult to visualize the role that mutual adaptation of two proteins plays in facilitating their interaction (Demchenko 2001).

When referring to the so-called native structure of a protein molecule, one tends to think of a unique tertiary structure determined solely by a particular nucleotide sequence. Such a view disregards the role played by the physico-chemical environment in controlling the folding process. The role played by the chemical environment on a molecule's conformation is clearly illustrated for instance in the case of cyclosporin which was found to have completely different conformations in different cell compartments (Altschuh et al. 1994; Zeder-Lutz et al. 1993). Differences in the cellular environment are also responsible for the common finding that when eukaryotic proteins are expressed in a recombinant bacterial system they often do not possess the anticipated 'correct' conformation.

When the quaternary structures of protein assemblies such as ribosomes or virus particles are viewed as static configurations it becomes impossible to account for their biological properties (Bamford et al. 2001). In viral capsids, for instance, it has been found that small changes in pH and temperature can lead to a rearrangement of parts of the protein subunits. As a result, internal regions of the subunits become transiently exposed to antibodies, cellular receptors and proteases (Quesniaux et al. 1983; Canady et al. 1996; Roivanen et al. 1993; Bothner et al. 1998). If these dynamic features of virus particles are not taken into account, it is impossible to explain some of the biochemical and immunological properties of a virus from its X-ray crystallographic structure. Since the analysis of biological systems must necessarily always include the dimensions of time, history and evolution, biological structure needs to be thought of as a dynamic rather than a static property (Edmundson et al. 1987).

# Functions

The term function as used in biology is highly ambiguous. Biochemists tend to regard the function of a protein simply as what the molecule does, i.e. its functioning or activity at the molecular level. According to this restrictive sense of function, the function of a protein corresponds to what it does and how it acts, as well as to when and where it acts (Murzin and Patthy 1999). The primary activity that is considered is usually binding activity and function then becomes synonymous with binding.

In this case the meaning of function is restricted to the level of the protein molecule itself, and functions that are meaningful only at the level of the cell or of the organism as a whole are ignored (Bork et al. 1998). In a biological context, however, it is possible to differentiate between functions at different levels of organization, for instance:

1. at the molecular level, there are functions like binding, catalysis, signalling;
2. at the cellular level, functions resulting from the integrated interactions of many individual proteins or assemblies like ribosomes or nucleosomes;
3. at the organism level, functions expressed as metabolic and physiological pathways that arise from the co-operative action of many gene products.

Since proteins are able to interact with a large number of other molecules in different cellular compartments, thereby contributing to a variety of integrated activities, it is to be expected that the functions of proteins at the cellular and organismic level will be very difficult to unravel. In the absence of information regarding local concentrations, the localization of specific ligands or the nature of the chemical environment in the cell at a given time, the disentangling of integrated functions becomes highly problematical. Fortunately, the recent development of proteomics and bioinformatics is giving biologists new tools for analysing the bewildering complexity that arise from the inumerable interactions that are possible among the thousands of proteins expressed in a cell (Pandey and Mann 2000).

## The Use of Functional Language in Biology

It is not possible to describe biological systems without referring constantly to the roles that organs, cells, organelles and molecules play in meeting the needs of an organism and keeping it alive. To a human observer, organisms appear to have goals, ends and purposes and it seems that their behaviour can be explained in terms of such end-directed or teleological features.

When biologists attribute a purpose to a process or explain it by citing its goal or function, they do not imply the type of human intentionality that is used to explain purposive human action. Biologists do not explain a plant's heliotropism, i.e. the movements that the plant does to maximize its exposure to the sun, by the plant's desire or wish to increase the number of photons landing on its leaves (Rosenberg 1985, p. 44). The presence of such a process is explained teleologically only by referring to its contribution to the reproductive fitness of the organism in which it occurs. Attributing a function to heliotropism helps to explain why such a behaviour is present and has the form that it does. Performance of a useful function is the reason why the process of natural selection retained a particular feature during evolution. Ultimately, a functional explanation for the presently observed structure is provided in terms of superior fitness in the past.

It is assumed that the performance of any function must confer some good to the biological system as a whole, for instance by contributing to its health, performance, survival or reproduction. To attribute a function to an item is to say that the item has that behaviour because it has a certain evolutionary history and produced certain beneficial consequences in the past; those consequences themselves had the effect of reproducing items with that behaviour (Macdonald 1992; Kitcher 1998). Evolution thus operates on the DNA sequence through feedback from its effects, similar to a pattern of backward causation (Rosenberg 1994).

It is an anthropomorphic fallacy to view evolution as a process that solves problems of adaptation within the context of certain goals and purposes. Evolution operates through blind variation and selective retention of useful functional features and it appears to be goal-directed only by analogy with the intentionality of human behaviour. Functional descriptions only give the appearance of goal directed activities and they have a heuristic value because they allow us to predict the future behaviour of a system from knowledge of its past behaviour. The usefulness of functional language is illustrated by the many metaphors that are constantly used by molecular biologists, e.g. molecular recognition, discrimination capacity, proofreading, replication errors etc. It seems that biological phenomena can only be made intelligible by ascribing functions to underlying structural features.

The binding site of an immunoglobulin molecule, for instance, is a functional rather than a structural concept. It is defined by the binding activity that arises when 10–15 residues (out of the 50 or so hypervariable residues of an immunoglobulin) participate directly in an interaction with an individual epitope. However, the full complement of all the residues of the six hypervariable loops is only a virtual site since it does not constitute an actualized structural binding site for any antigen. Its identity as an antibody binding site emerges only through the use of functional language.

## Causal Explanations Versus Functional Explanations

Causality is the relation of cause and effect, whereby event A is the cause of event B, and B is the effect of A, if and only if the following three conditions apply:

1. B follows A in time;
2. A produces B;
3. the relation between A and B is constant and necessary.

Causation is a mode of event generation. Causal relations are relations between successive events or processes and not between two material objects or between a structure and an event. A biological event such as a binding reaction can thus not be caused by something that is not an event, such as the structure of one or both interacting partners. There is also no unique causal relation between the structure and activity of a protein molecule since most proteins have a multiplicity of activities or functions, while a single activity can be generated by a variety of structures (Martin et al. 1998).

It is helpful to consider what we mean when we say that we understand the molecular basis of antibody specificity. Many scientists equate understanding with explanation and would say that whatever has been explained is also understood. The nature of explanation, however, is anything but simple (Achinstein 1983; Schaffner 1993) and it is not easy to decide if an explanation is adequate or not. According to some authors, the best explanations are based on causal relations since the search for the explanation of a phenomenon often involves finding its cause. However, the ontology of causality, i.e. what a cause actually consists of, is controversial and remains a central issue in the philosophy of science (Mahner and Bunge 1997; Salmon 1998).

The appeal to causes as an explanation of events and behaviour is reinforced by the human predisposition to look for the reasons (sometimes called causes) of human actions. However, reasons are not the causes of events and prior states of a thing are not causes of its posterior states. Since things are not causes, ancestors are not causally related to descendents (via reproduction) and DNA sequences or genes cannot be causes of phenotypic traits or of behaviour patterns (Mahner and Bunge 1997, p. 39; Rose and Rose 2000).

Most molecular biologists adhere to the view that 'structure determines function'. From this, they infer that there is a linear, causal pathway that goes from the nucleotide sequence to the protein sequence (step 1) then to the protein conformation (step 2), to specific binding activity (step 3) and finally to biological function (step 4). Such a pseudo-causal pathway appears to possess explanatory power because it is presented in a context-independent manner, i.e. as if the cellular context and environment did not matter. In step 1 of the pathway, the role played by the cellular machinery in bringing about the translation process is ignored and in step 2, it is assumed that the physico-chemical environment during the folding process is irrelevant. In step 3 no consideration is given to the fact that the relations between structure and activity are many–many, and that binding sites are relational entities defined by the interacting partners and not by intrinsic structural features that could be identified independently of the interaction with a particular ligand (Van Regenmortel 1989a). The structure of a binding site, in contradistinction with the structure of a molecule, cannot be described without considering the binding partner. It is indeed impossible to deduce binding activity from the structure of one of the interacting molecules without considering a particular relation with a partner. Since a specific binding site cannot be visualized if the partner has not been identified, structure and binding activity should be viewed in an integrated manner and analyzed in a four dimensional, spatio-temporal framework (Van Regenmortel 1996).

As explained above, causality is not a relation between a material object and an event. This means that the structure of an antibody cannot be the cause of its binding activity or of its capacity to neutralize an infectious agent such as a virus. When attempts are made to analyse structure–function relationships, correlations rather than causal relations should be investigated. Instead of looking for causal links between the structure and activity of an antibody, the aim should be to uncover correlations between the atomic structure of binding sites and biological activity and to analyse the multiplicity of factors that can influence the activity.

Causal explanations are reductive because one factor is singled out for attention and given undue explanatory weight on its own. Since biological systems are complex, any observed effect is always the result of a complex network of interactions and an analysis in terms of a single cause is rarely satisfactory. Instead of invoking causes, it is preferable to refer to the many factors that simultaneously influence the features of a biological system. In biology, network causality is a more useful explanatory category than linear causality of the push–pull variety relevant in physics (Van Regenmortel 2002c).

The often quoted assertion of the geneticist Theodosius Dobzhansky: 'Nothing in biology makes sense except in the light of evolution' is particularly relevant to one type of functional explanation. This type of explanation is based on a broad historical, evolutionary account of the selective advantage that the possession of a certain trait conferred to an organism in the past. Such a general evolutionary account, however, can be given for just about any biological function and it lacks power because it gives no information regarding the contingent constraints that favoured one particular adaptive solution and pathway as opposed to its many possible alternatives. It must also be pointed out that the selection mechanism is blind to structure: it selects for effects and cannot discriminate between different structures with identical effects (Rosenberg 1994, p. 25). The best an evolutionary account can do is to provide an answer to a 'why' type question, such as: why do vertebrates possess antibodies? The answer would list the known and conjectured advantages that the possession of antibodies confers to an organism.

There are, however, many other types of questions that demand functional explanations in biology. Some examples are: how does the immune system produce antibody diversity? How do antigens trigger an immune response? What is the basis of antibody specificity? How do antibodies neutralize the biological activity of an antigen?

Answers to such questions rely on functional explanations that employ concepts and assert relations that cannot be defined nor derived from the physical sciences (Nagel 1961, p. 431). Questions about the functions of antibodies cannot be answered in a mechanistic manner by analysing in great detail their internal structure. The answer as to why antibodies are organized the way they are does not lie inside but outside the antibody molecule. The answer, therefore cannot be obtained in a reductionist manner by analysing the antibody's chemical composition (Cohen and Stewart 1994, p. 243). Biologists favour functional explanations for a currently observed biological structure in terms of superior fitness in the past, rather than a structural explanation for a presently observed function.

All biological functions in a cell or organism are interdependent and internally regulated and since their occurrence is context-dependent, they cannot be understood in isolation. Functional explanations are thus more appropriate for understanding complex biological systems exhibiting many complex interactions than are causal explanations that focus on a single factor.

## Looking for Structural Correlates of Binding Sites

The biological function of a protein is always dependent on a first specific binding step and binding can thus be considered as the most basic activity of proteins. In recent years the quantitative measurement of protein binding activity has been greatly simplified by the availability of biosensor instruments based on surface plasmon resonance. These instruments make it possible to visualize the binding process as a function of time, allowing kinetic rate constants and equilibrium affinity constants of the interaction to be determined (Myszka 1999; Rich and Myszka 2000, 2001). The biosensor technology has therefore become the method of choice for measuring the binding affinity of biomolecular interactions and for establishing correlations between the structure and binding activity of proteins (Van Regenmortel 2001d).

To evaluate the contribution of individual residues to the interaction, the usual approach is to modify the proteins by site-directed mutagenesis or alanine scanning mutagenesis and to assess the effect of mutations on the binding affinity. Such studies invariably indicate the presence of hot spots in binding interfaces, i.e. regions made up of a small number of residues that contribute most of the binding energy (DeLano 2002). When these residues are mutated, the binding constant is decreased at least 10- or 100-fold. Out of the 50 or so residues that make up the six complementarity determining regions (CDR) of antibody molecules, it is usually found that about 5–10 residues contribute significantly to the binding energy (Laune et al. 1997). In view of the considerable flexibility of the CDR loops, there is often considerable mutual adaptation of both epitopes and paratopes. This makes it possible for a single antibody molecule to interact with a large number of related antigenic structures (Roberts et al. 1993). The resulting multi-specificity and polyreactivity of immunoglobulin binding sites gives rise to considerable variability in the atomic contacts that are possible and makes it difficult to derive general rules regarding the relative functional importance of most residue positions in CDR loops. The adaptation of paratopes to their binding function has resulted in the presence of a much larger number of tyrosine and tryptophan residues in the CDRs than is usual at the surface of proteins. These aromatic side chains can make large rotations with little entropic cost and they contribute significantly to the binding energy (Mian et al. 1991). In three lysozyme–antibody complexes studied by X-ray crystallography, it was found that tyrosine and tryptophan residues contributed 155 of the 302 interatomic contacts between these antibodies and the antigen (Padlan 1990).

Only a small fraction of the residues that are assigned to binding sites because they make contact with the ligand actually contribute to the binding free energy (Jin and Wells 1996). As a result, the so-called 'functional' binding site is always smaller than the 'structural' binding site defined on the basis of crystallographic coordinates of the complex. On the other hand, substitutions of residues that are not in contact at the interface are also frequently found to affect the binding affinity (Lavoie et al. 1992). A number of mutations in framework residues at the VH-VL interface of different Fabs have been shown, for instance, to affect the Fab-antigen interaction kinetics (Chatellier et al. 1996). The effect was larger on the off-rate than on the on-rate, but no clear relationship was found between the nature of the substitution and the observed functional effect (Ben Khalifa et al. 2000). The binding kinetics were affected by alteration of framework residues remote from the binding site and the effects were unpredictable and depended on the surrounding environment in the protein. Residue substitutions in a protein are able to produce small structural shifts and allosteric changes that propagate far beyond the mutated region; although these perturbations are not always detected at the current resolution of structural data, they are able to significantly affect binding affinity (Sturtevant 1994).

In a recent study, the influence on kinetic interaction parameters of substitutions in 14 conserved, surface—exposed charged residues located away from the paratope in two antibodies was investigated (Weidenhaupt et al. 2002). A few of the substitutions had a small effect on the kinetic association rate whereas the dissociation rate was not affected significantly. These results are in line with the general finding that electrostatic attractions play a major role in steering two proteins together (Sheinerman et al. 2000; Gabdoulline and Wade 1999) and influence on-rates more than off-rates.

When the binding properties of 15 engineered Fabs directed to an epitope of tobacco mosaic virus protein, which included four single, two double and three triple Fab mutants were compared, multiple substitutions were found to have non-additive, i.e. cooperative effects on the dissociation kinetics (Rauffer-Bruyère et al. 1997). The effect of any particular mutation depended on the molecular context and differed in the wild protein and in variants showing substitutions from previous mutations. The general conclusion is that the effects of multiple mutations on antibody function are difficult to predict, even when high resolution structural data are available.

Multiple mutations can lead to altered functional characteristics in the protein by new combinations of properties that are not manifested in the parents and which therefore cannot be part of a rational design strategy (Van Regenmortel 2000). This has led some authors to conclude that the engineering of improved protein function need not be based on structural information but is more likely to succeed by the combination of interactive rounds of random mutation, recombination and selection (Tobin et al. 2000; Oue et al. 1999; Ness et al. 1999).

## Predicting the Binding Properties of Antibodies

It is often said that the validity of an explanation can be tested by its predictive power, although this ignores the fact that successful predictions are possible even if the extent of understanding of a phenomenon is rather limited. This appears to be the case with structure–function relationships in antibodies since it is possible to predict the off-rate of an antigen–antibody interaction in the absence of in-depth knowledge of which physico-chemical and structural features of binding sites are responsible for the observed affinity. Recent studies have shown that it is possible to use a multivariate QSAR (quantitative structure-activity relationship) approach involving modifications in antigenic peptide sequence and buffer composition for predicting the kinetics of a particular interaction (Andersson et al. 2001; Choulier et al. 2002). By using a proper experimental design it was feasible to determine the influence of the chemical environment and of the chemical nature of the ligand on the off-rate kinetics and thus to derive mathematical models for predicting binding. By measuring the kinetic parameters in buffers containing various additives some information could be obtained regarding which non-covalent forces are involved in an interaction since different forces are targeted by various additives (Andersson et al. 1999). However, the predictions apply only to the particular antigen–antibody system that is being analyzed and they lack the generality that would be necessary to allow the successful engineering of antibodies with predetermined binding properties. Our knowledge of the links between the structure (when crystallographic data are available) and activity of binding sites is too rudimentary to allow a reliable calculation of the energetic contribution that all the atoms in a binding site make to the binding affinity. It is thus not possible to translate structural information into reliable binding energy predictions (Janin 1997). Although some progress has been made in understanding the pathway of protein-protein association, especially regarding the effects of mutation, viscosity, temperature, pH and salt (Roos et al. 1998), the calculation of absolute kinetic parameters from structural data is rarely feasible (Schreiber 2002).

The difficulties that are encountered when attempts are made to rationalize observed differences in binding constants are exemplified in studies of lysozyme–antibody interactions (Lipschultz et al. 2002). A single substitution at position 68 in hen egg lysozyme (Arg/Lys) was found for instance to result in a change in equilibrium constant of over three orders of magnitude mainly because of the presence of an additional water molecule at the epitope-paratope interface (Xavier et al. 1999). In general, it is found that residue substitutions affect dissociation rate parameters more than association rate parameters.

## Knowledge of Structure: Activity Correlates in Antibodies are of Little Use in Vaccine Design

A particular functional activity of antibodies that is of considerable importance is the capacity that some antibodies have to neutralize the infectivity of infectious agents such as viruses (Dimmock 1993). Many studies have been devoted to the structural analysis of monoclonal antibodies endowed with neutralizing activity with the aim of discovering which structural features differentiate antibodies that neutralize from those that do not. The rationale for such studies is the hope that if we understand the molecular basis of antigen–antibody recognition and of antibody neutralization, this would allow to the development of a new generation of vaccines based on structure-based design (Van Regenmortel 2001b). Such unwarranted expectations arise because a clear distinction is not drawn between antigenicity and immunogenicity.

While it is true that knowledge of the structure-activity correlates in a neutralizing antibody combining site may allow one, using molecular design principles, to improve the steric and chemical complementarity between a particular epitope and the corresponding paratope, such a strategy will only affect the binding parameters, i.e. the antigenicity of a molecule and not its immunogenicity. Whereas antigenicity can be reduced to the chemical level of an interaction between an epitope and a paratope, such a reduction is not feasible in the case of immunogenicity, i.e. the ability to induce an immune response, which depends on many complex biological interactions involving various components of the immune system. The immunogenic capacity of a molecule depends on the potentialities of the host being immunized, i.e. the immunoglobulin gene repertoire, self-tolerance, the production of cytokines and various cellular and regulatory mechanisms. These factors cannot be controlled or optimized because one understands the structural basis of a single epitope– paratope interaction, even if the paratope belongs to an antibody that is able to neutralize an infectious agent.

Some of the attempts to develop a vaccine against human immunodeficiency virus (HIV) have followed the classical reductionist approach of analysing the atomic structure of monoclonal antibodies that are directed against HIV and possess the capacity to neutralize the infectivity of the virus. The justification for such studies is the belief that knowledge of structure–function relationships in neutralizing anti-HIV antibodies will facilitate the design of an HIV vaccine. Such studies correspond to a reductionist approach because they focus on the antigenicity of single epitope– paratope pairs and do not take into account:

1. that infectivity neutralization by antibody is a ternary interaction between path-ogen, antibody and host cell;
2. that antigenic sites and antibody combining sites are fuzzy recognition sites, each one consisting of several individual epitopes and paratopes respectively (Van Regenmortel 1999b);
3. that neutralization antigenic sites often consist of several overlapping or nonoverlapping epitopes that are functionally active in a collective manner;
4. that cooperativity effects and synergy between individual antibodies (Zeder-Lutz et al. 2001) cannot be reduced to a simple additive effect of single molecules;
5. that there is no causal link between antigenicity and immunogenicity and no general chemical strategy for designing immunogens that elicit neutralizing rather than non-neutralizing antibodies.

As stated by Wong-Staal and Gallo (2002) 'There is perhaps no greater need in medical science at the inception of the twenty-first century than the development of a preventive vaccine against HIV'. At a time of unprecedented faith in the powers of structural analysis for solving biological problems it is understandable that structuralists and crystallographers would try to address the AIDS vaccine problem by elucidating the structure of antibodies that can protect against viral infection (Kwong et al. 1998; Saphire et al. 2001). An additional aim was to compare the structures of the HIV-1 gp 120 envelope glycoproteins of laboratory-adapted and primary isolates of HIV-1, since it is known that these two types of viral isolates differ in their ability to be neutralized by antibodies (Kwong et al. 2000). The results of such studies indicated that there was considerable flexibility in functionally important regions of the gp 120 protein and that neutralization resistance was specified by quaternary interactions that were not readily amenable to modification by structure-based design (Kwong et al. 2000; Tugarinov et al. 2000; Monaco-Malbet et al. 2000). Although such studies provide extensive information on structure–function correlates in individual neutralizing antibodies, they do not in any way facilitate vaccine design and the search for suitable immunogens capable of inducing protection against disease. On the other hand, the structural information derived from this type of study may help in the development of effective therapeutic inhibitors capable of controlling viral infection, for instance peptides or intracellularly expressed antibodies (Rezacova et al. 2001). A possible application of anti-HIV monoclonal antibodies, used singly or in various combinations is their use in passive immunotherapy, since there is evidence that they may be effective in preventing mother-to-infant transmission of HIV (Gauduin et al. 1997; Mascola et al. 2000). Classical structure-based drug design (Kuntz 1992; Gschwend et al. 1996; Amzel 1998) is feasible because the relationship between a drug and its receptor or target molecule is fairly unique, which is not the case for the polyspecific recognition typical of antigen–antibody interactions. In the absence of rules for designing immunogens that elicit a protective immune response, it seems that structure-based vaccine design is not a realistic enterprise.

**Acknowledgement**  This review was published in a Festschrift in the honour of Allen Edmundson. It has been a rewarding experience to interact with Allen Edmundson over the years, also in his capacity as JMR editor. Our own appreciation of the role of mobility in antigen-antibody interactions (Westhof et al. 1984) was in tune with his flexible keys and adjustable locks, at a time when few crystallographers favoured the mutual adaptation of interacting partners.

# Reductionism and Complexity in Molecular Biology: Scientists Now Have the Tools to Unravel Biological Complexity and Overcome the Limitations of Reductionism

The reductionist method of dissecting biological systems into their constituent parts has been effective in explaining the chemical basis of numerous living processes. However, many biologists now realize that this approach has reached its limit. Biological systems are extremely complex and have emergent properties that cannot be explained, or even predicted, by studying their individual parts. The reductionist approach—although successful in the early days of molecular biology—underestimates this complexity and therefore has an increasingly detrimental influence on many areas of biomedical research, including drug discovery and vaccine development.

The claim made by Francis Crick (1966) that "The ultimate aim of the modern movement in biology is to explain all biology in terms of physics and chemistry" epitomizes the reductionist mindset that has pervaded molecular biology for half a century. The theory is that because biological systems are composed solely of atoms and molecules, without the influence of 'alien' or 'spiritual' forces, it should be possible to explain them using the physicochemical properties of their individual components, down to the atomic level. The most extreme manifestation of the reductionist view is the belief that is held by some neuroscientists that consciousness and mental states can be reduced to chemical reactions that occur in the brain (Bickle 2003; Van Regenmortel 2004a).

Reductionists analyse a larger system by breaking it down into pieces and determining the connections between the parts. They assume that the isolated molecules and their structure have sufficient explanatory power to provide an understanding of the whole system. As the value of methodological reductionism has been particularly evident in molecular biology, it might seem odd that, in recent years, biologists have become increasingly critical of the idea that biological systems

EMBO Reports, 2004, 5, 1016–1020.
Marc H V Van Regenmortel

Marc H V Van Regenmortel, *HIV/AIDS: Immunochemistry, Reductionism and Vaccine Design*, https://doi.org/10.1007/978-3-030-32459-9_6

can be fully explained using physics and chemistry. Their situation is similar to that of an art student asking about the significance of Michelangelo's David and being told that it is just a piece of marble hewn into a statue in 1504. This is certainly true, but it evades pertinent questions about the anatomy of the statue, its creation at the beginning of the Florentine Renaissance, its significance in European art history, or even the scars on its left arm that were plastered after it was broken in three places during the anti-Medici revolt of 1527. In an analogous way, the biology, development, physiology, behaviour or fate of a human being cannot be adequately explained along reductionist lines that consider only chemical composition. Anti-reductionists therefore regard biology as an autonomous discipline that requires its own vocabulary and concepts that are not found in chemistry and physics. Both sides have discussed their standpoints at several recent international meetings (Bock and Goode 1998; Van Regenmortel and Hull 2002; Van Regenmortel 2004a) and the main disagreement between the protagonists is about what constitutes a good scientific explanation.

## The Most Extreme Manifestation of the Reductionist View Is the Belief that Is Held by Some Neuroscientists that Consciousness and Mental States Can Be Reduced to Chemical Reactions that Occur in the Brain

Today, it is clear that the specificity of a complex biological activity does not arise from the specificity of the individual molecules that are involved, as these components frequently function in many different processes. For instance, genes that affect memory formation in the fruit fly encode proteins in the cyclic AMP (cAMP) signalling pathway that are not specific to memory. It is the particular cellular compartment and environment in which a second messenger, such as cAMP, is released that allow a gene product to have a unique effect. Biological specificity results from the way in which these components assemble and function together (Morange 2001a). Interactions between the parts, as well as influences from the environment, give rise to new features, such as network behaviour (Alm and Arkin 2003), which are absent in the isolated components.

Consequently, 'emergence' has appeared as a new concept that complements 'reduction' when reduction fails (Van Regenmortel 2004a). Emergent properties resist any attempt at being predicted or deduced by explicit calculation or any other means. In this regard, emergent properties differ from resultant properties, which can be predicted from lower-level information. For instance, the resultant mass of a multi-component protein assembly is simply equal to the sum of the masses of each individual component. However, the way in which we taste the saltiness of sodium chloride is not reducible to the properties of sodium and chlorine gas. An important aspect of emergent properties is that they have their own causal powers, which are not reducible to the powers of their constituents. For instance, the experience of pain can alter human behaviour, but the lower-level chemical reactions in the neurons that are involved in the perception of pain are not the cause of the altered behaviour, as the pain itself has causal efficacy. According to the principles of emergence, the natural world is divided into hierarchies that have evolved over evolutionary time (Kim 1999; Morowitz 2002). Reductionists advocate the idea of 'upward causation' by which molecular states bring about higher-level phenomena, whereas proponents of emergence accept 'downward causation' by which higher-level systems influence lower-level configurations (Kim 1999).

## Anti-Reductionists Regard Biology as an Autonomous Discipline that Requires Its Own Vocabulary and Concepts that Are Not Found in Chemistry and Physics

Although biology has always been a science of complex systems, complexity itself has only recently acquired the status of a new concept, partly because of the advent of electronic computing and the possibility of simulating complex systems and biological networks using mathematical models (Emmeche 1997; Alm and Arkin 2003). Because complex systems have emergent properties, it should be clear from the preceding discussion that their behaviour cannot be understood or predicted simply by analysing the structure of their components. The constituents of a complex system interact in many ways, including negative feedback and feed-forward control, which lead to dynamic features that cannot be predicted satisfactorily by linear mathematical models that disregard cooperativity and non-additive effects. In view of the complexity of informational pathways and networks, new types of mathematics are required for modelling these systems (Aderem and Smith 2004).

Another essential property of complex biological systems is their robustness (Csete and Doyle 2002; Kitano 2002). Robust systems tend to be impervious to changes in the environment because they are able to adapt and have redundant components that can act as a backup if individual components fail. A further characteristic of complex systems is their modularity (Alm and Arkin 2003): subsystems are physically and functionally insulated so that failure in one module does not spread to other parts with possibly lethal consequences. This modularity, however, does not prevent different compartments from communicating with each other (Weng et al. 1999). An additional peculiarity of complex biological systems is that they are open—that is, they exchange matter and energy with their environment—and are therefore not in thermodynamic equilibrium. In the past, the reductionist agenda of molecular biologists has made them turn a blind eye to emergence, complexity and robustness, which has had a profound influence on biological and biomedical research during the past 50 years. In the following sections, I describe some of the harmful effects of reductionist thinking in drug-discovery programmes and vaccinology.

The number of new drugs that are approved by the US Food and Drug Administration has declined steadily from more than 50 drugs per annum 10 years ago to less than 20 drugs in 2002. This worrying trend has persisted despite continuous mergers and acquisitions in the industry and annual research and development expenditures of approximately US$30 billion. Commentators have attributed this poor performance to a range of institutional causes, such as inefficient project management, increased regulatory requirements, a decline in the clinical science that deals with whole organisms, an overemphasis on technology-driven research and an unwillingness to concentrate on products that are not likely to generate sales of at least US$0.5–1.0 billion per annum (Drews 2003; Gershell and Atkins 2003; Kubinyi 2003; Miska 2003). Furthermore, it seems that the new strategies of drug discovery, which are based on high-throughput screening, combinatorial chemistry, genomics, proteomics and bioinformatics, are not bringing forth the new products that were anticipated (Kubinyi 2003; Glassman and Sun 2004). Knowledge of the genome sequences of humans and various pathogenic agents has led to the identification of only a limited number of new drug targets (Drews 2003). Moreover, Glassman and Sun (2004) listed several biotechnological projects that have, so far, failed to live up to expectations, including gene therapy, stem-cell research, antisense technology and cancer vaccines. A common problem with many of these innovations is that their potential risks and unwanted side effects tended to be overlooked initially, as was the case for gene therapy (Williams and Baum 2003).

## It Remains True that Human Disease Is Best Studied in Human Subjects

However, there is probably a more fundamental reason for these failures: namely, that most of these approaches have been guided by unmitigated reductionism. As a result, the complexity of biological systems, whole organisms and patients tends to be underrated (Horrobin 2001). Most human diseases result from the interaction of many gene products, and we rarely know all of the genes and gene products that are involved in a particular biological function. Nevertheless, to achieve an understanding of complex genetic networks, biologists tend to rely on experiments that involve single gene deletions. Knockout experiments in mice, in which a gene that is considered to be essential is inactivated or removed, are widely used to infer the role of individual genes. In many such experiments, the knockout is found to have no effect whatsoever, despite the fact that the gene encodes a protein that is believed to be essential. In other cases, the knockout has a completely unexpected effect (Morange 2001a). Furthermore, disruption of the same gene can have diverse effects in different strains of mice (Pearson 2002). Such findings question the wisdom of extrapolating data that are obtained in mice to other species. In fact, there is little reason to assume that experiments with genetically modified mice will necessarily provide insights into the complex gene interactions that occur in humans (Horrobin 2003).

## Vaccination Is Firmly Anchored in the Biological Realm and Cannot Be Reduced to the Level of Chemistry

The disappointing results of knockout experiments are partly caused by gene redundancy and pleiotropy, and the fact that gene products are components of pathways and networks in which genes acting in parallel systems can compensate for missing ones (Morange 2001b). As many factors simultaneously influence the behaviour of a system, one part might function only in the presence of other components. The essential contribution of other genes in achieving a particular function will therefore be missed, which will further encourage the reductionist view that a single gene has adequate explanatory power (Van Regenmortel 2004a).

Another factor that is responsible for disappointing results in drug discovery is the excessive reliance on in vitro systems. Many researchers claim that in vitro cell cultures, or even computer models, might be able to reflect accurately and reliably the functioning of an intact human. There is considerable evidence, however, for a lack of congruence between in vitro assays and the in vivo systems that they attempt to model. There is no doubt that pharmaceutical research is hampered by insufficient whole-animal studies. Furthermore, even animal models of human disease are often inadequate and are a poor surrogate for clinical studies in humans. It remains true that human disease is best studied in human subjects (Horrobin 2003).

Another defect of reductionist thinking is that it analyses complex network interactions in terms of simple causal chains and mechanistic models. This overlooks the fact that any clinical state is the end result of many biochemical pathways and networks, and fails to appreciate that diseases result from alterations to complex systems of homeostasis. Reductionists favour causal explanations that give undue explanatory weight to a single factor. By contrast, many biologists favour functional explanations for a structure or cellular process, and emphasize the selective advantage of these features during evolutionary history—after all, evolution selects for function, not structure. Functional explanations are more useful for understanding complex biological systems with many interactions than are causal explanations that give unwarranted importance to a single factor (Van Regenmortel 2002a). Lewontin (2000) also stressed the reciprocal relationships between genes, organisms and their environment, in which all three elements act as both causes and effects.

Another area of biomedical research that has been strongly influenced by reductionist thinking is the so-called rational design of vaccines, which is based on the assumption that the principles of structure-based drug design are applicable to vaccines. However, this disregards the fact that the relationship between a drug and its receptor or target molecule is fairly specific, whereas the relationship between an antigen and an antibody is much less restricted. The binding site of an immunoglobulin molecule comprises around 50 hypervariable residues that together make up the complementarity determining regions (CDRs). Approximately 10–15 of these residues usually participate in the interaction with an individual epitope, but the full complement of all 50 hypervariable residues does not constitute an actual binding site for any epitope. This means that around 35 CDR residues can potentially bind to other epitopes that bear little or no resemblance to the first, which explains the extensive multispecificity of immunoglobulins and the occurrence of many different paratopes or binding sites in each molecule. The ability of an immunoglobulin molecule to bind various antigenic structures is further enhanced by the considerable flexibility of the CDRs, which allows the binding site to adopt various conformations (James et al. 2003). The binding reaction involves a combination of conformational selection and induced fit (Bosshard 2001; Goh et al. 2004), and entails a mutual adaptation of the two interacting partners (Westhof et al. 1984; Tainer et al. 1985).

In recent years, rational design has become fashionable in vaccine research as opposed to empirical discovery (Van Regenmortel 2000). The term 'rational' implies that research makes extensive use of molecular data and structural knowledge, whereas the term 'design' indicates that the biological activity of the developed products is predictable. Rational design is therefore presented as a more scientific approach than the empirical 'trial-and-error' screening and selection of molecules. The belief that a molecular-design strategy will be successful for developing new vaccines is typical of the reductionist mindset, as it assumes that a biological phenomenon, such as protection against infection, can be reduced to the level of chemistry. However, there are many reasons why a reductionist approach to vaccine development is unlikely to succeed.

## The Impossibility of Reducing Biology to Chemistry Is Responsible for the Lack of Success in Developing Structure-Based Vaccines

First, the antigenic determinants, or epitopes, of an infectious agent are emergent entities that are defined by their specific antibody partners and exist only in the context of the immune system. Epitopes and paratopes are not intrinsic features of an antigen and an immunoglobulin molecule, respectively, and cannot be identified independently of a binding reaction. Furthermore, antigenic and immunoglobulin-combining sites are fuzzy recognition sites that consist of several individual epitopes and paratopes (Van Regenmortel 1999a). Whereas a molecular-design strategy for improving antigenic reactivity is applicable to a single pair of interacting molecules-for instance, one epitope and a monoclonal antibody, it cannot be applied to the numerous epitopes that are involved in the protective immune response to a given pathogen.

Second, eliciting antibodies that simply bind to the pathogen is of little value in vaccine development. What are required are antibodies that have a functional activity, namely, the ability to neutralize the infectious agent in vivo. Our ability to predict the function of proteins is limited and our capacity to predict the neutralizing activity of an antibody from its chemical structure is practically nonexistent (Van Regenmortel 2000, 2002a). Vaccination and protective immunity have a meaning only at the level of the whole organism: molecules, tissues and organs cannot be vaccinated. Vaccination is therefore firmly anchored in the biological realm and cannot be reduced to the level of chemistry.

Third, despite an unprecedented global research effort, no vaccine against human immunodeficiency virus (HIV) is in sight (Burton and Moore 1998). Although a reductionist approach to HIV-vaccine development continues to be advocated (Burton et al. 2004a), there is no evidence that this will be effective. This approach involves determining the atomic structure of monoclonal antibodies against HIV antigens using X-ray crystallography, with the aim of elucidating the structure of the HIV epitopes. The justification for these studies is the assumption that knowledge of the structure of the epitopes that are recognized by neutralizing antibodies will help to design an effective HIV vaccine. The X-ray crystallographic analysis of broadly reactive HIV-neutralizing antibodies might indeed determine the structure of epitopes inside antibody-binding pockets, but it does not tell us how to use immunization to induce antibodies with the same specificity (Van Regenmortel 2002a; Burton et al. 2004a). The structures of epitopes and paratopes that are present in a complex represent the final conformation at the end of a dynamic process of conformational selection, induced fit and somatic mutation. It is not possible to infer which epitope conformation in the immunogen was ultimately responsible for the appearance of neutralizing antibodies.

In fact, immunogenicity depends on the biological potential of the host that is being immunized; in other words, on extrinsic factors, such as the immunoglobulin gene repertoire, self-tolerance, the production of cytokines, and various cellular and regulatory mechanisms. Unfortunately, we do not know how to control these properties of the immune system to produce neutralizing antibodies (Van Regenmortel 2001b). Further difficulties are that antibodies act in a collective manner and that the neutralizing synergy between various antibodies cannot be reduced to the simple additive effect of individual molecules (Zeder-Lutz et al. 2001). Sometimes the synergy occurs because the binding of one antibody leads to a conformational change in the antigen, which then increases its accessibility for other antibodies.

The reductionist approach of using peptide fragments of a virus protein for vaccination purposes has also achieved little success (Van Regenmortel 2001b). The peptides that are used are either short sequences, which are known as continuous epitopes of the viral protein (Van Regenmortel 1999a), or so-called mimotopes, which are peptides that show little or no sequence similarity to any of the viral proteins but are believed to mimic a discontinuous epitope of the virus (Meloen et al. 2000). Discontinuous epitopes are made up of amino-acid residues from distant regions of the viral protein, which are brought together by the folding of the peptide chain. As antibodies harbour many paratope subsites-each able to bind to related or unrelated epitopes-it remains possible that the mimotope binds to a different subsite from the one that interacts with the discontinuous epitope that induced the antibody. In fact, the extent of mimicry achieved by a mimotope peptide might be so limited that it might not be able to elicit antibodies that recognize and neutralize the virus (Van Regenmortel 1999a). Attempts have been made to reconstitute discontinuous epitopes by synthesis or by the selection of phage-displayed peptides. However, although such reconstituted epitopes might bind to viral antibodies, they are rarely able to elicit protective antibodies (Enshell-Seijffers et al. 2003; Oomen et al. 2003; Villen et al. 2004).

## Extreme Holism, According to Which Everything Is Connected, Certainly Does Not Provide a Methodological Alternative

Once more, it is the failure to distinguish antigenicity—that is, antigenic reactivity—from immunogenicity that leads to the unwarranted expectation that it should be relatively straightforward to design effective peptide-based synthetic vaccines. The impossibility of reducing biology to chemistry is responsible for the lack of success in developing structure-based vaccines. Moreover, it is safe to assume that vaccine development will continue to rely on the same empirical approaches that have been used successfully in the past (Van Regenmortel 2001b, 2002a).

In light of these failures, it has become popular to criticize the reductionist approach that is used in the study of biological systems (Lewontin 2000), although it is more difficult to determine what should be done instead. Extreme holism, according to which everything is connected, certainly does not provide a methodological alternative. What are needed are new experimental techniques for investigating the unique complexity of biological systems that results from the bewildering diversity of interactions and regulatory networks. Recent developments in high-throughput microarrays, nanotechnologies, bioinformatics and systems biology are providing data that molecular biologists might use to simulate the behaviour of complex biological networks and systems (Kitano 2002; Alm and Arkin 2003; Aderem and Smith 2004; Blake 2004). If these simulations make it possible to predict the reactions of a system, we will have achieved some degree of understanding, even if we cannot identify the innumerable causal interactions that are involved (Berger 1998).

Gene ontologies that provide a standardized vocabulary for data exploration (Blake 2004), and software programmes, such as Cytoscape (Aderem and Smith 2004), which create visual representations of biological systems, make it possible to handle enormous amounts of data and build useful models of complex systems. An important present limitation is the paucity of quantitative information about the kinetic parameters that underlie all protein–protein and protein–DNA interactions (Alm and Arkin 2003). However, it is undeniable that molecular biologists now have at their disposal tools that may help to unravel biological complexity and overcome the limitations of reductionism. Given our failures in developing drugs and vaccines against a wide range of debilitating diseases, this move away from the reductionist viewpoint and toolset is a high priority for both biological and biomedical research.

# Editorial: Biological Complexity Emerges from the Ashes of Genetic Reductionism

Fifty years after the elucidation of the double helix structure of DNA, heralded by some as providing the solution to the mystery of life, an increasing number of biologists have become critical of the reductionist view that biological systems can be fully explained by the physico-chemical properties of their constituent parts. Biology is increasingly regarded as an autonomous discipline requiring its own explanatory concepts not found in chemistry and physics, and it has become accepted that biological systems can only be understood in terms of their evolutionary history on Earth. The current rejection of reductionism as an acceptable approach for explaining biological phenomena may seem paradoxical since the value of reductionism as a research strategy for dissecting and analysing the constituents of biological systems has never been questioned. The outstanding success of molecular biology in unravelling the structural and chemical basis of living processes is indeed a clear testimony of the value of reductionism as an analytical methodology.

Dissatisfaction with reductionism has a different origin altogether and arises because scientists nowadays increasingly question the validity of reductive explanations for providing an understanding of what is causally relevant in bringing about biological phenomena. The debate between reductionists and antireductionists is thus very much a debate about what constitutes a good scientific explanation (Byerly 2003). The limits of reductionism for providing adequate biological explanations have been discussed at a number of international meetings (Bock and Goode 1998; Van Regenmortel and Hull 2002), the most recent meeting being the conference on Reduction and Emergence organized by Max Kistler at the Ecole Normale Supérieure in Paris on 12–15 November 2003. This conference brought together 25 philosophers and scientists who discussed in depth the nature and limits of reductive explanations in the physical and biological sciences. Many of the

Journal of Molecular Recognition, 2004, 17, 145–148.
Marc H V Van Regenmortel

© Springer Nature Switzerland AG 2019
Marc H V Van Regenmortel, *HIV/AIDS: Immunochemistry, Reductionism and Vaccine Design*, https://doi.org/10.1007/978-3-030-32459-9_7

presentations at the meeting are available on the web site: http://www.institutnicod. org/reduction.htm. Some of the issues that were discussed concerning reduction and emergence in molecular biology are summarized below.

## The Limits of Reductionist Explanations in Biology

According to William Wimsatt, a reductionist is interested in understanding the properties and behaviour of a biological system in terms of the properties of its parts and their interrelations and interactions. A reductionist starts by choosing, designating or constructing a system for analysis, and by so doing he will partition the world of study into that system and its environment. Assumptions will be made about which variables must be controlled or randomized and the description and observation of the environment will be greatly simplified compared with the description of the entities internal to the system of study. By fixing the environment, the reductionist makes it artificially disappear as a variable. Fitness, for example, will be described as if it were a property of phenotypes or genes and the fact that it is a relation between organism and environment will tend to be ignored. By keeping environmental variables constant, the reductionist will be unaware of how the system depends on them and he will tend to overlook the context-dependence of biological phenomena. When environmental variables are not monitored, the data necessary for detecting external interactions will simply not be available (Wimsatt 2007).

Reductionists believe that the behaviour of a biological system can be explained by the properties of its constituent parts and by an analysis of the multiple interactions that exist between these parts. They assume that, in principle, the activity of the whole can be inferred, deduced, calculated and predicted from the properties of the parts. Antireductionists disagree and maintain that the properties of the whole cannot be deduced from the properties of the parts. They argue that interactions between parts as well as certain inputs from the environment give rise to novel features that are absent in the parts taken in isolation. Antireductionists stress that biological systems possess emergent (also called relational or systemic) properties that are possessed only by the system as a whole and not by any isolated part of the system.

According to Achim Stephan, a property of a system is considered to be emergent (1) if it is irreducible to properties of the parts, i.e. if it is not reductively explainable, and (2) if it could not have been predicted before its first instantiation. The function of a protein, for instance, cannot be predicted de novo from its structure, although it can sometimes be predicted by comparison with a protein of similar structure whose function had previously been established empirically.

A binding site in a protein molecule is also a relational or emergent feature since it can only be defined by the relational nexus with a particular interacting partner. Intrinsic structural features of the protein cannot identify the binding site independently of the ligand. The specificity of a binding reaction is also an emergent property since it corresponds to the discriminating capacity that a molecule possesses when faced with two different ligands. Specificity is thus a ternary relational property that requires at least three potential interactors in order to be meaningful (Van Regenmortel 1998).

Since scientists tend to subscribe to a world view known as physicalism according to which nothing exists but the physical, they accept that systems that possess emergent properties nevertheless consist exclusively of physical entities. For biologists, living beings and cognitive systems consist of the same basic entities that make up inanimate nature.

Emergence is a concept that is complementary to reduction since emergence is said to occur when reduction fails. Emergent properties cannot be reduced to properties of the constituents parts of the system and they resist any attempt at being predicted or deduced by explicit calculation or any other means. Jaegwon Kim emphasized that emergent properties have their own distinctive causal powers and that causality is not restricted to lower level phenomena. For instance, the experience of pain can alter human behaviour and it is not necessary, in order to explain this, to invoke the causal efficacy of lower level chemical reactions occurring in neurons. Downward causation is equally acceptable as an explanatory scheme as is a bottom up explanation that moves from lower level to higher level phenomena.

Biological phenomena can be studied at different levels, depending on the question being asked, and the nature of the enquiry is in fact defined by the analytical approaches and methodology that are used. The muscle contraction that occurs when a frog jumps can be studied as a physiological phenomenon or biochemically as the interaction between actin and myosin. However, the biochemical event is not the cause of the physiological event since both occur simultaneously. According to Achinstein's classification of scientific explanations, the biochemical event is not a causal explanation of the physiological event but corresponds to a so-called identity explanation (Achinstein 1983). Instead of reducing a phenomenon described at a higher level to one at a lower level, this type of explanation simply replaces one reaction by another. Kenneth Schaffner incorporates this type of pseudo-reduction in his general reduction replacement model (Schaffner 1993). Instead of being an instance of a successful reduction of biology to chemistry, such an explanation amounts to a shift in subject matter whereby the attempt at explaining the biological phenomenon has been abandoned. When the cellular or organismic context is left out of the picture, the biological enquiry becomes a purely chemical investigation since it attempts to provide answers to chemical rather than to biological questions.

## Causal Explanations Versus Functional Explanation

A causal explanation is reductive in the sense that one factor is singled out for attention and is given undue explanatory weight on its own. In biological systems, any observed effect always results from a complex network of interactions and an explanation in terms of a single cause is never satisfactory. Instead of invoking causes, it is better to refer to the many factors that simultaneously influence the behaviour of a biological system (Van Regenmortel and Hull 2002, p. 50). It should also be stressed that the initial state of a thing is not the cause of its subsequent state. Since things are not causes, DNA sequences and genes cannot be causes of phenotypic traits or of behaviour patterns (Mahner and Bunge 1997, p. 39).

It is not possible to describe biological systems without referring constantly to the roles that molecules, organelles, cells and organs play in keeping an organism alive. To a human observer, organisms appear to have goals and purposes and it may be tempting to try to explain their behaviour in terms of such end-directed features. However, when biologists attribute a function to a cellular process and explain it by referring to its goal or purpose, they do not imply the type of human intentionality that is characteristic of purposive human action. Instead, the occurrence of biochemical and cellular processes is explained by the fact that they contribute to the reproductive fitness of the organism in which they are found. Biologists use functional explanations for a currently observed structure or cellular process and they point to the selective advantage that these features conferred to the organism during its evolutionary history. The strength of functional explanations in biology lies in the fact that functions have to pass through the sieve of natural selection.

It is unfortunate that the term biological function is highly ambiguous. Biochemists tend to regard the function of a protein simply as what the molecule does, i.e. its functioning or activity at the molecular level. However, functions become meaningful only when the biological context is taken into account and this implies that they must be analysed at the level of the cell and of the organism as a whole. Although the activity of an enzyme such as trypsin can be analysed purely at the chemical level, for instance in terms of which peptide bonds in a protein are cleaved, the biological function of trypsin emerges only at the cellular and physiological levels through its participation in protein degradation and digestion processes. Molecular biologists tend to emphasize so-called molecular functions analysed at the biochemical level although functions possess a biological connotation only when they are integrated at the cellular level and they are actually playing a role in ensuring the self-maintenance of a living organism.

## The Rejection of Genetic Reductionism and Determinism

Michel Morange analysed in detail why genetic determinism has lost its attraction as an adequate explanation for the appearance of specific phenotypic traits. In the wake of 50 years of highly successful research in molecular biology (Morange 1998), we have at present a very good understanding of the intricate mechanisms that allow genes to be translated into proteins. However, this knowledge has given us very little insight about the causal chains that link genes to the morphological and other phenotypic traits of organisms.

Many attempt have been made to characterize all the genes and gene products involved in particular biological functions. Knockout experiments, whereby a gene considered to be essential is inactivated or removed, have been widely used in an effort to confirm the hypothetical function of a gene established on the basis of in vitro experiments. In many such experiments, the knockout was found to have no effect whatsoever, despite the fact that the gene was known to code for a protein thought to be essential. In other cases, the knockout had an effect completely different from the one that was expected. According to Morange, such unexpected findings are in part due to the fact that gene products are components of pathways and networks. In a network it is always difficult to anticipate the modifications which may result from the alteration of one of the links.

Two other features also contribute to the puzzling results of knockout experiments, namely gene redundancy and gene pleiotropy (or multi-functionality). Gene redundancy dampens the effects of the inactivation of a single gene while gene pleiotropy is responsible for the fact that one genotype can lead to many different phenotypes (Morange 2001a). Wimsatt attributes the surprise generated by the unexpected results of knockout experiments to unwarranted assumptions on the part of the experimentalist. One fallacy is to assume that the function of a gene is to produce whatever the system fails to do when that gene is absent. This could lead to the view that spark plugs are 'sputter suppressors'. Another fallacy is to assume that the function of the gene is to produce whatever the system produces when that gene is activated or simulated. Since many factors always simultaneously influence the behaviour of a system, the function of a part may be realized only against a background of the activities of many other interacting components. The essential contribution of other genes in achieving a particular function may thus be missed because they are always there to provide it as background in a constant context. Furthermore, no consideration is given to the possibility that the influence of a gene deletion may not be revealed because of the testing conditions used. In most cases, therefore there is no justification for considering that the subtracted or added behaviour of the system represents the so-called function of the manipulated gene.

When it comes to human behaviour, genetic determinism has become totally discredited for providing plausible explanations that link traits to genes. Since genes never act in isolation and any trait is always the result of the activity of many gene products, recurrent reductionist claims that genes have been discovered for alcoholism, violence, depression, rape or criminal behaviour are nowadays rarely taken seriously, except in the popular press (Van Regenmortel and Hull 2002).

## Cancer as a Tissue Organization Problem

Carlos Sonnenschein and Ana Soto described their tissue organization field theory of neoplasia which postulates that carcinogens act by disrupting the normal interactions that take place between the cells of certain tissues and organs, in particular in epithelium and stroma tissues. According to this theory, carcinogenesis should be studied at the level where it is identified, i.e. at the tissue level of organization rather than at the cellular or molecular levels. In current cancer research, established cell lines are studied in vitro on the assumption that it is legitimate to reduce phenomena occurring at the tissue or organismic level to purely cellular phenomena. Cancers are reduced to transformed cells and carcinogenesis is reduced to enhanced proliferation of cells in a cultured dish. Sonnenschein and Soto believe that the somatic mutation theory (SMT), which has been for many years the prevailing paradigm in the field of carcinogenesis, is based on an unwarranted reductionist simplification (Sonnenschein and Soto 1999, 2000). According to the SMT, cancer arises when a single somatic cell becomes a neoplastic cell through the accumulation of multiple mutations in genes controlling cell proliferation and the cell cycle. The SMT encouraged biologists to search for oncogenes which were believed to correspond to a particular kind of gene responsible for excessive cell proliferation. Since cancer actually results from the deregulation of normal regulatory processes, oncogenes in fact do not correspond to a particular category of genes. Although more than a hundred oncogenes have been identified, the presumed association between a pattern of mutated oncogenes and cancer type has not been found. The effect of an oncogene depends on the type of cell in which it is expressed and overexpression of a given oncogene may enhance growth in one cell type and induce apoptosis in another (Weinstein 2002). Furthermore, the neoplastic phenotype can be normalized at a frequency much higher than when a mutant reverts to wild-type. It has also been observed that liver carcinoma cells can revert to normal cells when they are injected into normal livers and that normal cells can be made to behave like carcinoma when they are repositioned. It seems therefore that, instead of explaining cancer by the action of mutated genes operating at the level of single cells, it is possible to interpret the cancer phenotype as an emergent phenomenon of altered gene expression arising at the tissue level of biological organization.

## Molecular Neuroscience: The Last Bastion of Ruthless Reductionism

From the account I have presented so far of the discussions that took place at the Paris meeting on Reduction and Emergence, it should be clear that the general mood of the participants was very much a case of 'emergence is in, reductionism is out'. Genetic reductionism, in particular, has been abandoned as a useful explanatory scheme for understanding the phenotypic traits of complex biological systems. Genes are increasingly studied today because they are involved in the genetic programme that unfolds during development and embryogenesis rather than as agents responsible for the inheritance of traits from parents to offspring. It is now recognized that, during embryogenesis, morphogen gradients are formed by the diffusion of certain gene products away from a given source. Such gradients have been shown, for instance, to lead to the development of the anteroposterior and ventral–dorsal axes in Drosophila and in higher organisms. It can be said that the morphogens produce so-called positional information that allows cells to 'know' where they are (Fox-Keller 2002/2003).

Complex systems are defined as systems that possess emergent properties and which, therefore, cannot be explained by the properties of their component parts. Since the constituents of a complex system interact in a non-linear manner, the behaviour of the system cannot be analysed by classical mathematical methods that do not incorporate cooperativity and non-additive effects. Another peculiarity of complex biological systems is that they are open systems which exchange matter and energy with the environment and which, therefore, are not in thermodynamic equilibrium.

The acceptance of emergence and complexity as essential concepts for understanding the uniqueness of biological phenomena is responsible for the decreased popularity of 'Nothing but-ism' statements of the type: 'Genes are the only units of selection', 'Organisms are nothing but bags of genes', 'Biological entities are nothing but physico-chemical systems', 'The mind is nothing but neural activity'. One glaring exception to this general trend was the presentation by John Bickle entitled 'Reducing mind to molecular pathways'. In line with the title of his 2003 book, Philosophy and Neuroscience: a Ruthlessly Reductive Account, Bickle presented an ebullient and spirited defence of the reductionist view that molecular and cellular neuroscience can explain cognition. He described in detail the biochemical reactions that occur when thin slices of the hippocampus of rodents are submitted to electrical stimuli. Calcium channels open up, calcium flows into cells and binds to calmodulin, the resulting complex converts ATP to cAMP and the amount of glutamate in the tissue increases. Since all these molecular events also occur when the animal is remembering a visual 'input', Bickle claims that this establishes a 'link' between cognition and certain molecules and that it amounts to a successful reduction of mind to cellular and molecular pathways!

A phenomenon known as long-term potentiation (LTP) is a favourite topic of study by neurophysiologists. LTP is the conversion of labile, short-term memories into long-term memories lasting several days and can be brought about by short electrical pulses. Furthermore LTP can be abolished in knockout mice when a protein that binds to cAMP is deleted. This allows neuroscientists to intervene at the molecular level in specific neurons and to track the resulting change in memory consolidation. Bickle maintains that this approach which he calls 'intervene cellularly/molecularly and track behaviourally' provides a full reductive explanation of cognitive behaviour. During the ensuing discussion, Huub Looren de Jong pointed out that a link is not a reduction and that reduction is not the correct term for describing the experimental protocols used by neuroscientists. For Bickle, however, cognitive scientists who disagree with his ruthless reductionist stance simply demonstrate their ignorance of mainstream molecular neuroscience and its experimental protocols!

One could not help being reminded of the simplistic expectations of molecular biologists who, in the not too distant past, assumed that the whole of biology was going to be successfully reduced to chemistry and physics. Molecular neuroscientists are currently discovering that methodological reductionism is a powerful analytical approach to obtain information on how the components of nervous tissue work and interact. As molecular biologists do, they now publish their data in prestigious scientific journals like Nature, Science and Cell. No doubt they will discover in due course that the notions of emergence and complexity must also be applied to their field of investigation and that reductionism is not the panacea for understanding the mind.

# The Rational Design of Biological Complexity: A Deceptive Metaphor

## Reductionism, Emergence, and Complexity

The reductionist approach of decomposing biological systems into their constituent parts has dominated molecular biology for half a century. Since organisms are composed solely of atoms and molecules without the participation of extraneous forces, it has been assumed that it should be possible to explain biological systems on the basis of the physico–chemical properties of their individual components, down to the atomic level. However, despite the remarkable success of methodological reductionism in analyzing individual cellular components, it is now generally accepted that the behavior of complex biological systems cannot be understood by studying their individual parts in isolation (Dupré 1993; Rose 1997; Bhalla and Iyengar 1999; Van Regenmortel 2004b). The reason is that complex systems possess so-called emergent (also called relational or systemic) properties that are not present in the isolated components and cannot be predicted, deduced, or calculated from the properties of the parts (Holland 1994). Interactions between parts as well as certain inputs from the environment give rise to novel features such as network behavior that exist only in the system as an integrated whole (Alm and Arkin 2003). Emergence is an explanatory concept that is complementary to reduction since emergence is invoked when reductionist dissection is not able to explain the behavior of a complex system (Van Regenmortel 2004b).

Biology has always been a science of complex systems that are made up of many parts (Weng et al. 1999) but it is only recently that the notion of complexity has acquired the status of a new analytical concept, partly through the development of informatics and the possibility of simulating complex systems and networks using mathematical models (Emmeche 1997; Auffray et al. 2003; Barabasi and Oltvai

Proteomics, 2007, 7, 965–975.
Marc H V Van Regenmortel
Copyright © 2007 WILEY-VCH Verlag GmbH & Co. KGaA, Weinheim

© Springer Nature Switzerland AG 2019
Marc H V Van Regenmortel, *HIV/AIDS: Immunochemistry, Reductionism and Vaccine Design*, https://doi.org/10.1007/978-3-030-32459-9_8

2004). The constituents of complex systems interact in various ways such as negative feedback and feed-forward control and this leads to dynamic behavior which does not fit linear mathematical models that disregard cooperativity and synergy. In order to unravel the complexity underlying biological systems, it is necessary to investigate the rules and patterns of the multiple interactions that link all the components of such systems (Csete and Doyle 2002). This is one of the major goals of systems biology (Kitano 2002; O'Malley and Dupré 2005).

## The Nature of Biological Systems

A system is a complex object of which every part or component is related to at least one other component (Bunge 2003). Systems can be analyzed in terms of their composition, environment, structure, and dynamics. The environment of a system is the collection of things that act on, or are acted upon, the system's components. Scientists often find it difficult to partition the material they study into components internal or external to the core of the system. In their experiments, they tend to assign certain factors to the environment and they then usually assume that these are invariant. As a result, the environmental factors artificially disappear as variables and the biological system appears to be context-independent. Such a procedure conceals the fact that biological systems are open systems that exchange matter and energy with their environment and are not in thermodynamic equilibrium.

The structure or organization of a biological system is the collection of relations (i.e. bonds, links, and connections) existing among the system's internal components as well as among these and the environment. An important aspect of cellular organization is the existence of cycles, for instance, the cell cycle and various biochemical metabolic cycles. A system's dynamics arises from the many interactions that occur between the individual elements of pathways and networks. These give rise to characteristic features of living systems such as their self-organization, autonomy, homeostatic control, and self-repair potential.

Biological systems usually comprise a number of subsystems or modules that have a composition and structure that are included in those of the more encompassing system but which nevertheless are physically insulated and functionally semi-autonomous. This modularity (Alm and Arkin 2003; Weng et al. 1999; Emmeche 1997; Auffray et al. 2003; Barabasi and Oltvai 2004) ensures that failure in one subsystem does not necessarily spread to other parts with possibly lethal consequences. However, this does not prevent different subsystems from communicating and interacting synergistically with each other (Weng et al. 1999; Csete and Doyle 2002). Examples of subsystems are genetic networks, metabolic pathways, organelles, the circulatory system, and whole organs.

The development of high-throughput and "omics" methods for collecting data, for instance on protein–protein and protein–DNA interactions, has shown that biological systems can be described as molecular interaction networks (Alm and Arkin 2003; Barabasi and Oltvai 2004). Networks consist of a series of nodes representing genes, proteins, or metabolites that are connected to each other by links, each link representing the interactions between the components. Nodes are characterized by their connectivity or degree which indicates how many links the node has to other nodes. The links can have a selected direction which represents, for example, the direction of material flow from a transcription factor to the gene that it regulates. In undirected networks, the links do not have an assigned direction. This is the case for protein interaction networks where a link represents a mutual binding relationship (Barabasi and Oltvai 2004). Most biological networks have been found to be so-called small-world, scale-free networks characterized by a power-law connectivity distribution, where most of the nodes have only a few links but a few nodes, called hubs, have a large number of links. In metabolic networks, most substrates participate in only one or two reactions, while a few participate in dozens of reactions and function as metabolic hubs (Barabasi and Oltvai 2004).

One of the aims of systems biology is to determine how the various interaction networks (i.e. protein–protein interaction, metabolic, signaling, and transcription-regulatory networks) form a network of networks that leads to the characteristic integrated behavior of living systems (O'Malley and Dupré 2005). Bork and Serrano (2005) pointed out that systems biology will need to utilize four-dimensional analyses incorporating spatial and temporal data if it is to achieve its declared aim of successfully simulating, predicting and controlling biological systems. In particular, it will be necessary to reconcile its reliance on structural data with the dynamics inherent in the mobility of molecules (Uversky et al. 2005) and the mutual adaptation of interacting partners (Demchenko 2001; Goh et al. 2004).

Two other features of complex biological systems are their robustness and the fact that they possess many redundant components that are able to perform the same or similar tasks. Robustness is present at all levels of biological organization and means that systems continue to function in the face of major genetic or environmental perturbations, as demonstrated by the ability of organisms to survive and reproduce (Kitano 2002; Wagner 2005). Proteins can tolerate thousands of amino acid changes, metabolic networks continue to sustain life even after the removal of important chemical reactions, and gene regulation networks continue to function after alteration of key gene interactions (Wagner 2005, p. 2). The evolvability of organisms is actually facilitated by their robustness.

## Function and Functional Language in Biology

A further peculiarity of biological systems is that it seems difficult if not impossible to describe them without referring constantly to the functional roles that cellular and molecular constituents play in meeting the needs of organisms and keeping them alive. Although the use of functional language in biology is difficult to avoid, the notion of function itself is highly ambiguous (Van Regenmortel 2002a; Shrager 2003). Biochemists tend to regard the function of a protein simply according to what the molecule does, i.e. its function or activity at the molecular level (Murzin and

Patthy 1999) and they focus therefore on activities like binding, catalysis, or signaling. Since proteins are able, on average, to interact with as many as five partners through individual binding sites, it is to be expected that proteins will be able to serve different functions under different circumstances.

Functions can also be conceptualized at higher levels of cellular and organismic organization, when they are interpreted as contributing directly to the survival and reproduction of the organism. In fact, any functional explanation in biology follows the good consequence doctrine according to which the performance of a function must confer some good to the system (Achinstein 1983). Within the framework of Darwinian evolution, it is indeed possible to "explain" any biological function by saying that it exists because it contributes to the reproductive fitness of the organism in which it occurs. Instead of looking for a structural explanation for a currently observed function, biologists tend to favor evolutionary explanations for presently observed structures (Van Regenmortel 2001c). This type of explanation is based on the broad historical, evolutionary account of the selective advantage that the possession of a trait conferred to an organism in the past. However, such explanations can be provided for just about any conceivable biological function, and they lack power because they give no information regarding the contingent constraints that favored, in a given environment, one particular adaptive solution and pathway instead of its many possible alternatives (Van Regenmortel 2002a). Like history, evolution never repeats itself and all attempts to derive convincing explanatory accounts are doomed since every evolutionary event is a one-off, unrecovable phenomenon (Hanke 2004).

## Human Intentionality and the Design Metaphor

Design can be defined as the deliberate and intelligent conceiving of an artificial thing or process. The designer's task is to pose and solve an inverse problem: given a desired performance, he has to imagine what would execute it (Bunge 2003). Most inverse problems have multiple solutions although some problems may have none. Plausible solutions are tested by trial and error until the preset goal is attained.

When it tries to achieve certain ends, the human mind is probably the most efficient goal-directed system that exists. Humans act and make things with certain ends in mind and they look at the artifacts they created in terms of their intention to achieve a certain goal. Human intentions are experienced as the cause of purposive human action, and this leads to the anthropomorphic fallacy that the behavior and activities of all living organisms are explainable in terms of their goals and purposes, i.e. by end-directed or teleological features. Any functional part that contributes to the working of a living system is interpreted as achieving a certain goal, and this leads to the view that organisms were designed by a thinking mind to function in a preordained way. In the same manner as it is explained that a knife is sharp because it is intended for cutting, one tends to regard the parts of living organisms as if they were manufactured and designed to carry out particular functions. Organisms with their adaptations may indeed give the appearance of having been designed and this encouraged the use of the design metaphor when describing evolution by Darwinian natural selection (Ruse 2002, p. 276).

This is somewhat of a paradox since Darwin was responsible for the removal of the mythical figure of the designer or Creator as the agent responsible for the diversity of living forms on our planet. In an attempt to rescue design as an explanatory concept, it has been suggested that it is possible to have a "design without a designer" (Ayala 2004). Since organisms give the appearance of being designed, this encourages the use of metaphors of goal-directed teleology for describing biological processes in terms of design, purposes, and functions (Ruse 2002).

In reality, a biological function does not entail design for that function and functional descriptions need not be based on psychological notions of design, intention, and purpose (Allen and Bekoff 1995). Doing something by design is synonymous to doing it intentionally. It would be more appropriate to say that organisms are fashioned or shaped by selection pressures rather than invoke design and its corollary of intentionality as an explanatory concept.

Many biologists believe that the use of a design terminology is a harmless figure of speech, but others think that it perpetuates the unscientific mental habit of supposing that objects or events have purposes (Hanke 2004). Attributing a purpose to an object is entirely subjective since purpose has no real existence outside the mind thinking of it.

It has been argued that the word function implies a purpose and hinders scientific thinking because it encourages scientists to split any biological entity into sealed compartments with different functions, instead of seeing it as an integrated whole with many connections (Hanke 2004).

Natural selection has replaced the Creator as an explanatory concept although there is in fact no selection, only differential survival. Fitness, which corresponds to anything that increases the chances of survival in a given environment, is often regarded as the basis for selection, although fitness has been described as "another phantom of the human mind" (Hanke 2004). Fitness is not a property of phenotypes or genes: it is a relation between the organism and the environment.

Today it is generally accepted that Darwinian natural selection does not proceed according to the preconceived plan of a designer but that it occurs blindly through the increased survival and reproduction of adaptive random variations. Although the mutations that provide the hereditary variations arise at random, this randomness is counteracted by natural selection which preserves what is useful in their carriers in a given environment and eliminates what is harmful (Ayala 2004). Darwin's theory of natural selection provided an explanation for the observed adaptations of organisms and removed the need for an intelligent deity as an explanation for the evolution of life and the emergence of the human species. According to the principles of emergence, the natural world is divided into hierachies that have evolved over evolutionary time (Morowitz 2002). Since blind law is able to explain the diversity of organisms, invoking creationist miracles is no longer necessary and this has put another nail in the coffin of the argument for design (Ruse 2005, p. 265).

According to Dawkins (1995, p. 155), accepting Darwinian evolution entails abandoning the human predilection for intentions: "In a universe of blind physical forces and genetic replication, some people are going to get hurt, other people are going to get lucky and you would not find any rhyme or reason in it, nor any justice. The universe we observe has precisely the properties we should expect if there is, at the bottom, no design, no purpose, no evil, and no good, nothing but blind pitiless indifference". Dawkins' harsh scientific vision, however, does not mean that the potential for morality and ethical behavior in human societies is rendered meaningless and not worthy of serious consideration (Dembski and Ruse 2004; Kitcher 1982, p. 186).

## The Bogus Scientific Claims of the Intelligent Design Movement

The spurious explanatory value of design for understanding biological phenomena is nowhere more apparent than in the claims of adepts of the intelligent design movement (Dembski and Ruse 2004; Young and Edis 2004; Pennock 2004; Kitcher 1982). So-called intelligent design, which is actually a misnomer for divine or supernatural design, pretends to explain the occurrence of life and evolution on our planet by simply reinstating the creationist credo: "God did it". Such a claim which invokes an unknowable deity with inscrutable intentions cannot have the status of a scientific explanation since it requires the acceptance of unverifiable miracles and unintelligible supernatural causation. Those who make such claims stand resolutely outside the boundaries of science.

The evident complexity of biological systems has led some biologists and philosophers to claim that it is impossible to attribute the evolution of living systems on our planet to the blind action of Darwinian natural selection. The underlying idea is that complex systems must have appeared all at once, because gradual, small increments of complexity would not have been sufficiently functional to be selected. The usual argument is that complex systems made up of a large number of mutually adapted components, such as the bacterium flagellum, the blood coagulation cascade, the vertebrate eye, or the Krebs citric acid cycle, could not have arisen gradually since, in order to be selected, such systems would already have to be fully operational (Behe 1996). Superficial reasoning assumes that the absence of a single component would render the system nonfunctional, preventing it from being selected. Such reasoning also implies that biological systems are analogous to man-made artifacts or machines which evidently, do not assemble spontaneously by the random and unguided assembly of a large number of components.

It is well established, however, that evolutionary processes occur through parallel processing rather than by the sequential addition of specifically designed and perfected components one after another. Furthermore, for each of the systems that intelligent design could not claim to have arisen gradually, plausible evolutionary routes based on the ability of proteins to slowly acquire new functions have been proposed (Melendez-Hevia et al. 1996; Thornhill and Ussery 2000; Miller 2004; Weber and Depew 2004; Kirschner and Gerhart 2005).

There is also no need to assume that all the subsystems and modules of a complex system had to appear simultaneously in order to constitute a functional system. For instance, there is good evidence that the immune system which comprises inter alia—MHC molecules, T-cell receptors, antibodies, cytokines, and the complement cascade—arose by sequential addition and modification of individual elements from simpler, pre-existing systems (Bottaro et al. 2006).

There is, therefore, no need to follow the creationists and invoke the intervention of an intelligent deity able to conceive of and create simultaneously all the interlocking components that have to be present in a complex biological system. Recently, an attempt to have intelligent design taught in US schools alongside evolution failed when a Federal Court ruled that it would have meant allowing the unconstitutional teaching of religious creationism (Bottaro et al. 2006; Gross 2006).

## The Lure of Rational Design Claims

Another unfortunate consequence of the continued use of design phraseology is that it encourages scientists to claim that they follow a rational design approach in their research work. This claim implies that they are able to conceive a solution to the problem at hand by a rational process of deduction and prediction. However, this would require an extensive knowledge of the innumerable causal connections present between the components of the system, a knowledge which clearly does not exist.

Since scientists need to secure funding for their research, it is indeed tempting for them to suggest that their current understanding of complex biological networks is sufficient to allow them to predict the outcome of biochemical or genetic interventions aimed at restoring failing biological systems. By suggesting that rational design will succeed in reaching intended therapeutic goals, they accredit the view that currently available simplistic models of linear causality, which disregard the unpredictable behavior of complex networks, will allow them to achieve wanted therapies in a preordained and predictable manner. It is fashionable today to present experimental work in the field of drug design, gene therapy, or cancer vaccines as realistic endeavors based on sound molecular knowledge and devoid of unwanted side-effects (Kubinyi 2003; Glassman and Sun 2004). In many cases, such claims amount only to unjustified hype. Unfortunately, such practices may lead to subsequent disenchantment with science when the promised solutions do not materialize (Van Regenmortel 1999b, 2000).

Instead of presenting rational design strategies as the most effective approach for developing new therapeutics, it would be better to acknowledge that nearly all advances in biomedicine are derived from empirical observations based on trial and error experimentation (Van Regenmortel 2000). An empirical approach is always preferable since the outcome of experiments involving biological systems is usually not predictable from pre-existing knowledge. Furthermore, opposing rational and empirical approaches in scientific research is actually misleading since most scientific knowledge is derived empirically in the course of experiments that are necessarily planned and analyzed in a rational manner (Van Regenmortel 1999b). Any suggestion that our limited understanding of complex biological systems will allow us to do away with empirical screening of compounds for biological activity and unwanted side effects can only lead to unrealistic expectations.

Another striking metaphor used to describe genetic engineering experiments is to say that scientists doing this type of work are "playing God." This metaphor, which is often utilized by those who otherwise have no use for theological language, implies that these scientists disregard the precautionary principle and are guilty of genetic meddling or trespassing that could have unforeseen and harmful consequences (Graham 2002). Those who today oppose such things as designer babies or stem cell experimentation tend to have a religious motivation and they use the playing God metaphor which is based on preDarwinian beliefs that credited God with having designed all living creatures. Given the complexity of living systems and our limited ability to control them by rational design, the playing God metaphor is a salutary reminder that humans are unlikely to equal the extraordinary design capacities previously attributed to the Creator.

## Complex Causal Connections Are the Rule in Biological Systems

When scientists claim that they follow a rational design strategy, it implies that their knowledge of the many causal connections that exist in the system they study is sufficient to allow them to predict the outcome of experimental interventions or manipulations. This in turn suggests that our everyday notion of causation is an adequate guide for unraveling the multiple complex connections present in biological systems. However, there is no evidence that this is actually the case and it is thus relevant to examine whether our common sense ideas of causality derived from observations on simple physicochemical systems are applicable to the study of complex biological systems.

The appeal to causes as an explanation of events and processes arises from the human predisposition to look for the reasons, often called causes, of human actions. However, human intentions and reasons are not the causes of events, nor is the prior state of a thing the cause of its subsequent state. The caterpillar is not the cause of the butterfly.

Causation is a mode of event generation (Bunge 2003). Causal relations are relations between successive events and not between two material objects or between a structure and an event. A biological event such as a binding reaction can thus not be caused by the structure of one or both interacting partners and structures possess no causal efficacy in bringing about a biological activity. The analysis of so-called structure–function relationships should therefore consist of uncovering correlations rather than causal relations (Van Regenmortel 2002b). Since things are not causes, ancestors are not causally related to descendents via reproduction, and DNA sequences or genes cannot be causes of phenotypic traits or behavior (Mahner and Bunge 1997, p. 39).

The ontology of causality, i.e. what a cause actually consists of, remains a controversial field in the philosophy of science. According to one influential school of thought, causal interactions involve the exchange and conservation of a quantity, such as energy or momentum, which corresponds to the linear causality of the push–pull variety relevant in physics (Salmon 1998). In everyday life, our conception of causality is linked to the experience of being able to achieve a wanted result by acting in a certain way, which then leads to the belief that we caused the wanted result to happen (Menzies and Price 1993). This ability of "making things happen" is central to the manipulationist or interventionist account of causation and causal explanations developed by Woodward (2003). According to this interpretation, causal relationships are relationships that are potentially exploitable by humans for purposes of manipulation and control. It is undeniable that causal knowledge is valuable to humans because of its practical utility arising from the ability to intervene and control.

As long as biology remained a descriptive science, there was little opportunity for developing causal explanations of biological phenomena. It is only with the advent of molecular biology that causal mechanisms could be investigated because it became possible to manipulate and control biological systems and to observe the results in ways that were not previously possible (Woodward 2003, p. 9).

The reason why establishing causal relationships is more important than establishing correlations is that causal links provide an opportunity to modify outcomes in a desired manner. It is well-known, for example, that many up-or down-regulated proteins identified by proteomics analysis are not causally linked to the parameter that was allowed to vary in the experimental set up. It is therefore always necessary to establish that a causal connection exists by subsequent experimentation since this is the only way to establish that the hypothetical cause elicits the predicted effect. In the same way, causal links in a biological network can only be established experimentally by manipulating and intervening in the system.

No single event in biology can be said to be the cause of another event on the basis that it was necessary and sufficient on its own for the effect to occur. This means that simple linear causality can never provide a relevant explanation in biology. Nancy Cartwright (1983) pointed out that even in physics, simple causal links derived from Newtonian mechanics usually cannot account for the behavior of actual objects since they are subject to forces other than gravitation, for instance electromagnetic forces. The reason why the law of universal gravitation applies to celestial bodies is that they move in an interstellar vacuum, free of external interference. However, on Earth, the laws of Newtonian mechanics are true only if one adds an unrealistic *ceteris paribus* (other things being equal) clause in order to rule out interfering forces (Cartwright 1983). To believe that these physical laws somehow apply in spite of the many interfering or counteracting forces that prevent objects to behave in accord with them is very much a metaphysical assumption.

In biology, only contributory causes exist in the sense that a multiplicity of conditions are necessary but not sufficient to bring about an effect. Any observed effect is always the result of a complex network of interactions and internal regulations and it does not help to focus on a single factor and assume that the clause other things being equal is realistic when hundreds of background conditions are involved. Instead of giving undue explanatory weight to one causal factor, it is always necessary to take into account the numerous factors that simultaneously influence the system and bring about a particular end result in a given biological context (Van Regenmortel 2001c). Because of synergy as well as interference phenomena, there is often no linear relationship between the magnitude of a causal factor and the size of the effect. Many causal factors can take a range of values and it may only be above a certain threshold that an effect will be observed. When it is said that smoking causes lung cancer, it is not clear whether smoking 20 or 60 cigarettes a day provides an acceptable explanatory causal relation. Furthermore, even if smoking causes an increase in the probability of contracting cancer, it is also true that some lung cancer victims never smoked and that some heavy smokers never develop cancer (Rosenberg 2000, p. 52).

Although computer simulations sometimes make it possible to predict the outcome of multiple causal interactions in a biological network, features like negative feedback, feed-forward control, interference, synergy, cooperativity as well as the causal efficacy of emergent phenomena (Van Regenmortel 2004a) usually do not make it possible to identify the innumerable causal links that are involved (Berger 1998). A final comment is in order regarding the irrelevance of so-called Aristotelian final causes. The belief that biological objects have intrinsic natural goals that are masquerading as final causes is an anthropomorphic relic of intentional and teleological thinking whereby human actions are analyzed in terms of their intended goals. Regarding an intention as the cause of a biological goal that has been attained implies that organisms have been designed to function in a certain way. This is no longer an acceptable interpretation of how causality is at work in evolutionary biology.

# A Case Study: The Rational Design of Synthetic Peptide Vaccines

In immunology, reductionist thinking has led to the view that the development of peptide-based synthetic vaccines is a realistic enterprise. Since the effectiveness of existing vaccines has been shown in many cases to depend on their capacity to elicit protective antibodies in vaccine recipients, most research efforts have concentrated on stimulating the B-cell arm of the immune response. The antigenic properties of viruses reside in B-cell epitopes of viral proteins which correspond to regions of the virus particles that are recognized by the binding sites (or paratopes) of soluble and membrane-bound antibody molecules. These epitopes are classified as either continuous or discontinuous depending on whether the amino acid residues included in the epitope are contiguous in the polypeptide chain or not. This terminology suggests that the units of recognition operative in epitope–paratope interactions are individual amino acid residues whereas it is, in fact, at the level of individual atoms that the immunological recognition takes place (Van Regenmortel 1999a). Since discontinuous epitopes are made up of residues distant from each other in the sequence but brought together by the folding of the peptide chain, their immunological reactivity obviously depends on the conformation of the protein. It may be less obvious that the zreactivity of continuous epitopes is also conformation-dependent as shown by the fact that an antibody will not recognize in the same manner a peptide antigen when it is free in solution, conjugated to a carrier or adsorbed to a solid-phase (Van Regenmortel 1999a).

B-cell epitopes are relational, emergent biological entities that can be defined only by their ability to be recognized by complementary paratopes in the context of an immune system (Van Regenmortel 1999a, 2004b). Epitopes and paratopes are thus not intrinsic features of antigen and immunoglobulin molecules, respectively, and they cannot be structurally analyzed on their own, independently of the relationship with their partners.

It is generally accepted that crystallographic analysis of antigen–antibody complexes gives the most reliable information on the structure of epitopes. However, the structures revealed in the complex are usually different from the structure of the respective binding sites in the free antigen and antibody molecules, *i.e.* before they have been altered by the mutual adaptation of the two partners that occurs during the interaction (Wilson and Stanfield 1994; Goh et al. 2004). Since it is the conformation of the epitope before complexation that is important for inducing the wanted immune response, structural information obtained from the complex may be of less value for guiding the construction of a vaccine.

X-ray crystallographic data show that the majority of protein epitopes are discontinuous which explains why most antibodies raised against a native protein will not react with short peptide fragments of the protein. Since it is extremely difficult to reconstitute discontinuous epitopes by chemical synthesis for use as potential vaccines (Enshell-Seijffers et al. 2003; Villen et al. 2004), efforts have focused on the use of continuous epitopes corresponding to linear peptide sequences of 6–20 residues. However, there is nowadays considerable skepticism regarding the potential use of such peptides as effective vaccines (Van Regenmortel 1999c). Many investigators take the view that the majority of continuous epitopes described in the literature and listed in databases probably correspond to unfolded regions of denatured protein molecules rather than to epitopes of native proteins (Van Regenmortel 2006). Such linear peptides are easy to synthesize but are unlikely to react with antibodies directed to the native structure of the protein antigen. It is also unlikely that these peptides will elicit the type of neutralizing antibodies reacting with the native proteins of infectious agents that are needed for immune protection.

Most of our knowledge of protein epitopes is obtained by measuring the ability of anti-protein antibodies to cross-react with short peptide fragments that have a different conformation from the one present in the corresponding region of the cognate, native protein. Such studies of the antigenic cross-reactivity of peptides correspond to a chemical investigation of epitope–paratope interactions and do not provide information on the immunogenicity of peptides, i.e. their ability to induce an immune response in a competent host. Immunogenicity, which is the property required from a vaccine, depends on extrinsic factors such as the host immunoglobulin repertoire, self-tolerance, the production of cytokines, and various cellular and regulatory mechanisms. These properties have a meaning only in the context of a particular host, which makes immunogenicity a biological property (Van Regenmortel 2001a).

Most peptides are immunogenic in the sense that they readily induce antibodies that react with the peptide immunogen. However, this type of immunogenicity is irrelevant for vaccination purposes. What is required is the induction of antibodies, which on the one hand recognize the cognate, native antigen (so-called cross-reactive immunogenicity) and, in addition, also neutralize the infectivity of the pathogen harboring the antigen (so-called cross-protective immunogenicity (Van Regenmortel 2006). For most continuous epitopes of viruses described in the literature, there is no evidence that they possess the required cross-reactive and cross-protective immunogenicity. Although it is sometimes possible to increase the cross-protective immunogenicity of peptides, for instance by peptide cyclization, this strategy of constraining the conformation of peptides is not necessarily effective. It is plausible that the intrinsic disorder in certain loop regions at the surface of virus particles is responsible for the finding that peptides corresponding to such disordered regions often tend to give the best results as vaccine immunogens (Langeveld et al. 1994; Fischer et al. 2003; Uversky et al. 2005).

A further difficulty arises from the relatively weak immunogenic activity of peptides, which usually necessitates that peptides be conjugated to carrier molecules. When several peptide moieties are coupled to a protein carrier, the microenvironment at each point of attachment of the peptide is likely to be different and this can lead to different conformations in the peptide (Friede et al. 1994). In addition, certain residues of the carrier protein may be recognized by the immune system together with residues of the peptide moieties and this may lead to an heterogeneous, ineffective immune response.

A good example of the difficulties that must be overcome to transform a continuous epitope into an effective vaccine immunogen is provided by the peptide ELDKWAS present in the conserved membrane-proximal external region (MPER) of the gp41 protein of human immunodeficiency virus 1 (HIV-1). This peptide, which is recognized by the anti-HIV-1 broadly crossreactive neutralizing mAb 2F5, has for a long time been regarded as a promising vaccine candidate because it is present in a conserved region necessary for envelope-mediated fusion of the virus. Several groups incorporated this peptide into a variety of immunogenic constructs in an attempt to have it elicit antibodies with the same neutralizing capacity as mAb 2F5. When additional gp41-derived flanking residues were added to the peptide

(Tian et al. 2002) or when the ELDKWAS sequence was constrained in an a-helical or b-turn like conformation, it was found to have up to tenfold higher affinity for the 2F5 antibody than that exhibited by the free, unconstrained peptide. However, in spite of this improved antigenicity, the peptide constructs, when used as immunogens, were still unable to induce antibodies with detectable neutralizing capacity (Joyce et al. 2002; Ho et al. 2005).

In an attempt to identify substitutions that could lead either to increased or decreased neutralization by the 2F5 antibody, infectious mutants of HIV-1 were constructed by introducing substitutions in the MPER of gp41 (Zwick et al. 2005). Again, the results were disappointing since it was not possible to ascertain which structural elements in spatial proximity to the ELDKWAS residues controlled the neutralization capacity. Attempts to obtain this type of information were made by determining the crystal structure of the 2F5 antibody in complex with 7-mer, 11-mer, and 17-mer peptides of the gp41 MPER which all contained the ELDKWAS sequence (Ofek et al. 2004). The 3-D structure of the bound peptides was found to differ significantly from the corresponding region in the postfusion six-helix bundle conformation of gp41. This underscores the fact that the conformation of the epitope region is flexible and may also differ in the prefusion gp41 viral spike which is the relevant immunogenic structure in the virus. It seems that this epitope is able to assume different conformations depending on the fusogenic state of gp41 and that its accessibility may vary during the course of infection. It is, therefore, not clear which conformation should be stabilized in peptide constructs intended for vaccination. Epitopes with a variable conformation have been called transitional epitopes (Zolla-Pazner 2004) and they illustrate the need to include the fourth dimension of time in the description of antigenic specificity (Van Regenmortel 1996).

Another important observation made by Ofek et al. (2005) is that the hydrophobic membrane which is close to the MPER at the viral surface, strongly determines the type of antibody such as 2F5 that is produced. The absence of such an hydrophobic environment in experimental gp41 peptide immunogens may have contributed to the failure of eliciting 2F5- like neutralizing antibodies.

These findings suggest that the ELDKWAS sequence actually corresponds to only a part of the more complex discontinuous epitope that elicited the neutralizing mAb 2F5 (Van Regenmortel 2006; Ofek et al. 2004). Unfortunately, it is extremely unlikely that the precise configuration of atoms in the residues making up a discontinuous epitope structure could be reconstituted by a short linear peptide sequence (Enshell-Seijffers et al. 2003; Villen et al. 2004).

The belief that it should be possible to develop synthetic peptide vaccines is based on the assumption that the principles of structure-based drug design are applicable to vaccines (Van Regenmortel 1999c; Nicholson 1994). Such an assumption overlooks the fact that the relationship between a drug and its receptor or target molecule is fairly unique, whereas the relationship between an antigen and an antibody is less restricted and can be described as promiscuous.

A further difficulty is that antigenic recognition and discrimination are typically fuzzy processes, which do not allow absolutely clear-cut demarcation lines to be drawn between the presence or absence of a reaction (Van Regenmortel 1999b) and even between immunological self and nonself (Leng and Bentwich 2002).

Only some of the antibodies that recognize an antigenic site in a virus possess neutralizing activity and it is not known which structural features differentiate antibodies that possess such activity from those that do not. In the past, attempts to find out have relied on the analysis of the atomic structure of neutralizing mAbs able to protect against infection (Kwong et al. 1998; Saphire et al. 2001). Although such crystallographic studies reveal the structure of viral epitopes recognized by neutralizing antibodies, they do not provide information on which synthetic immunogens will be able to elicit antibodies with the same neutralizing capacity (Van Regenmortel 1999b). Although a molecular design strategy could be used to improve the binding capacity of a mAb destined for passive immunotherapy of viral infection (Parren and Burton 2001; Presta 2003), this approach is not feasible for modulating immunogenicity and optimizing the numerous cellular interactions in the host that lead to a protective immune response. In order to obtain an effective peptide vaccine by design, one would need to know which features of the peptide control cross-reactive immunogenicity, whether helper T-cell or cytotoxic T-cell epitopes must be incorporated in the peptide construct and how this could be done most effectively, what would be the best adjuvant and delivery route, etc. It is now increasingly appreciated that this type of information will be obtained only by an empirical approach in which numerous antigens are used to immunize animals and the sera are characterized for specificity and neutralizing activity (Van Regenmortel 1999b; Zwick et al. 2005). There is thus little doubt that attempts to develop peptide-based vaccines will continue to be driven by trial and error experimentation rather than by rational design.

## How Can Systems Biology Help to Unravel the Immune System?

One of the aims of systems biology is to understand the emergent biological phenomena that arise from the multiple interactions occurring between the individual components of a system (Kitano 2002). The immune system which involves the dynamic interaction of a wide array of tissues, cells, and molecules, can lead to a variety of different biological outcomes and this complexity makes the study of immune responses particularly apt to a systems biology approach (Aderem and Smith 2004). A system—level unraveling of the immune system could be expected to involve the following steps:

i. Define all the components of the immune system using proteomics approaches.
ii. Map the network of interactions among the components and display/visualize them by means of a suitable software such as Cytoscape (http://www.cytoscape.org).
iii. Describe the kinetic parameters that underlie the various interactions including the kinetics controlling competitive binding phenomena exhibited by T-cells and antibodies.

iv. Elucidate the regulatory pathways that control the system.
v. Describe the dynamic behavior of the system in terms of positive and negative feedback loops and of connectivities between pathways and modules.
vi. Develop mathematical models to accurately represent and predict the overall behavior of the system.

Currently, there is a considerable amount of research activity to achieve steps (i) and (ii). Multiplexed protein micro-array systems, allowing the simultaneous measurement of hundreds of proteins in complex mixtures, have been developed for characterizing different types of humoral immune responses (Kingsmore 2006). These assays utilize protein chips containing allergens, autoantigens, or microbial antigens arrayed as microscopic spots, and they allow the characterization of humoral immunity in allergies (Lebrun et al. 2005), autoimmune disorders (Robinson et al. 2003), tumors (Haab 2005), and infectious diseases (Bacarese-Hamilton et al. 2004). Such multiplexed microarray immunoassays involve picoliter antibody spots and require only microliter sample volumes and their miniaturization has the added benefit of decreased reaction times resulting from short diffusion distances (Kingsmore 2006). Multiplexed arrays are also particularly useful for measuring simultaneously the large number of agonist and antagonist cytokines that regulate immune responses (Nagorsen et al. 2004). Since a network of cytokines is involved, the measurement of a single cytokine is mostly of little value.

Proteome-based approaches have also been used to identify candidate antigens of pathogens for several diagnostic, therapeutic, and vaccine applications. This is usually done by using human sera from infected subjects who have mounted an antibody response to the pathogen (Haas et al. 2002; Havlasova et al. 2002; Bahk et al. 2004; Kowalczewska et al. 2006).

Protein microarray systems mostly require the use of mAbs as the immobilizing reagents to be deposited on microchip surfaces (Kusnezow and Hoheisel 2003). Tens of thousands of specific mAbs have been developed, and these are used to analyze the entire proteome of different organisms using protein microarrays. These arrays are also useful for investigating the frequent occurrence of nonspecific antibody binding, a phenomenon which needs to be controlled if reliable conclusions are to be drawn from immunoassays (Michaud et al. 2003).

Since traditional animal-based antibody generation is laborious and time-consuming, other approaches have been developed to obtain the required mAb. Recombinant antibody technology using phage display libraries can shorten the time of antibody generation from months to days and is likely to become increasingly used in proteomics research (Ohara et al. 2006). Recently, another improved method for cloning cells secreting specific antibodies, based on a microengraving process, has been described. This method allows about 80,000 cells to be analyzed simultaneously and decreases the antibody screening time by a factor of about 20 (Love et al. 2006).

Our understanding of T-cell immune responses has improved considerably in recent years, partly through the application of proteomics techniques. Tens of thousands of different peptides are presented at the surface of antigen—presenting cells, and some of these alert the immune system to changes in the intracellular environment that may be associated with infection, oncogenesis, or other abnormal cellular processes (Williamson and Purcell 2005). This pool of peptides, bound to MHC molecules, has been coined the immunoproteome or immunopeptidome, and its analysis by affinity chromatography and MS has helped to elucidate some of the molecular events involved in antigen processing (Klade 2002; Purcell and Gorman 2004). By comparing the immunoproteomes of healthy and cancerous cells, it is hoped that tumor-specific antigens may be identified which could in future be targets for novel cancer-specific immunotherapeutic agents (Shoshan and Admon 2005).

There is currently very little information available on the kinetic parameters that control the interactions occurring in biological systems (step (iii) above) and this is a major short-coming of today's proteomics techniques. This is unfortunate since the on-and off rates of a biomolecular interaction are crucial parameters that determine whether a biochemical reaction is biologically significant or not (Markgren et al. 2000). Surface plasmon resonance (SPR)-based biosensors which allow interaction kinetics to be measured easily have been available since the early 1990s (Fagerstam and Karlsson 1994) but it is only in the last 2 years that biosensors instruments capable of handling several hundred samples a day have been commercialized. SPR technology in conjunction with MS has tremendous potential in the area of interaction proteomics (Larsericsdotter et al. 2006) and there is no doubt that this technique will be increasingly used in systems biology (Schuck 1997; Karlsson and Fält 1997; Van Regenmortel 2004c; Rich and Myszka 2005a, b).

Mathematical models have been used to analyze the molecular mechanisms that underlie the key aspects of T-cell immune responses such as T-cell activation and thymocyte selection (steps (v) and (vi) above). Using in silico models, it is possible to discriminate between the various possible ways a molecule can influence a cascade of events since its effect on different cellular responses can be modeled separately (Chakraborty et al. 2003). It has also been possible to show that a mathematical model of the immune network system proposed by Jerne in 1974 was able to account for the memory features and dynamics that were predicted for that system (Bernardes and Zorzenon Dos 1997).

It is now generally accepted that computer simulation may be the only way to assess the relative magnitude of the many factors that contribute to an effective immune response (Bock and Goode 1998). In silico modeling may make it possible in future to predict cooperative and synergy effects between phenomena such as T-cell and B-cell activation (Zwick et al. 2001b; Silverstein and Rose 2003), regulation by cytokines, and immunological memory (Crotty and Ahmed 2004).

Although traditional reductionist approaches have been very successful in the past for analyzing components of the immune system individually, it seems likely that a more integrative systems biology approach will be required to account for many of the still unexplained features of the immune system. It remains to be seen whether more integrative approaches to vaccine development will be more successful than currently used strategies based on rational design and manipulation of individual components of the immune system, one at a time.

# Basic Research in HIV Vaccinology Is Hampered by Reductionist Thinking

## Introduction

Before it was recognized that the vast majority of epitopes in proteins are discontinuous, i.e., composed of surface residues originating from distant parts of the protein sequence, short peptide segments of viral proteins able to react with antiprotein antibodies were considered to be continuous epitopes mimicking the antigenicity of the protein. This led to the expectation that if such peptides could be made to adopt the 3D structure observed when the corresponding regions of the viral protein are bound to neutralizing monoclonal antibodies (nMabs), the peptides would be able to act as effective vaccine immunogens. More than a thousand peptides were tested over the years as potential synthetic peptide vaccines against a variety of pathogens but not a single peptide passed phase III clinical trials nor was marketed for use in humans (Hans et al. 2006). These attempts to develop synthetic peptide vaccines were based on the premise that continuous epitopes reacting with antibodies specific for a viral protein may be able to induce antibodies that recognized the virus and neutralized its infectivity. It was usually found, however, that few continuous epitopes of viral proteins were able to elicit antibodies that recognized the native protein although most of them readily induced antibodies that reacted with the peptide immunogen (Van Regenmortel 2009a).

In order to be useful as a vaccine immunogen, a peptide must not only induce antipeptide antibodies but must also possess so-called cross-reactive immunogenicity, i.e., the ability to induce antibodies that recognize the cognate protein, as well as

Frontiers in Immunology, 2012, 3, 194.
Marc H V Van Regenmortel

© Springer Nature Switzerland AG 2019
Marc H V Van Regenmortel, *HIV/AIDS: Immunochemistry, Reductionism and Vaccine Design*, https://doi.org/10.1007/978-3-030-32459-9_9

cross-protective immunogenicity, i.e., the ability to induce antibodies that neutralize the infectivity of the pathogen (Van Regenmortel 2006). Since very few linear peptides were found to possess the required cross-reactive and cross-protective immunogenicity, it became generally accepted that the prospects of developing effective synthetic peptide vaccines were very poor. In the present review, it will be argued that many of the impediments that prevented the development of synthetic peptide vaccines in the past are responsible for the lack of success observed in current attempts to design HIV-1 vaccines based on the 3D structure of more complex discontinuous epitopes.

Arguments will be presented that the failure to develop an HIV-1 vaccine by rational design is mainly due to the underlying reductionist thinking that pervades much basic research in immunology and vaccinology. A reductionist mindset obscures the fundamental divide between antigenicity and immunogenicity, i.e., between the chemical nature of antigen–antibody recognition processes and the biological nature of the immunogenic processes that allow a viral antigen to give rise to a protective immune response in a competent host.

## Antibody Polyspecificity and the Relational Nature of Epitopes and Paratopes

Epitopes and paratopes are usually identified by solving the 3D structure of antigen–antibody complexes and determining which amino acids in the two partners make contact with each other (Sundberg and Mariuzza 2002). When a paratope is defined solely in terms of residues that make contact with an epitope, it is difficult to account for the observation that the binding activity of an antibody often depends on structural features distant from the paratope itself (Schildbach et al. 1993; Chatellier et al. 1996). Similarly, residues in the antigen that are not in contact with the antibody may be able to affect the binding process, for instance by influencing the conformation or stability of the free form of the protein or by participating in long-range allosteric effects. Complex formation is driven by the free energy change associated with the binding process, and residues away from the contact regions are often able to contribute to this free energy change as demonstrated by the changes in affinity and specificity that result when these residues are mutated (Greenspan and Di Cera 1999). Mutational studies have also shown that not all contact residues contribute to the interaction energy, confirming that epitopes defined structurally differ from epitopes identified in functional assays (Cunningham and Wells 1993). Furthermore, the structures visualized in antigen–antibody complexes may differ from the structures of the binding sites in the free molecules, before the processes of mutual adaptation and induced fit that usually occur when the two partners interact (Wilson and Stanfield 1994; Bosshard 2001; Kim et al. 2011). As a result the structure of an epitope bound to an nMab may not correspond to the structure that is recognized by B-cell receptors (BCRs) during the immunization process and is presumed to be required in a vaccine. It is also known that residues in the antigen that are not in contact with paratope residues may be able to modulate the immunogenic activity of epitopes (Moudgil et al. 1998). It was also found that the sensitivity to neutralization of HIV-1 strains that harbor the core epitope recognized by nMab 4E10 was modulated by amino acid substitutions elsewhere in the viral envelope (Gray et al. 2008). All these observations are consistent with the view that epitopes and paratopes are fuzzy binding sites devoid of clear-cut structural boundaries (Van Regenmortel 1998).

Specificity has been defined by Medawar and Medawar (1978) as the comple-mentary relationship existing between an agent and something acted on, which arises from the stereochemical complementarity found between such partners as antigen and antibody, enzyme and substrate, or receptor and ligand. In addition to being defined as a measure of goodness of fit between paratope and epitope, antibody specificity is better viewed as resulting from the capacity of an antibody to discrim-inate between two or more antigens (Day 1990, p. 291; Frank 2002, p. 42).

Antibodies recognize complementary antigens through the extremely versatile binding sites of immunoglobulin (Ig) molecules which are able to recognize virtually every molecular structure (Nezlin 1994; Wucherpfennig et al. 2007). The Ig binding site consists of 50–70 hypervariable residues distributed over the six complemen-tarity determining regions (CDRs) of the variable domains of the heavy and light Ig chains. Each Ig binding site contains numerous overlapping and non-overlapping paratope subsites of 10–20 residues, each approximately 2800 $A^2$ in area, that are able to bind to different antigens. The surface of one paratope corresponds to only 20–35% of the total surface encompassed by the CDRs of an Ig molecule (Denisova et al. 2010). An Ig molecule may sometimes harbor two non-overlapping paratopes, which will allow it to bind simultaneously to two small antigens (Richards et al. 1975; Bhattacharjee and Glaudemans 1978; Eisen and Chakraborty 2010). A more common situation is that paratope subsites present in an Ig molecule at least partly overlap which prevents two different antigens from binding simultaneously to the same Ig. The presence of multiple paratopes in Ig molecules means that antibody molecules are always polyspecific (Frank 2002).

The polyspecificity of antibodies is also demonstrated by their ability to bind large numbers of small peptides possessing limited sequence similarity and by the fact that many residues of an epitope can be replaced by any other amino acid without impairing the epitope's antigenic reactivity (Getzoff et al. 1988; Geysen et al. 1988). When peptide libraries are tested for their ability to bind Mabs raised against a protein, it is usually found that many peptides that bind Ig residues situated outside the paratope region show little sequence similarity with the target antigen. Such peptides are therefore poor mimics of any epitope of the native protein although they are often referred to as mimotopes (Van Regenmortel 2009a; Denisova et al. 2010; Irving et al. 2001, 2010).

The fact that every antibody molecule always harbors numerous paratopes, allowing it to bind many related or unrelated epitopes, was recognized as soon as myeloma proteins and Mabs became available and their specificity could be ana-lyzed (for a review, see Eisen and Chakraborty 2010). It is therefore astonishing that the implications of antibody polyspecificity for vaccine development were pointed out only recently (Van Regenmortel 2011a, 2012a) since the degeneracy of the immune system has been known for many years (Sperling et al. 1983; Parnes 2004; Wucherpfennig et al. 2007).

The entire surface of proteins harbors a large number of overlapping discontinuous epitopes which can be identified only when many Mabs reacting with the protein have been isolated. Discontinuous epitopes consist of two to five short linear stretches of residues that are distant in the protein sequence but are brought together by the folding of the peptide chain. Since the number of Mabs available for each protein is usually limited, it may seem that antigenicity is located in discrete epitope regions rather than forming an antigenic continuum at the surface of the protein (Van Regenmortel 2009b). Since discontinuous epitopes cannot be isolated in active form from the protein in which they are embedded, it is impossible to study their capacity to act as effective vaccine immunogens on their own. The intact, native protein must always be used as immunogen and this inevitably produces a heterogeneous response against the numerous epitopes present in the protein (Van Regenmortel 2012a).

A second type of protein epitope called a continuous epitope is defined as any linear peptide region of the protein, usually 5–8 residues long, that is able to cross-react with antibodies raised against the protein. Continuous epitopes usually do not exist as individual, discrete binding or immunogenic sites in the native protein, and only some of their residues may be present at the surface of the protein where they are usually part of discontinuous epitopes (Chen et al. 2009).

Antibodies are often given names that suggest they are specific for particular proteins or antigenic sites which always harbor numerous epitopes. This practice can lead to confusion since an antibody cannot be specific for a multiepitopic antigen as a whole but only for one of its epitopes. There is therefore no clear answer to the question whether an Mab or a polyclonal antiserum is the more specific reagent since it depends on which antigens the investigator is trying to differentiate (Van Regenmortel 1998). An individual Mab is usually better able to discriminate between two cross-reactive epitopes than a polyclonal antiserum, although the antiserum tends to have greater specificity for a multiepitopic protein. The reason for this is that the many antibodies in an antiserum which recognize the same protein through separate epitopes give rise to an additive specificity effect. Since each polyspecific Ab in the polyclonal serum cross-reacts with a large number of epitopes that are different for each Ab, the collective cross-reactive potential of the antiserum is diluted out which leads to the separate cross-reactivities of the different antibodies being masked (Talmage 1959; Richards et al. 1975; Al Moudallal et al. 1982; Parnes 2004). The ability of the immune system to specifically recognize a huge number of multiepitopic antigens is therefore not due to the existence of myriads of antibodies, each one recognizing a unique epitope present in only one antigen, but arises from the combinatorial effect of several polyspecific antibodies recognizing separate epitopes on the same antigen (Wucherpfennig et al. 2007; Van Regenmortel 2012a).

# Causation and Explanation in Biology and Immunology

Philosophers usually describe the process of scientific explanation by the so-called deductive-nomological model which presents explanations as being logically deduced from one or other relevant law of nature (Klee 1997; Psillos 2002). For instance, the occurrence of an eclipse will be explained using Newton's laws of universal gravitation together with certain initial conditions. Newton's laws of motion apply to celestial bodies because they move in an interstellar vacuum, free from external interference. On Earth, the laws of Newtonian mechanics explain phenomena only if one adds a *ceteris paribus* (i.e., other things being equal) clause to rule out interfering forces such as electromagnetic forces (Cartwright 1983). Newton's laws have sometimes been blamed for leading scientists to expect that all explanations in science should be able to provide the precise predictions achievable in astronomy. In physics, there are many well-established laws that do lead to extremely reliable predictions and have given us a remarkable understanding of the physical universe. Unfortunately, this is not possible in biology for the simple reason that there are no universal laws in biology (Dupré 1993). Causal explanations in terms of a single cause acting according to a law of nature are therefore not possible in the biological sciences.

Causation is a mode of event generation (Bunge 2003) and causal relations are relations between successive events and not between two material objects or between a structure and an event. A biological event such as the binding reaction between an antibody and an antigen is thus not caused by the structure of the reactants. An antibody on its own possesses no causal efficacy in bringing about a biological activity such as infectivity neutralization which involves a ternary relationship between antibody, pathogen, and host and always depends on numerous immunological and pathophysiological factors.

In biology, only contributory causes can be identified because a multiplicity of background conditions or factors are always involved in bringing about an effect. Because of synergy and various interference phenomena, there is also no linear relationship between the magnitude of one causal factor and the magnitude of a biological effect. Since any observed effect always results from the complex network of interactions and internal regulations that exist in every biological system, a single causal factor can never be presented as an explanation since it is not realistic to assume that the clause "other things being equal" is relevant when hundreds of background conditions contribute to an effect. In non-linear dynamic systems, the notion of causality has very little explanatory value (Berger 1998; Wagner 1999).

The immunogenicity of an epitope is a biological property, partly determined by its intrinsic chemical structure, which mainly depends on numerous extrinsic factors such as the host Ig repertoire, the presence of appropriate BCRs and T cell help, the use of adjuvants, the process of antibody affinity maturation, self-tolerance, and various cellular and regulatory mechanisms that exist only in the biological context of the immunized host. Explaining immunogenicity by referring to its multiple

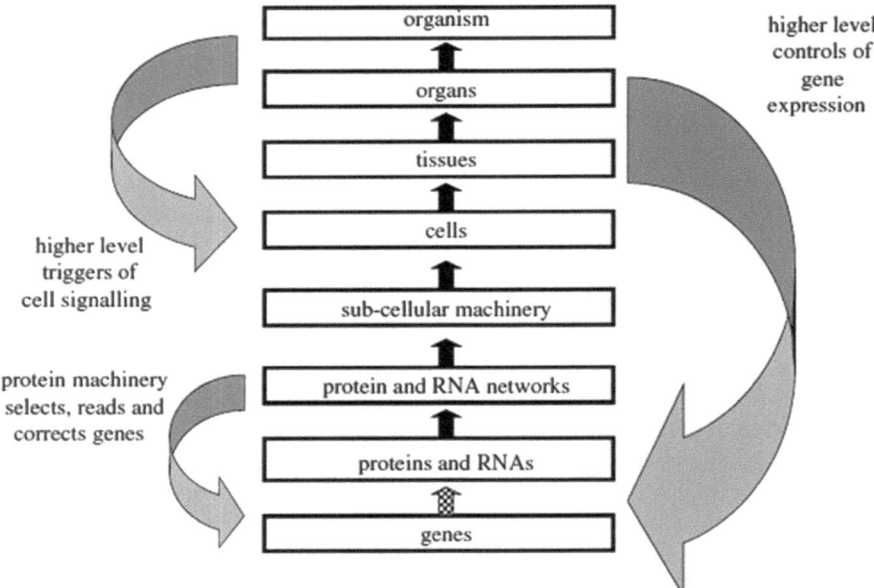

**Fig. 1** The reductionist upward causal chain from genes to organisms, and various forms of downward causation that regulates lower level components in biological systems. This representation does not take into account the influence of the environment at the various levels (Noble 2012)

"causes" is therefore not helpful for controlling a process that is always influenced by a large number of interdependent and cooperative biological interactions.

The dictum "*structure determines function*" has many adepts among molecular biologists and often leads to the assumption that there is a linear causal pathway operating in biomolecules which links gene sequence to protein sequence and then to protein conformation, binding, and function. Such a pseudo-causal pathway appears to possess explanatory power because it is presented in a context-independent manner as if the chemical and cellular environment did not contribute to the process of protein folding or that the natural Darwinian selection process that led to the existence of a functional binding partner could be ignored (Van Regenmortel 2002c). Such a hypothesized linear pathway is often presented as shown in Fig. 1, which suggests that the unidirectional flow of genetic information that occurs from DNA to protein can be extended to the upper levels of biological complexity present in tissues, organs, and organisms. However, genes do not cause phenotypic characters and although they contribute to determining phenotypic traits, they do not act upon them. The prior state of a thing is also not the cause of its subsequent state, the caterpillar not being the cause of the butterfly (Mahner and Bunge 1997, p. 39). Although it has been said that to explain an event is to provide some information about its causal history (Psillos 2002, p. 217), this does not mean that it is possible in immunology to demonstrate a single chain of successive causes and effects where each effect is itself the cause of a subsequent effect. The arrows pointing upwards in Fig. 1 should thus not be interpreted as representing a single causal relation linking events occurring at different levels and providing an ultimate explanation of an immunological mechanism in terms of the genome. In the case of antibodies, Ig genes are randomly spliced together from gene segments and they undergo extensive somatic hypermutation following activation with antigen; antibody specificity is therefore determined by genetic mechanisms that are unique to the immune system.

There is no single upward causality chain linking all the levels of biological organization since the phenotypes observed at each level result from innumerable interactions between genetic, epigenetic, biochemical, and environmental factors. Although the misleading metaphors of codes, instructions, blueprints, and genetic programs continue to be used to describe the transmission of information from genes to proteins and phenotypes, the misconceptions that arose from the use of such terms are now generally recognized (Kay 2000; Noble 2008a; Shapiro 2009). The correspondence between a particular codon and the amino acid it codes for is usually described as being arbitrary or symbolic (Maynard-Smith 2000). However, it is equally plausible that the code is not arbitrary but originated from the need to have a pattern of hydropathic complementarity between the peptides specified by the sense and antisense strands of DNA molecules (Blalok 1990; Tropsha et al. 1992). It is, indeed, remarkable that there is no exception to the rule that codons and anticodons always code for amino acids of opposite hydropathicity, i.e., either hydrophilic or hydrophobic residues. This leads to peptides coded by the two DNA strands that tend to bind to each other. This phenomenon explains why peptide analogs that retain the original hydropathic profile found in a continuous epitope are usually able to bind the same antibody although they possess little or no sequence similarity with the original epitope (Hanin et al. 1997; Van Regenmortel 1998).

The earlier confusion between the concepts of gene as a predictor of phenotypic characters and a developmental factor in ontogeny has been resolved (Bradie 2003; Moss 2003) and it is now evident that our considerable understanding of the mechanisms that allow genes to be translated into proteins has in fact given us very little insight on how phenotypic traits are produced in organisms (Oyama et al. 2001; Oyama 2009). Although a gene-centric approach to understanding biological systems is still sometimes advocated (Brenner 2010) it is generally accepted that there is no privileged level of determination in biology at any of the levels illustrated in Fig. 1 and that both bottom up and top down determination can occur (Mahner and Bunge 1997; Noble 2012). Since the hierarchy of levels labeled "up" and "down" are actually metaphors (Noble 2012), one can also refer to horizontal causation for describing the internal causal relations that operate within the boundaries of each of the different levels.

Any level in the hierarchy of biological complexity can be the starting point for a causal analysis, provided a certain initial state of affairs is considered to be in need of an explanation. However, the explanations must be framed in terms of network causality and they always have to consider the numerous factors that simultaneously influence the system in a given biological context. The upward and downward arrows in Fig. 1 therefore represent only one factor among many others that contribute to the observed effect. In all cases, a causal link attributed to one particular factor or mechanism will have to be confirmed by subsequent experimentation which alters that factor since this is the only way to establish that the hypothetical cause elicits the predicted effect in a particular context (Van Regenmortel 2007). This means that in immunology, explanations in terms of causal links must be validated by experimental interventions and manipulations in order to satisfy Woodward's (2003) requirement of "making things happen".

## Reductionism and Emergence in Immunology

The reductionist mindset is epitomized by the assertion of Crick (1966): "The ultimate aim of the modern movement in biology is to explain all biology in terms of physics and chemistry". Most biologists believe that organisms are composed solely of atoms and molecules without the participation of extraneous or spiritual forces, and they tend to take the view that biological systems can therefore be fully described and understood in terms of the physico-chemical properties of their constituent parts. In viral immunology, reductionist thinking leads to the expectation that both cell-mediated immunity (Zajac and Harrington 2008) and antibody-mediated immunity to viruses (Neurath 2008) will eventually be fully understood in terms of the molecular properties of T-cell receptors (TCRs), BCRs, antibodies, MHC molecules, cytokines, and various other cellular constituents.

It is true that methodological reductionism which dissects a biological system into its constituent parts has been extremely successful as a research strategy in molecular biology and immunology and this seems to accredit the view that biology is reducible to physics and chemistry. It is, therefore, somewhat paradoxical that in recent years biologists have become increasingly dissatisfied with the view that biological systems can be fully explained in a reductionist manner using physico-chemical principles. Misgivings about the validity of reductionist explanations in biology have been expressed at several international meetings (Bock and Goode 1998; Van Regenmortel and Hull 2002; Byerly 2003; Kistler 2003; Van Regenmortel 2004a) and the claims of genetic reductionism which link human traits to genes have become totally discredited.

Reductionists assume that the behavior of a biological system can be understood by analyzing the multiple interactions that exist between its constituents and that the activity of the whole can be inferred, deduced, calculated, and predicted from the properties of the parts. Antireductionists disagree and claim that this is not feasible because interconnections and interactions between the parts as well as certain inputs from the environment give rise to novel, emergent properties which are absent in the parts taken in isolation and cannot be predicted or deduced from the properties of the parts (Holland 1994). Since dissecting the immune system into its components severs the connections that link the various parts together in a functionally integrated manner, essential features that regulate the system's behavior are destroyed and it is no longer possible to account for the workings of the system as a whole

(Van Regenmortel 2002a). Negative feedback and feed-forward control lead to dynamic behavior that cannot be predicted satisfactorily by linear mathematical models that disregard cooperativity, synergy, and non-additive effects. Emergent properties have their own distinctive causal powers exemplified by the downward causation that occurs when an organism controls biochemical activities that take place at a lower level.

Proponents of reductionism often claim that explanations of biological phenomena can be provided without invoking biological concepts at all. When the interaction between a viral epitope and a paratope is described at the molecular level, it is debatable whether the analysis belongs to the biological field of viral immunology or to the chemical field of protein science. Although such an analysis could be presented as a reduction of biology to chemistry, it can also be interpreted as an analytical shift that no longer addresses a biological question, especially if no attempt is made to study the interaction as a biological recognition phenomenon between pathogen and antibody or as part of an infectivity neutralization process (Van Regenmortel 2002c).

Biochemists and structuralists tend to regard the function of a protein simply as what the molecule does, i.e., its functioning or activity at the molecular level. However, the function of an enzyme or an antibody may become meaningful only when the biological context is taken into account and the study takes place at the cellular or organismic level. The antigenic and immunogenic activities of a viral discontinuous epitope may become actualized only if the epitope is studied when embedded in a native protein at the surface of a virion or when it interacts with a BCR during the immunization process. When the activity of an epitope is reduced to a chemical interaction with an antibody, the functional implications of such a recognition event cannot be predicted especially if the binding occurs simultaneously with numerous reactions of other antibodies present in a polyclonal antiserum. Extrapolating from the neutralizing activity observed with one Mab to the collective neutralization potential of an antiserum is always hazardous (Mascola et al. 1997) since the initial binding of an antibody often leads to conformational changes in the antigen that modify its ability to interact with other antibodies.

The structural parameters of an epitope bound to an Mab relate only to its binding activity as an antigen since the epitope structure is unlikely to be identical with the immunogenic structure that initiated the immunization process and gave rise to the Mab (Wilson and Stanfield 1994; Bosshard 2001; Kim et al. 2011). Recent studies involving deep sequencing of antibodies in human serum have shown that the initial immunogen which triggers the affinity maturation process leading to a particular nMab usually recognizes a germline-like version of the BCR that differs considerably from the one corresponding to the mature Mab, implying that different epitopes are involved (Xiao et al. 2009).

Predicting the binding and functional activity of a protein de novo solely on the basis of its structure remains an impossible task. What is possible is to predict the probable activity of a protein by comparison with a structurally similar protein of known activity. In the case of antibodies, their multiple binding activities depend on the existence of a relational nexus with several unknown and unpredictable partners.

Attempts to predict the neutralizing capacity of an antibody from its structure is even more unrealistic since it involves a ternary interaction between antibody, antigen, and host which is entirely context-dependent. Furthermore, it is impossible to instruct the immune system to produce polyclonal antibodies endowed with a synergistic neutralizing capacity, regardless of whether the structural correlates of that activity are known or unknown.

Vaccination as an immunological intervention is meaningful only at the level of an entire organism, since organs, tissues, or molecules cannot be vaccinated. The protection against disease which is the goal of all vaccination procedures therefore remains firmly anchored in the complexities of biological systems and cannot be reduced to the chemical level of a molecular interaction (Van Regenmortel 2004b).

## Human Intentionality and Rational Design Strategies

The term "design" can be defined as the deliberate conceiving of an artificial, novel object or process by an intelligent being. The designer's task is to pose and solve an inverse problem namely to imagine, using available knowledge, what would bring about a desired outcome (Bunge 2003). In most cases, possible solutions must be tested by trial-and-error experiments until the preset goal is attained. This means that opposing rational design and empirical approaches in vaccine research (Karlsson-Hedestam et al. 2008) is fallacious since all scientific knowledge in the experimental sciences is derived from empirical testing that is necessarily planned and analyzed in a rational manner. One empirical research area that may justify additional investigations is the use of chemically inactivated HIV-1 immunogens. Several new methods are currently available to chemically inactivate the virus while retaining the functional integrity of the Env protein (Van Regenmortel 2011b and references therein). It may thus be worthwhile to re-examine the potential value of a killed, genetically modified HIV-1 vaccine, provided the safety of such material has been established.

Doing something by design is synonymous to doing it intentionally, and human intentions tend to be perceived as the cause of all human behavior and actions. This often leads to the anthropomorphic fallacy that the behavior and activities of all living organisms can be understood in terms of intentions and purposes. Since all the individual components of a biological system contribute in an integrated manner to its functioning and survival, living organisms may give the impression of having been designed. However, biologists no longer use psychological notions of intentions, design, and purposes to explain biological functions since it is universally accepted that organisms were fashioned and shaped by the filter and pressure of Darwinian natural selection (Ruse 2002; Hanke 2004; Van Regenmortel 2007).

It is nowadays commonly believed that rational design offers the best prospects for developing new drugs and vaccines (Bramwell and Perrie 2005; D'Argenio and Wilson 2010) and that this approach is vastly superior to the empirical screening and trial-and-error strategies used in the past. It has been claimed, for instance, that: "One of the goals of research in biotechnology is to transform the process of developing a drug from a trial-and-error empirical operation into a rational, structure-based process" (Amzel 1998). This denigration of empiricism contradicts the fact that even if a molecule or a vaccine has been designed following structure-based predictions, it is still necessary to screen and verify its activity in the biological context in which it is to be used. The label "rational" applied to modern vaccine design is similar to the concept of "rational drug design" which describes the structure-based strategy used in drug development (Kuntz 1992). This strategy relies on knowledge of the 3D structure of a biological target for predicting and designing candidate molecules that will bind with high affinity and selectivity to the target and inhibit its biological activity. This computer-assisted strategy based on structural bioinformatics and molecular docking has been highly successful, for instance, for developing the antiretroviral drugs that are used to inhibit various HIV enzymes (Wlodawer 2002; De Clercq 2009; LaFemina 2009).

In the biochemical and biomedical literature, the term rational is used to refer to any procedure that is based on the common sense decision to focus on elements of the system under study for which molecular information is available (Obeid et al. 1995; Van Regenmortel 1999b). "Rational" is then synonymous with "reasonable" or "sensible" and does not correspond to the notion of rationality used in the philosophy of science and which is used in statements such as "scientific judgment is guided by rationality" (Newton-Smith 1981; Giere 1988). It has been suggested that rationality should no longer be used as the ultimate justification for scientific procedures and decision making and that it may be preferable to replace it by the concept of "bounded rationality" (Gigerenzer and Selten 2002, p. 147–171; Lucas and Roosen 2010). Bounded rationality recognizes that when scientists have to reach decisions on how to proceed, they operate under three unavoidable constraints caused by (1) the limited information they have, (2) the limited capacity of their minds to process huge amounts of data, and (3) the limited amount of time they have to make a decision. One consequence of these limitations is that it is then no longer feasible for them to achieve the optimization process that in principle is required from a genuinely "rational" design procedure that must take into account all the factors that could play a role.

There is one additional reason for being skeptical of the claim that the rational design of vaccines is a realistic endeavor. Although it is possible to rationally design an epitope or antigen so that it will have an improved structural complementarity to one particular nMab, this only represents antigen design in the context of a single epitope–paratope pair and it should not be called immunogen design. When authors discuss the rational design of an HIV-1 vaccine (Douek et al. 2006; Walker and Burton 2010; Nabel et al. 2011), they only refer to studies that improve the degree of complementarity in one epitope–Mab pair and they do not clarify how an improved antigen could actually be "designed" to become an immunogen capable of generating protective antibodies. Presumably this would require an investigation of the numerous factors, extrinsic to epitope–paratope recognition, that originate in the immunized host and control the type of protective immune response that a vaccine immunogen is expected to elicit. Optimizing the binding activity of an antigen by structure-based design is clearly not equivalent to controlling or improving by trial-and-error the numerous chemical and cellular interactions involved in the generation of a neutralizing immune response. Antigen design is actually masquerading as immunogen design because of the unacknowleged, implicit assumption that antigenic reactivity necessarily entails an immunogenic capacity to produce antibodies similar to the one used as template for designing the antigen. Although some authors acknowledge the difference between antigenicity and immunogenicity, they still tend to adhere to the view that an antigen designed to fit an nMab will also be an efficient vaccine immunogen able to elicit neutralizing and protective antibodies. All the experimental immunization results obtained so far (Van Regenmortel 2011b) as well as the nature of the biological processes and mechanisms underlying antibody synthesis and maturation speak against such an assumption (Verkoczy et al. 2011; Chen et al. 2012). It is therefore not astonishing that all the studies reporting the successful rational design of a viral antigen have failed to demonstrate that the engineered antigen is also an effective vaccine immunogen (Van Regenmortel 2012a).

## Reverse Vaccinology and the Rational Design of HIV-1 Antigens

The concept of reverse vaccinology (RV) was introduced in the field of bacterial vaccines by Rappuoli (2001) about 10 years ago. RV does not attempt to develop vaccine candidates by the usual approach of fractionating bacterial extracts and determining which antigens are able to induce a protective immune response but by predicting potential vaccine immunogens using bioinformatics analyses of entire bacterial genomes. In silico analysis of the genome provides a list of all the surface-exposed proteins that the pathogen is able to express and these proteins are then produced by high-throughput technologies and tested for their immunoreactivity with patient sera as well as for their ability to induce protective antibodies (Rappuoli and Bagnoli 2011). This strategy was called "reverse" vaccinology because the investigators operate in a so-called reverse manner, i.e., starting from the genome rather than from the organism, to discover which proteins are potential vaccine immunogens. This genome-based strategy has the advantage that hundreds of bacterial proteins can be identified as candidate immunogens even if the bacteria cannot be cultivated.

In virology, RV has a different meaning altogether and refers to the strategy of generating a vaccine from the known crystallographic structure of nMabs bound to viral epitopes. The term "reverse" is used metaphorically in the sense that the investigator is trying to generate a vaccine starting from neutralizing antibodies instead of trying to generate such antibodies by immunization (Burton 2002; Walker and Burton 2010). It is not clear, however, which vaccine discovery process is being reversed since the bnMab is simply used as a template to reconstruct its epitope outside the context of the natural antigen, using structure-based design technology.

The assumption is simply made that the reconstructed antigen designed to fit the bnMab will possess the immunogenic capacity of inducing a polyclonal antibody response with the same neutralizing capacity as the Mab.

Since the RV approaches used for developing bacterial and viral vaccines are completely different it would have been preferable to call them genome-based and structure-based RV, respectively (Van Regenmortel 2011a). Both strategies require trial-and-error experimentation to ascertain which candidate bacterial or viral proteins are able to act as effective vaccine immunogens, and it is therefore misleading to present RV as a rational design procedure that overcomes the empirical nature of all vaccine science (Mascola 1999; Karlsson-Hedestam et al. 2008). It has also been argued that the development and discovery of vaccine immunogens by rational design is actually a misnomer. The ensuing discussion will therefore only consider attempts to rationally design antigens able to bind Abs without making the assumption that what is being designed is also a vaccine immunogen (Sattentau and McMichael 2010).

In a recent comprehensive review paper entitled: "Structure-based vaccine design in HIV: blind men and the elephant," Pejchal and Wilson (2010) referred to the Indian tale of three men trying to describe an elephant and reaching different conclusions because they touch different parts of the animal. They compared this unsatisfactory procedure to the way investigators try to identify potential HIV vaccine targets by determining which Env regions are targeted by individual nMabs. An amusing illustration of this approach is shown in Fig. 2 where one vulnerable "epitopic" region in the animal has been hit by a single Mab arrow.

The question of whether an effective vaccine can be developed by focusing the immune response on a single vulnerable site or epitope is a controversial issue (Scheid et al. 2009; Walker et al. 2009; Walker and Burton 2010; Lynch et al. 2011; Moore et al. 2011). An alternative approach is depicted in the painting of

**Fig. 2** This cartoon illustrates how one vulnerable epitope region in an elephant is successfully hit by a single nMab arrow

**Fig. 3** The painting of St Sebastian by Mantegna in the Ca d'Oro museum in Venice. The arrows symbolize a polyclonal antibody response which is not directed to a single vaccine target

Mantegna (Fig. 3) which illustrates symbolically that an effective vaccine may require multiple vulnerable regions to be targeted by a polyclonal Ab response (Doria-Rose et al. 2012).

Although the advent of Mabs has revolutionized our understanding of immuno-chemistry by making it possible for investigators to successfully dissect polyclonal Ab responses in terms of single epitope–paratope interactions, the use of Mabs also had the unfortunate consequence that vaccine hunters have tended to focus on individual epitopes as elicitors of neutralizing Abs. This led them to neglect the synergistic effects that result from the collective neutralizing activities of antibodies directed to different epitopes (Laal et al. 1994; Mascola et al. 1997; Zwick et al. 2001b; Van Regenmortel 2012a).

Attempts to design improved HIV-1 antigens have used as templates a small number of nMabs that recognize different antigenic sites of the Env protein such as the conserved CD4-binding site (CD4-bs) (Burton et al. 1994; Kwong et al. 1998; Zwick et al. 2003a, b; Li et al. 2007), the CD4-induced (CD4i) antigenic site that becomes accessible after gp120 interacts with CD4 (Labrijn et al. 2003), the semiconserved V3 loop (Javaherian et al. 1989; Zolla-Pazner 2004; Zolla-Pazner and Cardozo 2010), the membrane-proximal external region (MPER) antigenic site

(Zwick 2005), and the glycan antigenic site (Scanlan et al. 2002; Pantophlet 2010). It must be emphasized that each antigenic site harbours numerous different epitopes. This means that if an immune response directed to one HIV-1 antigenic site is considered to represent a single specificity, this does not exclude that a large number of different Abs will recognize overlapping targets within the same antigenic region (Walker and Burton 2010).

Various strategies have been used to improve the antigenic reactivity of the epitopes recognized by Mabs by reconstructing the epitopes outside the viral protein context in which they occur. Such strategies include (1) hyperglycosylation (Pantophlet et al. 2003a), (2) amino acid substitutions (Pantophlet et al. 2003b), (3) stabilization of the CD4-bound state by introducing cross-links or deleting variable loops (Dey et al. 2007; Zhou et al. 2007c; Wu et al. 2010), and (4) immunofocusing, i.e., decreasing the ability of the CD4bs to bind to non-neutralizing antibodies while retaining the capacity to bind to bnMabs (Pantophlet and Burton 2003). Although some of the engineered antigenic sites that were obtained reacted better with bnMabs, none of them were found to be effective immunogens able to induce broadly neutralizing antibodies (Pejchal and Wilson 2010; Kong and Sattentau 2012; Van Regenmortel 2012a).

Attempts were also made to improve the antigenic reactivity of the MPER continuous epitopes recognized by Mabs 2F5, 4E10 (Muster et al. 1993; Julien et al. 2008) and Z13 (Nelson et al. 2007). This was done by adding flanking residues to the core epitopes (Tian et al. 2002), constraining the epitopes in various conformations (Joyce et al. 2002; Ho et al. 2005), mimicking the conformation of the epitopes when they are bound to bnMabs (Ofek et al. 2004), and grafting the epitopes into various protein scaffolds (Arnold et al. 2009; Burton 2010; Correia et al. 2010; Ofek et al. 2010a; Guenaga et al. 2011). Some of these modifications increased the binding capacity of the peptides but they did not endow them with the capacity to elicit bnAbs (Law et al. 2007; Guenaga et al. 2011). This may be due to the fact that the short core peptide regions recognized by 2F5, 4E10, and Z13 do not constitute the full epitopes which are likely to be chemically heterogeneous and include lipid moieties that participate in the binding reaction through hydrophobic interactions (Scherer et al. 2010; Ofek et al. 2010b). Other composite HIV epitopes have been shown to include glycan moieties (McLellan et al. 2011). Hydrophobic residues of the long CDR H3 loop of Mab 2F5 are able to insert into the viral membrane, producing major conformational rearrangements in both epitope and paratope which illustrates the role of induced fit dynamics in antigen–antibody interactions (Alam et al. 2009; Kim et al. 2011).

Epitopes in certain HIV-1 strains may become inaccessible to antibodies following hyperglycosylation, mutations or conformational changes, a phenomenon called antigenic masking (Krachmarov et al. 2005). This makes it impossible for such HIV-1 strains to be neutralized by certain nAbs since epitope exposure is usually a prerequisite for neutralization by antibody molecules (for a review, see Pantophlet 2010). Epitopes in the V3 loop that remain unmasked in many HIV-1 strains and are recognized by bnMabs have been identified using signature motifs of 2–4 residues and it has been suggested that cocktails of such epitopes may be good vaccine candidates (Agarwal et al. 2011). Recently, it was shown that cross-clade nAbs could be induced by using V3-scaffold immunogens in a DNA prime/boost regimen (Zolla-Pazner et al. 2011).

Many attempts have been made to develop vaccine immunogens by expressing surface loops containing continuous epitopes of different viruses as recombinant proteins (Hofnung and Charbit 1993) but even this fairly straight-forward approach did not produce any effective viral vaccine (Van Regenmortel and Muller 1999). Compared to simple loop structures, reconstructing HIV-1 discontinuous epitopes (Moore and Ho 1993; Van Cott et al. 1995a) and presenting them in the required conformation at the surface of a carrier or scaffold protein is a much more difficult task and all attempts to produce effective HIV-1 vaccine immunogens in this way have so far been unsuccessful (Burton 2010; Azoitei et al. 2011).

An even more difficult task is to reconstruct epitopes that arise from the quaternary structure of viral proteins. It has been known for more than 40 years that such epitopes which were initially called neotopes (see Van Regenmortel 1966, 2009b, 2012a) are present in capsid and membrane proteins and are easily detectable by appropriate immunoassays (Neurath and Rubin 1971; Van Regenmortel and Neurath 1985). These epitopes are nowadays commonly referred to as quaternary epitopes which is a somewhat odd terminology since it would correspond to calling continuous and discontinuous epitopes, primary and tertiary epitopes, respectively. Neotopes arise from the juxtaposition of residues in neighboring protein subunits or from conformational changes induced by intersubunit interactions (Van Regenmortel 2009b). They were shown to be present in HIV-1 gp120 trimers only fairly recently although they appear to be very common (Broder et al. 1994; Cho et al. 2000; Walker et al. 2009; Pantophlet 2010; Robinson et al. 2010; Zolla-Pazner and Cardozo 2010; Pancera et al. 2010; Wu et al. 2011a).

Mab 2902 is the first antibody directed to a neotope of the Env protein which had its structure elucidated (Gorny et al. 2005; Kimura et al. 2009). This antibody which is highly strain-specific has a 21-residue long CDR H3 loop which protrudes from the paratope surface and recognizes a neotope comprising portions of the V2 and V3 loops (Honnen et al. 2007; Changela et al. 2011). The two somatically related Mabs, PG9 and PG16 which neutralize about 80% of primary HIV-1 isolates also recognize a V2 V3 neotope (Walker et al. 2009; Pejchal and Wilson 2010; Pancera et al. 2010). The PG9 neotope contains a single $\beta$ strand and two glycans that form a canyon structure into which the long CDR H3 loop of Mab PG9 inserts itself (McLellan et al. 2011). Many anti-HIV-1 antibodies have been found to possess long CDR H3 regions (Stanfield et al. 2004; Huang et al. 2005; Burke et al. 2009; Changela et al. 2011) although they do not use their protruding loops to reach recessed epitopes in the same way as Mabs PG9 and 2909. Unfortunately, it is not known how antibodies with long CDR H3 loops can be induced by immunization.

Reconstructing HIV-1 neotopes by structure-based design may turn out to be an impossible task, partly because of the unstable and transient conformation of Env trimers (Du et al. 2009) which can alternate between open and closed quaternary conformations (Harris et al. 2011). It remains unclear whether such transient neotopes are advantageous for inducing neutralizing antibodies because their conformational variability is able to facilitate induced fit adjustments and BCR recognition. In studies with other HIV-1 epitopes, there are conflicting reports on whether immunogenicity is enhanced by increasing or decreasing epitope flexibility (Dey et al. 2009; Moseri et al. 2010; Ofek et al. 2010a; Guenaga et al. 2011).

In recent years, large numbers of additional bnMabs have been isolated from HIV-1 infected individuals (Scheid et al. 2009; Simek et al. 2009; Stamatatos et al. 2009; Corti et al. 2010) which demonstrates that the human immune system is able to induce such neutralizing antibodies more frequently than previously thought (Wu et al. 2010; Scheid et al. 2011; Tomaras et al. 2011; Overbaugh and Morris 2012). This has led to the expectation that additional HIV-1 epitopes and potential vaccine targets will continue to be discovered and that this should facilitate and guide the design of effective vaccine immunogens (Sattentau and McMichael 2010). Such optimism assumes that the identification of additional HIV-1 epitopes some-how entails that there is then an increased probability that the rational design of HIV-1 vaccine immunogens will become more feasible. It would be equally unwarranted to claim that discovering additional bnMabs useful for passive immu-notherapy is likely to improve our ability to determine which immunogens are able to elicit such protective antibodies by active immunization. Although the passive administration of rare human anti-Env bnMabs to non-human primates (NHPs) can protect the animals from virus infection, this type of successful immunotherapy does not in any way tell us how such protective antibodies can be elicited by vaccination. It is of course possible to improve the paratope binding efficacy of a particular Mab intended for passive immunotherapy, using structure-based design. However, as discussed above, what is not feasible is to rationally design an immunogen that will elicit a protective polyclonal antibody response of predetermined efficacy (Van Regenmortel 2012a).

## The Bottleneck of Antibody Affinity Maturation

Considerable knowledge is available regarding the genetic basis of antibody diver-sity. It has been known for many years that initial exposure to most antigens elicits a primary immune response characterized by low-affinity germline-encoded anti-bodies (Marks 1994). In naïve human individuals who have not yet encountered a viral antigen, a primary response is initiated when the antigen binds to germline BCRs derived by random genetic recombination from a small pool of variable (V), diversity (D), and joining (J) gene segments present in the human genome (Arnaout et al. 2011). These BCRs which arise from rearranged heterodimers of light and heavy chains are highly polyspecific and react weakly with a large number of different antigens, allowing the immune system to initiate a response against virtu-ally any antigen. Structural analyses have shown that the polyreactivity of germline Abs is enhanced by the flexibility and conformational versatility of the CDRs (Manivel et al. 2000, 2002; James et al. 2003; Yin et al. 2003; Khan and Salunke 2012). The polyreactivity of germline Abs and of the initial response to HIV-1 Env antigens is a general property of the human immune system and is not a special feature of HIV immune responses. HIV-1 antibodies have been shown to cross-react with cardiolipin (Haynes et al. 2005a), host cell and gut flora antigens (Liao et al. 2011), as well as with histones, ribonucleoprotein, centromere B, and Ro/Sjogren syndrome autoantigen (Morris et al. 2011). It has been postulated that B cells that produce autoreactive Abs may trigger tolerance mechanisms which could eliminate bnAb precursors (Haynes et al. 2005b).

After initial binding of the antigen to germline BCRs, a stochastic process of somatic hypermutation occurs in the variable regions of Ig genes at rates several logs higher than in other genes. This leads to the selective expansion of B cell clones that possess high affinity BCRs for the immunizing antigen and to the proliferation of plasma cells that secrete antibodies with increased affinity for the antigen.

Advances in large scale pyrosequencing of the antibodyome, i.e., of the complete set of antibody sequences present in an organism at a given time, has allowed deep sequencing of V (D) J gene segments and Ig genes (Fischer 2011). This made it possible to identify which germline Ig genes are used in a given B cell maturation process and to follow the mutational pathway from naïve, low affinity B cells to high affinity mature BCRs and antibodies (Dimitrov 2010).

Early studies of antigen-driven maturation in HIV-1 antibodies showed that extensive somatic hypermutation occurred in their variable regions amounting to an average mutation frequency of 10–20% (Andris et al. 1991; Moran et al. 1993; Huang et al. 2004). When the germline predecessors of the affinity-matured HIV-1 Mabs b12, 2G12, and 2F5 were examined, it was found that the antibodies derived from naïve B cells did not bind to the epitopes that were recognized by mature Mabs (Xiao et al. 2009). Subsequent studies revealed that most anti-HIV-1 bnMabs were highly mutated antibodies which had undergone a prolonged affinity maturation process, thereby acquiring a high neutralization potency (Mouquet et al. 2011). The affinity maturation observed in HIV-1 antibodies was much more extensive than the 5–10% mutation frequency usually observed with antibodies directed to other viruses (Zhu et al. 2008a; Chen et al. 2012). In Mab VRCO1 which neutralizes about 90% of the HIV-1 strains tested, more than 60 amino acids in the variable region were found to differ from the germline sequence, which amounts to a 30% mutation frequency (Zhou et al. 2010; Scheid et al. 2011; Wu et al. 2011b). In the bnMabs PG9 and PG16, about 20% affinity maturation was present (Pancera et al. 2010) whereas in the less broadly neutralizing Mab2909 the degree of affinity maturation was considerably less. There is now abundant evidence that the degree of affinity maturation in bnMabs correlates with their neutralizing potency.

The germline-like versions of all these Mabs showed little or no measurable binding to HIV-1 Env, indicating that the immunogens which initiated the affinity maturation process are unlikely to have been the epitopes recognized by the mature bnAbs used as templates in the RV experiments. These findings therefore invalidate one of the assumptions underlying attempts to develop an HIV-1 vaccine by rational design (Pantophlet 2010), namely that the epitopes recognized by matured bnMabs can be used to trigger the affinity maturation process required to obtain protective antibodies.

Some investigators have been unwilling to admit that these findings have led to an impasse in structure-based RV. If somatic hypermutation is indeed a prerequisite for obtaining anti-HIV-1 bnAbs, it may be mandatory to use vaccine immunogens able to bind germline BCRs rather than immunogens designed to mimic the epitopes recognized by mature bnMabs. It may also be necessary to administer vaccine immunogens over several months or years if large numbers of amino acid changes must accumulate in germline precursor genes to obtain antibodies of sufficiently high affinity and neutralization breadth (Gray et al. 2011; Euler et al. 2012). On the other hand, it is not known if all the mutations observed in matured antibodies are required for achieving an adequate binding and neutralization potency (Burton et al. 2005), nor if the B cells expressing BCRs that bind with low affinity to certain viral antigens will be able to outcompete B cells expressing BCRs that bind to other epitopes with higher affinity.

It has been argued by some that a low affinity interaction between a naïve B cell and an HIV-1 epitope may be sufficient for initiating the affinity maturation process required for developing a protective immune response (for a review, see Van Bubnoff 2010). However, an even greater hurdle may have to be overcome if it turns out that a stochastic somatic hypermutation process which could follow millions of different pathways (Zhu et al. 2011) and may have to be guided and controlled in every vaccinated individual, is a requirement for developing a protective immune response. This new appreciation of the importance of somatic hypermutation in BCRs has led to the realization that the strategy of using as vaccine immunogens epitopes designed to fit mature bnMabs had little chance of succeeding since the required affinity-matured BCRs do not exist in naïve, vaccinated individuals (Chen et al. 2012).

Attempts are currently being made to try to modify Env molecules so that they become able to bind germline BCRs or maturation intermediates (Wu et al. 2011b). Glycan-depleted Env molecules have been obtained that bind to unmutated precursors of Mab 2F5 and 4E10 but these were unable to elicit neutralizing antibodies (Ma et al. 2011).

# Systems Biology and the Identification of Immune Correlates of Protection

Systems biology is sometimes presented as a research strategy that grew out of the realization that studying separately the individual components of a complex biological system would not lead to an understanding of how the whole system works (Cornish-Bowden 2011). Since biological systems possess emergent, functional properties that are absent in their constituents parts and are characterized by dynamic networks, regulatory mechanisms, robustness, and modularity, it seemed impossible to explain their integrated behavior by considering only events occurring at the level of individual macromolecules and cells (Van Regenmortel 2004b; O'Malley and Dupré 2005; Mazzocchi 2008).

Systems biology is an attempt to describe the multiple interactions between all the parts of a biological system by focusing on the dynamics of the entire system (Ideker et al. 2001; Kitano 2002). Since the immune system utilizes the dynamic interaction of a wide array of biomolecules, cells, and tissues, the resulting complexity makes the study of immune responses particularly apt to a systems biology approach (Aderem and Smith 2004). Genomics, proteomics, and other "omics" technologies based on multiplex microarrays, high-throughput sequencing, and bioinformatics are the methods used for this purpose which have made it possible, for instance, to identify candidate antigens for diagnostic, therapeutic, and vaccine applications (Van Regenmortel 2007). Systems biology is sometimes criticized for being a data-driven approach that cannot lead to the scientific discoveries that follow from hypothesis-driven research (Allen 2001; Wilkins 2001) although some authors argue that both approaches are complementary (Kell and Oliver 2004). It has been suggested that because living organisms have evolved, in response to changing environments, through the accumulation of random, error-prone changes, they developed a type of complexity different from the complex law-like behavior of their underlying physico-chemical constituents (Kelley and Scott 2008). Biological science therefore does not exhibit the regularities found in the physical sciences and biologists are only able to explain phenomena by positing mechanisms and making predictions without actually offering explanations of how these mechanisms function. This has led many biologists to adhere to a viewpoint called "epistemological antireductionism" (Nagel 1998) and to opt for some form of holism, emergentism (Gatherer 2010), or relational biology (Rosen 1991; Cornish-Bowden 2006), while at the same time accepting that many biological phenomena are too complex to be comprehended by human intelligence (Gannon 2007).

In vaccinology, systems biology approaches have been used mainly for predicting vaccine-induced immunity (Pulendran et al. 2010; Oberg et al. 2011). A proof-of-concept study was done by identifying patterns of upregulated gene expression, called molecular signatures, that were induced in humans vaccinated with the yellow fever vaccine YF-17D (Querec et al. 2009). This vaccine is one of the most successful vaccines ever developed and it was possible to correlate parameters such as magnitude of antigen-specific $CD8^+$ T cell responses and several molecular signatures with the development of protective immunity against yellow fever. A major shortcoming of such studies is that it is very difficult to establish if any correlations revealed by genomics or proteomics analyses are causally linked to protection. Genomics analyses may reveal, for instance, that following an immunization protocol, hundreds of genes are up- or downregulated. However, since numerous gene products always act in combination to generate biological functions, the resulting functional diversity is truly staggering, leading to numbers of potential combinations of causal factors that could be larger than the total number of atoms ($10^{80}$) in the Universe (Feytmans et al. 2005; Noble 2006).

It should also be evident that the presence of immune correlates of protection in vaccines can only be identified retrospectively after an effective vaccine has been developed empirically. In individuals who have survived a natural infection and are protected against reinfection, it could also be feasible to look for immune correlates of protection. However, this is not possible in the case of HIV infection since there are no individuals who have cleared infection and are subsequently immune to reinfection. Studies with long-term non-progressors and elite controllers of HIV-1 infection (Okulicz 2012) are also of little value since it is not possible with such individuals to exclude an innate or genetic predisposition to non-infection nor to predict which effector functions would be mediated by an adaptive vaccine-induced immunity (Koup et al. 2011). Since it is also impossible to determine immune correlates of protection from failed HIV-1 vaccine trials, it is not astonishing that past attempts to identify such correlates have been unsuccessful. The recent RV144 vaccine trial which showed a modest level of efficacy (Rerks-Ngarm et al. 2009) is the first instance when it became possible to test if immunological assays could serve as surrogate endpoints for HIV-1 infection (Nakaya and Pulendran 2012; Rolland and Gilbert 2012). However, the RV144 trial was not designed to identify immune correlates of protection and a retrospective analysis of the small number of samples available from this low efficacy trial was unlikely to reveal which aspects of the antibody response correlated with immune protection (Koup et al. 2011). Only the presence of IgG antibodies to the V1/V2 loops was found to show some correlation with a limited protection against HIV infection in the first year of the trial.

The use of NHP models of HIV infection such as simian immunodeficiency virus (SIV) and simian-HIV virus (SHIV) for evaluating immune correlates is also unsatisfactory because of the differences in disease pathogenesis and immune responses in HIV, SIV, and SHIV infections (Koup et al. 2011). It is also no longer generally accepted that showing protection in NHPs should be a gatekeeper for advancing a particular vaccine product into human efficacy trials (Shedlock et al. 2009; Thomas 2009; Greek 2012). A particular product that works in macaques may not work in humans and a strategy that shows no efficacy in NHPs could nevertheless work in humans. Animal models are poor predictors for human responses partly because results depend on the viral strains and doses used for challenge experiments and on the routes of infection (Thomas 2009). Although it may theoretically be justified to pursue exploratory small-scale human trials when no evidence is available to show that the approach being tested works in NHPs, ethics committees continue to recommend that NHP studies should always be used for the preclinical evaluation of HIV-1 vaccines (Morgan et al. 2008).

## Scientific Understanding Arises from the Ability to Successfully Manipulate the Immune System

In recent years, many authors have stressed the need to embark on large scale basic research programmes in order to increase our understanding of the human immune system and of HIV-1 antigenic structure and pathogenicity (Thomas 2009; McElrath and Haynes 2010; Virgin and Walker 2010). Such recommendations arise from the realization that our current scientific knowledge is too limited to enable us to instruct the human immune system to generate a protective response against HIV-1 infection. It is not clear, however, whether an increase in our basic knowledge of immunology and of the way bnAbs develop in certain HIV-1 infected individuals will enable us to elicit such antibodies by vaccination.

Instead of emphasizing the importance of theories and basic knowledge for achieving scientific understanding and solving problems in applied science, Hacking (1983) in his influential book *Representing and Intervening*, suggested that it is the ability of experimentalists to successfully manipulate a given system that gives them the confidence that their scientific constructs and entities are real and that a particular scientific explanation may be adequate. Whereas observation alone may not justify a belief in the reality of an invisible fuzzy entity such as an epitope, it is our ability to successfully manipulate it in an experiment that convinces us that the entity exists. In other words, we need to interfere with the material world in order to obtain knowledge about it and our scientific understanding increases because we are able to intervene in a system and successfully manipulate the phenomenon under investigation (Kelley and Scott 2008). This means that we "understand" the immune system when we can manipulate and control it and are able, for instance, to achieve protective immunity by vaccination (Kumar et al. 2012). An understanding of the immune system is thus achieved because of a prior successful intervention. This is different from the usual assumption that we need to increase our understanding of basic immunology in order to be able to manipulate and control the immune system.

In his book *Making Things Happen*, Woodword (2003, p. 9) emphasized that causal relations and explanations are important to human beings because of their interest in manipulating and controlling nature. Although the complexity of the immune system may prevent us from identifying all its internal regulatory mechanisms, it is by trial-and-error experimentation that we discover if the system can be successfully manipulated to achieve protective immunity. Unfortunately, there is no guarantee that increasing our knowledge of viral immunology and pathogenicity will necessarily give us that capacity.

Philosophers of science have used the so-called convergence argument to explain why scientists are confident that their experimental findings justify a belief in the reality of the entities and phenomena they study (Klee 1997, p. 212). If a variety of experimental results all converge on establishing what is the cause of AIDS, then it would be perverse to follow AIDS denialists and doubt that HIV causes AIDS. Conversely, it seems odd to believe that the failure of hundreds of attempts to transform HIV-1 epitopes recognized by bnMabs into effective vaccine immunogens is simply a remarkable coincidence which does not demand that we modify our underlying assumptions. It certainly is more realistic to accept that the convergence of so many negative experimental outcomes justifies the conclusion that the RV approach that has been used is not appropriate for developing a preventive HIV-1 vaccine.

# Concluding Remarks

Vaccinologists are well aware that vaccine research is an empirical science (Mascola et al. 1997) and that effective viral vaccines have always been discovered by trial-and-error immunization trials rather than by rational design using the 3D structure of viral antigens. The reasons why successful strategies used in the past failed in the case of HIV are well documented (Hilleman 1992; Kusters and Almond 2010; Virgin and Walker 2010; Kong and Sattentau 2012; Van Regenmortel 2012a). (1) The natural immune response in HIV-1 infected individuals does not clear the infection and there is therefore no natural immunological mechanism that a vaccine could mimic; (2) during HIV-1 infection, antibodies are mostly elicited against variable and accessible Env loops rather than against functionally important but less accessible conserved domains such as the receptor and co-receptor binding sites; (3) HIV-1 integrates into the host genome and establishes a latent pool of infected cells which conceal the virus from immune recognition; (4) the virus progressively destroys the immune system; (5) HIV-1 isolates exhibit an enormous antigenic variability; (6) the immune system does not readily elicit bnAbs against cryptic and transient HIV-1 epitopes; (7) the degree of antibody affinity maturation required to obtain antibodies that neutralize HIV-1 is much higher than what is needed in the case of antibodies directed to other viruses.

The present review discussed why the reductionist nature of structure-based RV is unlikely to lead to an effective HIV vaccine. Reductionist thinking has been prevalent in molecular biology for half a century and still has a strong hold on investigators who aim to develop a preventive HIV-1 vaccine. The reductionist mindset, for instance, leads them to accept that the biological activities and functions of Abs can be reduced to their physico-chemical structures and that the immunogenic potential of a protein molecule can be deduced from its antigenic properties. Chemical antigenicity is thereby confused with biological immunogenicity. This leads reductionists to assume that it should be possible to control an immune system and have it produce neutralizing antibodies simply by vaccinating an individual with a viral epitope that has been engineered to fit an nMab using structure-based design technology. This assumption overlooks the fact that every anti-HIV-1 bnMab is polyspecific and can bind viral epitopes different from the one identified when the structure of the bnMab–HIV complex was solved. There is therefore no reason why the particular HIV-1 epitope identified by crystallography should be the one that triggered the immune response that gave rise to the Mab.

Since the surface of a viral antigen is an antigenic continuum, the dissection of antigens using individual Mabs tends to give a biased view of immunological specificity and protection against infection since these properties usually result from the collective activities of several different antibodies reacting with the same antigen molecule. By focusing rational vaccine design on single epitope–paratope pairs, the neutralization synergy that tends to occur with polyclonal antibody responses will be missed.

The structural parameters of effective HIV vaccine immunogens have not been elucidated and it is therefore unfortunate that an empirical approach to vaccine development is often denigrated since trial-and-error experimentation remains the best strategy for developing any vaccine.

# Commentary: Basic Research in HIV Vaccinology Is Hampered by Reductionist Thinking

Structure-based reverse vaccinology (RV) (Van Regenmortel 2011a) attempts to develop an HIV-1 vaccine by determining the 3-D structure of complexes of HIV-1 Env epitopes bound to broadly neutralizing monoclonal antibodies (bnMabs) (Burton 2002). For more than 10 years, hundreds of investigators have tried, without success, to transform Env epitopes of known structure into effective vaccine immunogens, because they thought that epitopes that bind to affinity-matured bnMabs would also be able to induce similar antibodies when used as immunogens (Van Regenmortel 2012a).

Using the convergence argument (Klee 1997), I suggested that all the independently obtained negative results justified the conclusion that RV was unlikely to lead to a successful vaccine because it is based on invalid reductionist thinking (Van Regenmortel 2012b). King (2016) disagreed with this conclusion because it was based on inductive reasoning (i.e., generalizing from a limited body of evidence), and there was therefore no "proof" that the conclusion was correct (Chalmers 1999). Indeed, experimental science never leads to absolute certainty, since certainties are only achieved by logical, deductive reasoning and are not derived from empirical experimentation. When Einstein declared "It makes no sense to do the same thing over and over again and expect a different result," he did not contest that scientific conclusions are always reached by inductive inferences that may have only a reasonable probability of being correct.

Frontiers in Immunology, 2016, 7, 266.
Marc H V Van Regenmortel
To view the original article, visit https://www.frontiersin.org/articles/10.3389/fimmu.2016.00266/full
A commentary on
"Basic Research in HIV Vaccinology Is Hampered by Reductionist Thinking"
by King MR. Front Immunol (2016) 7:42. doi: 10.3389/fimmu.2016.00042

My convergence argument that RV is inappropriate for developing an HIV vaccine, however, was only a back-up argument (Van Regenmortel 2012b), since the actual reason why RV failed is that it did not have a sound theoretical basis corresponding to our current knowledge of immunological specificity and anti-HIV immune responses. For instance, RV does not take into account that the immune system is degenerate (Van Regenmortel 2012a; Wucherpfennig et al. 2007) and that antibodies and paratopes are never monospecific for a single epitope but are always polyspecific (Eisen and Chakraborty 2010) or even heterospecific (Van Regenmortel 2014a) for a large number of epitopes. This means that a single antibody is always able to bind several epitopes, besides the one observed by X-ray crystallography of one paratope–epitope complex. Thus, there is no reason to believe that this epitope of known structure is necessarily the one that induced the antibody and could be expected to elicit bnAbs when used as vaccine immunogen.

Another theoretical misunderstanding by many proponents of RV is that they believe that when they improve the antigenic reactivity of one Env epitope with respect to a single bnMab, using molecular engineering, this amounts to "designing" an improved vaccine immunogen capable of eliciting protective antibodies (Van Regenmortel 2015a). In so doing, they confuse antigenicity, which is a chemical property that allows a molecule to bind to an antibody, with immunogenicity, which is a biological property involving an appropriate immune system. This is typical of reductionist thinking, which assumes that biology can be reduced to chemistry and that an antigen is necessarily able to elicit the antibodies that it can react with. In fact, many factors that determine which antibodies will be produced are external to epitope–paratope recognition and originate in the immunized host. RV is also ineffective in the case of HIV, because neutralizing anti-HIV Abs are only obtained after a lengthy process of Ab affinity maturation, which is usually not the case for immune responses to other viruses. As a result, RV is not applicable to the epitopes of known structure that are recognized only by affinity-matured antibodies. Current attempts to develop vaccine immunogens from Env epitopes that do not bind germline B cell receptors or maturation intermediates present in naive individuals actually depart from the original RV strategy, which does not require the unraveling of antibody maturation pathways (Van Regenmortel 2015a).

Another reductionist limitation of RV is that it makes use only of epitopes recognized by a limited number of bnMabs, thereby neglecting the fact that the entire surface of a protein contains a very large number of overlapping epitopes and potentially immunogenic regions (Sela-Culang et al. 2013). However, it is well-known that effective vaccine-induced antibody responses are always polyclonal and recognize a wide variety of epitopes (Mascola and Haynes 2013).

King (2016) also pointed out that HIV vaccine development and related efficacy trials in humans present numerous ethical constraints that are particularly challenging. He suggested that it may be unethical to pursue attempts to develop an HIV-1 vaccine by RV, if scarce resources could be used more effectively to combat the AIDS epidemic and its huge societal problems by other means. Such a conclusion is reinforced by the theoretical shortcomings of RV outlined above, which suggest that other vaccine approaches should rather be investigated and funded (Van Regenmortel et al. 2014).

It could also be argued that the unwillingness of regulatory authorities to allow small-scale human vaccine trials, unless positive results have been obtained earlier with non-human primates (NHPs), may also be ethically questionable. It is widely accepted today that a vaccine response in NHPs (whether positive or negative) is not at all predictive of what is likely to happen in humans (Shedlock et al. 2009; Shanks et al. 2009). For instance, this means that a vaccine that shows no efficacy in NHPs may never be tested in humans, and thus that its possible efficacy would not be discovered because of ethical considerations. For instance, it is obvious today that the thalidomide disaster could not have been avoided if pregnant NHPs had first been tested in toxicity trials, since the drug is only teratogenic in humans (Greek et al. 2011). It should in fact be accepted that the only reliable model system for a human vaccine are human subjects (Horrobin 2003; Van Regenmortel 2004b). Small-scale human trials (Gandhi et al. 2009; Palma et al. 2014), using, for instance, HIV-infected individuals with temporarily interrupted ART, may be one approach that could be used to evaluate potential therapeutic HIV vaccines.

# Nature and Consequences of Biological Reductionism for the Immunological Study of Infectious Diseases

## Reductionism and its Historical Background

Numerous calls have asked for new methods applicable to infectious disease research. They are motivated by: (i) insufficient information on host-microbial interactions; (ii) obsolete microbial classifications—including "pathogenic" and "non-pathogenic" species; (iii) the need to distinguish "infectiveness" from "virulence"; (iv) the apparent end of the antibiotic era; and (v) requests for more reliable medical diagnoses than those based on research involving a single factor (Cohen et al. 2015; Van Regenmortel 2014b, 2015a, b; Casadevall and Pirofski 1999, 2000, 2015; Anuforom et al. 2015; Pomorska-Mól and Pejsak 2012; Pirofski and Casadevall 2015; Wainwright et al. 2017; Esparza 2015; Conti et al. 2007). Hoping to foster biologically grounded methods, this mini review describes the properties of infectious disease-related data, as well as *reductionism*—the belief that biology can be reduced to few and simple variables.

Three types of reductionism (ontological, epistemological, and methodological) have been described (Mazzocchi 2012). While the first two types involve *abstract* (non-measurable) concepts, reductionist methods utilize *concrete* (measurable) operations. Because conceptualizations precede operationalizations, invalid concepts

Frontiers in Immunology, 2017, 8, 612.
Rivas, A.L., Leitner, G., Jankowski, M.D., Hoogesteijn, A.L., Iandiorio, M.J., Chatzipanagiotou, S., Ioannidis, A., Blum, S.E., Piccinini, R., Antoniades, A., Fazio, J.C., Apidianakis, Y., Fair, J.M. and Van Regenmortel, M.H.V.

may promote invalid methods. Therefore, the validity of methods already applied or expected to be used in infectious diseases, in the future, should be determined.

Biology has adopted methods used in Physics (Ghilarov 2001; Mazzocchi 2008). Most notably, reductionist approaches have been followed in the field of molecular biology (Conti et al. 2007; Mazzocchi 2008; Van Regenmortel 2004b, 2012b). While such a fact should not be construed to imply that physics is reductionist *per se*, the opposite can be emphasized: in contrast to many physical systems, biological systems are generally complex, requiring approaches that far exceed the study of isolated component parts (Macklem and Seely 2010).

While biological *reductionism* has been successful, it has also been associated with failure and cognitive stagnation (Conti et al. 2007). For instance, after 20,000 publications on sepsis, only one new drug has been legally approved (Deutschman and Tracey 2014). At least two facts suggest that reductionism has hampered vaccine development: (i) more than a thousand synthetic peptide vaccines have been generated but none has been approved, and (ii) reverse vaccinology has not yet produced effective HIV vaccines (Esparza 2015; Mazzocchi 2012; Van Regenmortel 2012b). The high percentage (up to 42%) of research funding reported to be wasted may be due to inadequate methods, which include reductionism (Mazzocchi 2008; Yordanov et al. 2015).

Reductionism has prevailed since Descartes published "The discourse on the method" (Mazzocchi 2008). It is based on *deductions*, as when Halley *predicted*, in 1705, that a comet would be seen in 1758 (Wallis 1964). In contrast, Biology thrives on *inductions* made *after* data are collected (Van Regenmortel 2012b).

While Descartes has been viewed as the founder of reductionism (Movie S1 in Supplementary Material), that is not what he proposed: in 1637, he described *four rules*, reductionism being only the second rule of a method that also included (i) data analysis (first rule), (ii) integration (the third rule), and (iii) comprehensive assessments (the fourth rule). Descartes' third and fourth rules have not yet been applied in Biology (Margineanu 2016).

Two centuries later, Claude Bernard championed biomedically grounded methods (Noble 2008b). He proposed to study the *internal milieu*—today known as *homeostasis* or feedback processes. Later, von Bertalanffy showed that biological systems are not closed, but open (von Bertalanffy 1950). Thus, "internal" and "external" factors—e.g., host–microbial interactions—should be investigated.

## Reductionism-Related Errors and Information Loss

The difference between *immunogenicity* and *antigenicity* illustrates why reductionism, in Biology, is failure prone (Van Regenmortel 2012b). *Antigenicity* is simply the chemical capacity of a protein (e.g., a viral protein) to bind some preexisting antibodies. In contrast, *immunogenicity* is the in vivo capacity of the immune system to respond against an immunogen (e.g., a viral antigen) when it is introduced into an animal with the purpose of producing antibodies directed against the antigen. While the *complex* immune system elicits poly-reactive antibodies that recognize numerous antigens, only some antibodies may neutralize the infectivity of the pathogen (Van Regenmortel 2014a).

The previous concepts explain why reductionist attempts to design vaccine immunogens by molecular engineering usually fail (Van Regenmortel 2012b). Two errors explain such failures: (i) because the neutralization capacity of a poly-clonal antiserum depends on many and different antibodies, *outcomes cannot be predicted from the structure of any one antibody*; and (ii) because in vivo interactions involve the pathogen, antibodies, and some but not all host cells, *outcomes depend on multifactor,* in vivo *relationships*, which are not considered by synthetic approaches (Van Regenmortel 2002a, 2016b; Talmage 1959; Richards and Konigsberg 1973; Schubert 2014; Berzofsky 1985).

Reductionism is unintentionally practiced in many fields. For example, computer sciences are influenced by the *"curse of dimensionality"*—a term that refers to the large number of calculations that computers may need to perform (Robson 2005). To avoid millions of calculations, the number of *dimensions* to be analyzed may be *reduced* (Binder and Blettner 2015). Fields that *reduce dimensions* lose valuable information, e.g., in epidemiology, controlled trials do not assess comorbidities, even though they play major roles in infectious diseases (Greenhalgh et al. 2014; Esper et al. 2006; Ronacher et al. 2015; Ford et al. 2015).

Some quantitative traditions also limit the analysis of host–microbial interactions, e.g., *correlation analysis* neither explains nor predicts (Kitano 2002). *Network analysis* (a static method) cannot capture dynamics (Tieri et al. 2010). While classic statistics assume *linearity, independence*, and also regard as *constant* the *meaning* of any numerical assessment, these beliefs do not apply to immunomicrobial data: leukocytes are neither linearly distributed nor independent, and numbers derived from immune cells may have different interpretations at different times. That is, leukocyte data can be non-informative or *ambiguous* (Amarasingham et al. 2015; Iandiorio et al. 2016).

Errors also happen due to inadequate *procedures*—such as those commonly used with *"compositional"* data (e.g., leukocyte percentages). Because the same ratio value may be found in different biological conditions, simple leukocyte ratios induce ambiguity (Katz and King 1999; Rivas et al. 2013; Chatzipanagiotou et al. 2016). Errors are also generated by *dichotomization*: when a cutoff divides continuous data (e.g., leukocyte percentages) into two subsets and discontinuous labels—e.g., "infection-negative" and "-positive"—are assigned to each subset, false-positive and -negative errors invariably occur (Cohen 1983).

## Toward Remedial Strategies (I): The Properties of Infectious Disease-Related Data

Infectious disease-related data reveal, at least, four properties: (i) *circularity*, (ii) *heterogeneous temporal scales*, (iii) *ambiguity*, and (iv) *hidden structures* (Iandiorio et al. 2016; Rivas et al. 2013; Chatzipanagiotou et al. 2016). Understanding their features or consequences may prevent errors and information loss.

Data *circularity* is detected when three-dimensional (3D) interactions are explored—which become four-dimensional (4D) when time is also measured (Rivas et al. 2013). The analysis of dynamics matters because *what has occurred in the past will*—or may—*be repeated in the future* (Bertuglia and Vaio 2005). Because the circularity of temporal data shows neither beginning nor end, dynamics cannot be studied with approaches that utilize confidence intervals (Rivas et al. 2013; Cohen 1983; Gill and Hangartner 2010).

Because some processes occur within minutes or hours (e.g., early antimicrobial responses), while other responses—e.g., healing—take place over days or weeks (Qu et al. 2011; Myers et al. 2007), the use of identical chronological units promotes *information loss*: any one unit may be too large or too small to detect all immune functions. To capture *heterogeneous temporal scales*, "biological" (not chronological) units may be needed. Two examples of "biological" units include: (i) the increased neutrophil values that characterize early inflammatory responses (expressed as higher neutrophil/lymphocyte [N/L] ratio values), and (ii) the augmented mononuclear cell/neutrophil [MC/N] values (typically observed in the resolution phase). Such well-conserved immune profiles could act as the biological equivalents of "early and late hours" (Rivas et al. 2013).

*Ambiguity* results when the same numerical value of the same variable is found in different biological conditions (Iandiorio et al. 2016). Also known as *spatial relativity*, it occurs when data collected over short time frames (e.g., 1 day before and 1 day after a new infection develops) occupy a large portion of the space under analysis, and *vice versa* (Auffrey and Nottale 2008).

Because, in 3D/4D space, the number of data combinations may approach infinity, some data structures may be "compressed", i.e., unobservable (Fair and Rivas 2015; Leitner et al. 2015). Hence, *hidden information* is a common consequence of the combinatorial properties that characterize Biology.

## Toward Remedial Strategies (II): Methodological Foundations

Three traditions facilitate method development: (i) those grounded on *theory*, (ii) methods expressed with a mathematical language ("*modeling*"), and (iii) approaches that do not consider theories or models, but "*mechanistic*" (i.e., limited) explanations (Ghilarov 2001). Thus, methods that capture a major biological theory in their operations can be more explanatory than alternatives.

Accordingly, methods centered on "organizing principles" have been proposed (Conti et al. 2007; Wolkenhauer and Green 2013). New methods could capture critical (system-level) *biological properties*—not features derived from convenience or borrowed from other fields—e.g.: (i) *"one-to-many/many-to-one"* combinatorial features (Tieri et al. 2010), (ii) *complexity* (Mazzocchi 2008; Van Regenmortel 2004b), and (iii) *three—/four-dimensional dynamics* (Iandiorio et al. 2016; Chatzipanagiotou et al. 2016). These properties are not necessarily different: they may express the same phenomena.

The *"one-to-many/many-to-one"* feature has two presentations: (i) any one element (e.g., a cell type) can participate in two or more functions, and (ii) to be performed, any one function requires two or more elements. For instance, macrophages promote or destroy neutrophils and, together with lymphocytes, conduct *complex* functions—for instance, antigen activation (Knowlton et al. 2013).

While *complexity* may be indefinable and defy human understanding (Leitner et al. 2015; Burggren and Monticino 2005; Gannon 2007), four features describe it: (i) *emergence*, (ii) *irreducibility*, (iii) *unpredictability*, and (iv) *autonomy*. *Autonomy* means non-linearity: effects are *not proportional* or *linear* (Van Regenmortel 2007). *Emergent* features—e.g., those of virulence—are observed when a highly complex structure is assembled (Casadevall et al. 2011). Emergence (distinct, non-random patterns) may be detected using dimensionless numbers derived from leukocyte data, which create complex (although hypothetical) data structures (Leitner et al. 2015; Klinke 2009). Because *emergence* can neither be *reduced* to, nor *predicted* from isolated variables, to detect it, "top-down" (not only "bottom-up") approaches are needed (Conti et al. 2007; Mazzocchi 2012; Van Regenmortel 2004b; Casadevall et al. 2011).

While reductionism measures *variables* in isolation, combinatorial (non-reductionist) approaches capture *spatial–temporal relationships*. Distinct patterns emerge when, in 3D/4D space, dimensionless indicators *converge*, not when a single variable changes (Osinga et al. 2012; Andreopoulos and Tsotsos 2013). Because multidimensional pattern recognition does not require numerical cut-offs, it prevents errors associated with *dichotomization*, *"compositional"* data, *circularity*, and *ambiguity* (Tieri et al. 2010; Amarasingham et al. 2015; Chatzipanagiotou et al. 2016).

While "organizing properties" are necessary, they are not sufficient to prevent two problems: (i) *data variability* and (ii) the *multiple scales* of *temporal data* (Qu et al. 2011). Both problems may be addressed with structures that reveal a *single* (one data point-wide) *line* of observations. Such structures eliminate variability from all dimensions—except along the line—and detect temporal changes that occur along the line, even when such changes are numerically small and/or the individuals being tested include "slow" and "fast" responders (Iandiorio et al. 2016; Fair and Rivas 2015).

## Visualization of Reductionist and Non-Reductionist Paradigms

Figure 1 outlines both reductionist and non-reductionist paradigms. It shows how combinations of few elements (cell types) can create numerous structures. Discrimination depends on pattern recognition—which, in turn, depends on complexity, i.e., the more *spatial–temporal* relationships captured, the higher the chances of differentiating data subsets. These concepts are explained with a mundane example: written language. While any "letter", alone, lacks information (the left side of Fig. 1), *combinations* of increasing complexity ("words", "sentences", "paragraphs",

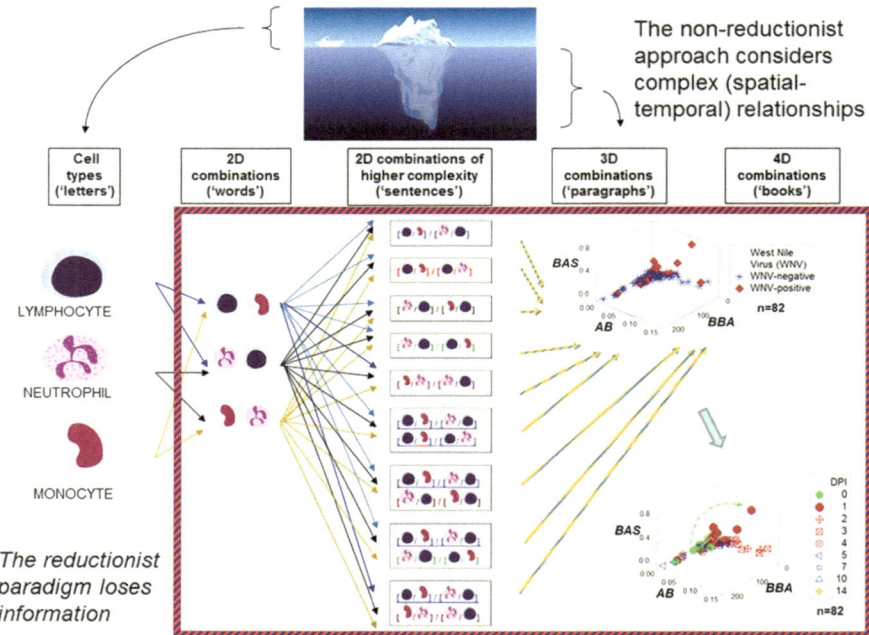

**Fig. 1** Reductionist and non-reductionist views. An iceberg is used to describe (i) reductionism (the "tip of the iceberg", i.e., an easily measured entity that does not express all the available information), and (ii) non-reductionism (a combinatorial and spatial–temporal analysis of biological complexity and dynamics, i.e., the area "below the surface"). These concepts are illustrated with an analogy that refers to written language. While simple elements ("letters") lack meaning, combinations of increasing complexity ("words", "sentences", "paragraphs", "books") exhibit distinct patterns that facilitate the partitioning of the data into subsets. The hypothetical indicators measured in the three-dimensional (3D)/four-dimensional (4D) plots shown on the right side in the figure—a set taken from the large group of dimensionless indicators shown in the central column—are identified with descriptors that lack any known biological meaning: "*BAS*", "*AB*", and "*BBA*." One example of a dimensionless indicator is the result from calculating: [M/L * N/M]/[N/L * L/M] over [M + L/N] * [L + N/M]/[N + M]/L * [M/N]. DPI: day(s) postinoculation with West Nile virus. Data source: Ref. (Rivas et al. 2013)

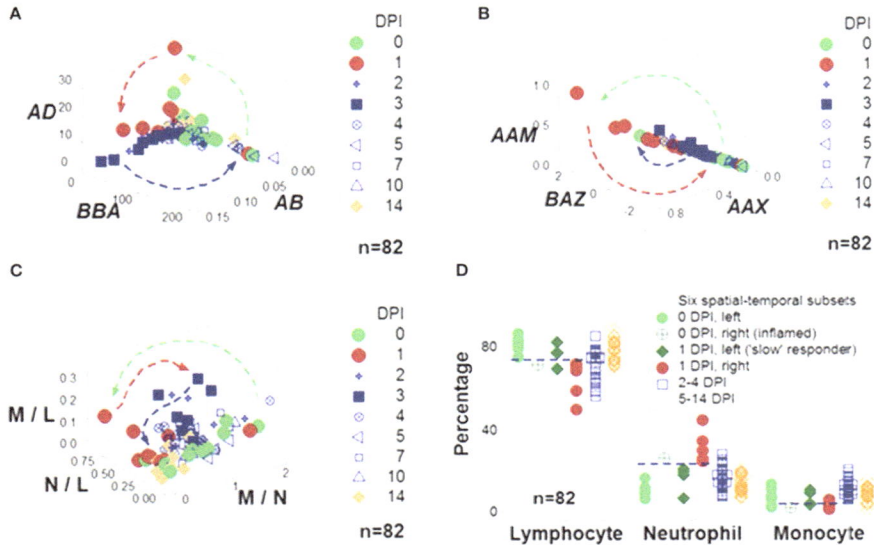

**Fig. 2** Integration of non-reductionism and reductionism. To both validate and interpret the non-reductionist graphic patterns (described in Fig. 1), additional non-reductionist data analyses and reductionist (cell type-based) operations may be required. Highly complex data structures can demonstrate both discrimination and robustness (**a, b**). In contrast, data structures of lower complexity may fail to distinguish changes that occur within 2 weeks (**c**). Based on spatial–temporal patterns, numerous data subsets may be identified and interpreted. For instance, in this example, before challenge [0 day(s) postinoculation (DPI)], all birds but one were located on the left side of the plots displayed in Fig. 1 [light green circles (**d**)]. In contrast, 24 h later (at 1 DPI), most challenged birds were on the right side [red symbols (**d**)]. However, some birds appeared to be "slow" responders: even at 1 DPI, they exhibited the profile of 0 DPI birds [dark green diamonds (**d**)]. The opposite profile was displayed by one 0 DPI animal, which revealed high neutrophil and low lymphocyte percentages [e.g., a profile indicative of an inflammation not due to the experimental challenge, dark, green circle with inserted cross (**d**)]. Inferences are facilitated by arrows that denote temporal data directionality (**a–c**) as well as non-overlapping data distributions [indicated by the horizontal lines (**d**)]. Because most data combinations have identical contents—except the three "words" [L and M, N and M, and L and N, shown in (**c**)], any other combination includes all data points of all three cell types (**a, b**), information does not depend on data inputs (identical for all but three indicators) but relationships, e.g., three-dimensional/four-dimensional (spatial–temporal) data "shapes", which can be rapidly validated and analyzed—as shown in the Movie S1 in Supplementary Material. Data source: Ref. (Rivas et al. 2013)

"books") possess meaning. When distinct *spatial* patterns emerge—such as the two perpendicular subsets exhibited by the upper 3D plot of Fig. 1—and *temporal* data are considered (the lower plot shown on the right side of Fig. 1, which displays 4D information), inferences can be based on arrows (*spatial–temporal* data *directionality*). While some inferences are visually obvious (Fig. 2a, b), not all 4D patterns are interpretable, e.g., the use of the three basic "words" (M–L, N–L, and M–N interactions, expressed as ratios) cannot distinguish dynamics that take place within

2 weeks (Fig. 2c). Yet, the spatial–temporal patterns shown in Figs. 1 and 2a, b support the detection of, at least, six immunological data subsets (Fig. 2d).

When emergent patterns are observed (which are not detected when reductionist approaches are utilized), one plausible inference is that they express *immunological functions* not previously recognized. As described in the Movie S1 in Supplementary Material, that hypothesis can be rapidly assessed.

Thus, non-reductionist data structures help discover preexisting functions (*propositional* knowledge). To validate such propositions, new tools or methods (*prescriptive* knowledge) may be required to conduct operations previously unfeasible (Van Regenmortel 2016b; Mokyr 2002a, b; Hacking 1983).

## Non-Reductionist Applications

The postulates described in Fig. 1 have been abundantly demonstrated (Iandiorio et al. 2016; Chatzipanagiotou et al. 2016; Fair and Rivas 2015; Leitner et al. 2015: Rivas et al. 2015). As shown in Fig. 2c, data *ambiguity* may occur when structures of low complexity are used (Iandiorio et al. 2016).

In contrast, new information emerges when highly complex data structures are utilized (Fig. 2a, b; Movie S1 in Supplementary Material). The discriminant process follows the geometric criteria described by Gestalt psychologists 80 years ago, including *similarity, proximity, continuity, closure, common fate, parallelism*, and *symmetry* (Andreopoulos and Tsotsos 2013).

Non-reductionist, combinatorial approaches can both detect false-negative and -positive errors and differentiate early from late immune stages (Fair and Rivas 2015). They also distinguish subsets of septic patients that differ in mortality rates and immunological profiles (Chatzipanagiotou et al. 2016).

Furthermore, non-reductionism can inform on patients empirically treated with antibiotics (Iandiorio et al. 2016; Chatzipanagiotou et al. 2016). While reductionist tests do not evaluate *antimicrobial potency* and only provide in vitro (*antimicrobial susceptibility test-based*) data (Anuforom et al. 2015), non-reductionist methods can provide *earlier* (within 24 h) and in vivo information on antibiotic–immuno–microbial–temporal interactions (Iandiorio et al. 2016). Because they may capture emergent (system-level) properties, non-reductionist analyses can yield more reliable results than those based on any one single factor (Conti et al. 2007).

Non-reductionist approaches can reveal interactions that involve cellular, supra-, and/or subcellular levels. Such approaches can simultaneously assess numerous *functions*, including (i) leukocyte activation, (ii) diapedesis, (iii) phagocytosis, (iv) early inflammation, and (v) the resolution phase of inflammation (Rivas et al. 2015).

# The Future: Integration of Non-Reductionist and Reductionist Operations

The *one-to-many/many-to-one* "organizing principle" is ubiquitous: all vertebrates are protected from thousands of microbes by up to five leukocyte types. Even if all cell types-estimated to be approximately 210 (Macklem and Seely 2010) -performed antimicrobial functions, they could not fend off tens of thousands of microbes should only "one-to-one" (immunomicrobial) relationships exist. Clearly, the reductionist "single structure/single sequence/single function" theory is implausible (Uversky et al. 2014).

In contrast, *multilevel functionality* seems to be one of Biology's "first principles" (Noble 2008b, 2010). Because it increases the complexity of the data—and, therefore, extracts more information—*multilevel functionality* may be operationalized by one-to-many/many-to-one constructs (Robson 2005; Cedersund and Roll 2009).

However, non-reductionist approaches may generate artifacts. To validate such methods, reductionist *operations*—e.g., statistical analyses that focus on individual cell types—may be required (Iandiorio et al. 2016; Chatzipanagiotou et al. 2016; Fair and Rivas 2015; Leitner et al. 2015; Rivas et al. 2015).

Given the problems associated with reductionist concepts, the previous statement seems contradictory. Yet, it is not: a non-reductionist paradigm (an abstract entity) may be partially implemented by operations (concrete entities) that include reductionist procedures.

Integrated (non-reductionist and reductionist) constructs may improve experimental designs (Roberts et al. 2015). Because experimental reductionism is inherently closed (Pilgrim 2015), it usually misses valuable information. In contrast, experiments conducted as a double (non-reductionist and reductionist) series of studies could circumvent the limitations of experimental reductionism.

# Conclusion

Because some properties of infectious disease-related data may possess undesirable consequences (e.g., data ambiguity prevents discrimination) and, in personalized medicine, decisions should be made even when the number of subjects $n = 1$, to diagnose and treat infectious diseases what is needed is not more data points (impossible when $n = 1$) but *temporal* data of greater *complexity*. To that end, immune profiles may be considered.

Because their repeatability can be easily determined—just a couple of studies can elucidate whether an observed immune pattern has been conserved across populations or species—immune profile-based inferences can measure *emergence*, i.e., patterns not shown by simple data structures that do not measure interactions—such as neutrophil percentages or counts—which may be revealed by 4D configurations of greater complexity (Iandiorio et al. 2016).

A two-step procedure may detect and validate "emergence." The first step is a non-reductionist, "top-down", hypothesis-free, combinatorial process that creates numerous and complex indicators with the purpose of generating distinct (non-randomly distributed) data subsets (Fig. 1). The second step is a reductionist (cell type-based) description of subsets meant to reveal, partially or totally, *non-overlapping* leukocyte data distributions which may also *differ temporally*. This double (spatial and temporal) data partitioning process is likely to be both *immunologically interpretable* and *statistically analyzable* (Fig. 2d).

Integrated (non-reductionist and reductionist) approaches may fill the gap of the Oslerian bio-medical paradigm—which looks for correlations but does not investigate pathogenesis—and merge disciplines and technologies (Schubert 2014; Loscalzo and Barabasi 2011). Complex and dynamic (combinatorial) methods may be more predictable than classic (reductionist or linear) models (Skinner 1994; Lecca et al. 2016).

Hence, the major message of this review refers to the *generation and interpretation of biological information*. Because most combinations of immunological data include exactly the same contents (Figs. 1 and 2; Movie S1 in Supplementary Material), *information does not depend on data inputs* but on procedures that include non-reductionist and reductionist steps: (i) detection of distinct patterns, followed by (ii) biological validation and statistical analysis of the data subsets identified in the first step.

# Part III

# The Search for a Preventive HIV Vaccine

# Limitations to the Structure-Based Design of HIV-1 Vaccine Immunogens

## Introduction

Many viral vaccines are effective because they elicit neutralizing antibodies that either block initial infection or help to eradicate the disease after infection has taken place. Human immunodeficiency virus type 1 (HIV-1) is probably the virus we know most about (Klatt et al. 2010) and its antigenic structure has been dissected in considerable detail (Zwick and Burton 2007). However, in spite of 25 years of intensive research, no vaccine capable of inducing adequate levels of neutralizing and protective antibodies has been developed (Hoxie 2010). Many investigators who report HIV-1 antigenicity studies tend to claim that the structural data they collect will most likely help in the design of an effective HIV-1 vaccine. Such statements which often amount to an article of faith assume that an effective vaccine immunogen can be deduced from the structure of a viral antigen bound to a neutralizing monoclonal antibody (Mab). This belief is so widespread that nothing short of a paradigm shift will have to occur to convince investigators that what is actually required are experimental immunogenicity trials since this is the only way to discover which HIV-1 immunogens are likely to be effective in a vaccine. The current exaggerated reliance on structural studies of HIV-1 antigens for the purpose of vaccine development has three main causes that will be analysed in this review, namely (1) misconceptions regarding the nature of protein epitopes, (2) a disregard for the analytical bias that arises when Mabs are used for dissecting the structure of antigens and (3) the belief that vaccine-relevant immunogens can be identified and predicted by analysing the structure of viral antigens.

Journal of Molecular Recognition, 2011, 24, 741–753.
Marc H V Van Regenmortel

© Springer Nature Switzerland AG 2019
Marc H V Van Regenmortel, *HIV/AIDS: Immunochemistry, Reductionism and Vaccine Design*, https://doi.org/10.1007/978-3-030-32459-9_12

# The Nature of Protein Epitopes

The B cell epitopes of proteins are the regions that are recognized by the binding sites or paratopes of antibody molecules. They will be referred to simply as epitopes since the T cell epitopes of proteins which are proteolytically cleaved peptides of the antigen that interact with the receptors of T cells will not be discussed. It is customary to classify protein epitopes as continuous or discontinuous depending on whether the amino acids that are included in the epitope are contiguous in the peptide chain or not (Atassi and Smith 1978). However, this distinction is rather theoretical since X-ray analysis of protein epitopes has revealed that the vast majority of epitopes is discontinuous and made up of surface residues distributed on two to five separate segments of the protein brought together by the folding of the chain. If some of these segments are able on their own to bind to antiprotein antibodies they will be given the label continuous epitope (Van Regenmortel 2009b).

   Continuous epitopes always have fuzzy boundaries since there are many ways to interpret the effect that removing or adding residues will have on the antigenic cross-reactivity of a peptide (Van Regenmortel 1998, 1999a). Few attempts are ever made to establish how many residues of a continuous epitope are actually exposed at the surface of the corresponding region of the cognate protein which would allow them to be recognized when an antibody or B cell receptor binds to the native antigen. When the structure of several peptide epitopes bound to antibodies was compared to that of the corresponding regions in the 3D structures of the cognate proteins, it was found that the antibodies were unable to bind to the parent proteins because of the existence of disruptive bumps and clashes that prevented antibody binding (Chen et al. 2009). The label "core epitope" is often given to short peptides of 5–6 residues identified in glycoproteins of HIV-1 and this may give the impression that such an epitope corresponds to the whole binding element that the antibody recognizes in the native protein. In fact these short peptides are only small fragments of the larger immunogenic structures that gave rise to antibodies able to cross-react with the peptides (Van Regenmortel 2006; Saphire et al. 2007).

## *Viruses Possess Quaternary Structure-Dependent Epitopes*

The quaternary structure of the viral subunits assembled in virions gives rise to epitopes that are not present in monomeric viral proteins. The term neotope was coined in 1966 (Van Regenmortel 1966) to describe epitopes that arise either from the juxtaposition of residues in neighbouring subunits or from the conformational changes that are induced in the subunits by intersubunit interactions (Van Regenmortel 1992a). Neotopes have been shown to be present in the capsids and membrane proteins of many viruses (Neurath and Rubin 1971) including HIV-1 (Broder et al. 1994; Gorny et al. 2005). When the glycoproteins of HIV-1 associate to form trimers, a portion of their surface is masked while other regions are conformationally altered and give rise to neotopes. The masked surfaces harbour the so-called cryptotopes which are epitopes that are only accessible to antibodies when the subunits are dissociated (Van Regenmortel 1992a; Broder et al. 1994). Obviously, both neotopes and discontinuous epitopes cannot easily be mimicked with synthetic linear peptides (Timmerman et al. 2007).

Since the quaternary structure of virions is sensitive to small changes in the chemical environment (Bothner et al. 1998) neotopes are often transient epitopes that can assume different conformations. In foot-and-mouth disease virus particles, two flexible regions corresponding to residues 141–160 and 200–213 of the VP1 protein were initially considered to be distinct epitopes but were later found to constitute a single neutralization neotope recognized by the same Mab (Parry et al. 1989). The loop corresponding to residues 141–160 of one viral subunit is located close to the C-terminal residues 200–213 of an adjacent subunit and a hybrid peptide immunogenic construct comprising both regions was able to elicit a stronger neutralizing antibody response than the loop 141–160 on its own (DiMarchi et al. 1986). In the case of some HIV-1 neotopes, it has been suggested that the epitope may have arisen from residues of two adjacent glycoprotein subunits since it involved loops located on opposite sides of one subunit monomer (Cho et al. 2000).

## Epitopes and Paratopes Are Relational Entities

The epitope nature of a set of amino acids can only be established when an immunoglobulin that binds to it has been found. Both epitopes and paratopes are relational entities defined by their mutual complementarity and they depend on each other to acquire a recognizable identity (Van Regenmortel 1998). They cannot be defined by any intrinsic feature of each partner that exists independently of the relational nexus. This relational dependence means that analysing the antigenicity of a protein amounts to analysing the size of the immunological repertoire of the host immunized with that protein. The number of epitopes present in a protein is thus equal to the number of different Mabs that can be raised against it. In this way, the insulin molecule was shown to possess 115 epitopes (Schroer et al. 1983) and the BLysS molecule more than a thousand (Edwards et al. 2003). Antibodies are often said to be specific for the antigen that was used to elicit them in the immunized host. This can lead to confusion since an antibody cannot be specific for the entire multiepitopic antigen as a whole but only for one of its numerous epitopes. If an investigator wants to distinguish between two related antigens using Mabs, a Mab that recognizes an epitope present in both antigens will be called unspecific while a Mab recognizing an epitope present in only one antigen will be called specific. Instead of referring to antibody specificity it is therefore preferable to speak of the discriminating potential of antibodies since it is the wish of the investigator to discriminate between two antigens that is relevant (Van Regenmortel 1998) It is now accepted that the entire accessible surface of a protein harbours a considerable number of overlapping epitopes that are identifiable only if a sufficiently large number of Mabs is available (Berzofsky 1985). The same residues at the surface of a protein can be part of neighbouring overlapping epitopes and even if the majority of residues recognized by two different Mabs are the same, the chemical bonding patterns between epitope and paratope in the two complexes may be completely different and involve entirely unrelated sequences of antibody complementarity determining regions (CDRs). In view of their relational nature, as soon as an epitope is slightly modified and binding to the antibody is affected, both the epitope and paratope are no longer the same. The size of the antigen surface in contact with a paratope has been called the antibody footprint and covers an area of about 900Å$^2$. Although it is not possible to draw clear boundaries between overlapping epitopes, one often refers to the existence of separate antigenic sites in proteins. Each antigenic site is delimitated at the surface of the protein on the basis of a particular spatial location as well as by certain structural features or functional activities. Each antigenic site always harbours many overlapping or non-overlapping epitopes (Van Regenmortel 1998).

## Mabs Introduce a Bias in Antigenic Analysis

The advent of Mabs has completely transformed our ability to dissect immune responses to proteins since they allow the analysis of single epitope–paratope pairs. However, the use of Mabs for dissecting and identifying epitopes leads investigators to focus on artificial boundaries between overlapping epitopes and to overlook the fact that the surface of a protein is an antigenic and functional continuum (Berzofsky 1985). The situation is somewhat similar to the fuzzy boundaries that exist between the colours in a rainbow which do not prevent us from recognizing separate colours.

When immune responses are dissected with Mabs, investigators tend to concentrate on a single epitope recognized by one neutralizing Mab in the hope that this will allow them to infer the structure of an immunogen able to elicit similar neutralizing antibodies. Although it should be possible, in principle, to isolate a very large number of Mabs specific for any HIV-1 antigenic site, the number that is currently available for any site is small and very much the result of chance events. When a rare neutralizing Mab has been isolated, it becomes conceptually associated with a unique, discrete epitope and this leads to the expectation that a vaccine containing this epitope should be able to induce similar antibodies and emulate the neutralization process that occurs when the Mab interacts with its epitope in the intact virion. In reality, most protective immune responses against pathogens are polyclonal and involve the collective neutralizing activities of antibodies directed to several different epitopes. Such immune responses are particularly effective because different neutralizing antibodies act in synergy, a phenomenon that is also observed when mixtures of anti-HIV-1 neutralizing Mab are utilized (for a review of early reports on such synergy effects with HIV-1, see Van Cott et al. 1995a, b). Different mechanisms have been proposed to explain the synergy between various neutralizing Mabs (Zwick et al. 2001b) one of them being that the binding of a first antibody induces a conformational change in the viral protein that exposes new epitopes and leads to enhanced binding by other Mabs. In this way, combinations of different antibodies achieve a higher degree of neutralization than would arise from the additive effects of each antibody.

Another bias introduced when neutralizing Mabs are used to characterize epitopes believed to be relevant for vaccine development is that the binding specificity of the Mab is very much dependent on the selection process that was used to obtain it. Mabs are often selected by using a library of linear peptides that may mimic an epitope of the virus and it is therefore not surprising that such Mabs often bind more strongly to peptides than to the immunizing virions used for eliciting the Mabs (Van Regenmortel 2009a) Since the structures of only few HIV-1 epitopes bound to neutralizing Mabs have been solved, a considerable amount of research has been devoted to a small number of peptide epitopes that may have little relevance for developing HIV-1 vaccines (Van Regenmortel 2009a; Montero et al. 2008). In view of the polyclonal nature of protective immune responses, it would probably be more relevant to investigate which combinations of protein immunogens are able to elicit synergistic neutralization effects in animal models. A major advance took place in recent years when international collaborative teams started to screen large panels of chronically HIV-1 infected individuals showing broad HIV-1-specific neutralizing antibody responses, in an effort to obtain many novel Mabs possessing broad neutralizing activities (Binley et al. 2008; Doria-Rose et al. 2009; Li et al. 2009, Simek et al. 2009; Gray et al. 2009). Using a new method of cloning antigen-specific memory B cells from HIV-1 infected subjects (Scheid et al. 2009) many antibodies specific for different antigenic sites were obtained that were able to neutralize numerous HIV-1 subtypes (Stamatatos et al. 2009; Wu et al. 2010). These results demonstrate that during natural HIV-1 infection, many individuals are able, after prolonged viral exposure, to elicit broadly reactive neutralizing antibody responses against HIV-1 epitopes not previously recognized. Since the immune response against HIV-1 appears to be more diversified than previously thought, these new data will certainly stimulate the study of additional immunogens. However, it should be emphasized that the structural delineation of additional HIV-1 epitopes bound to neutralizing Mabs will not necessarily allow investigators to deduce which immunogens will be successful in an HIV-1 vaccine.

## Antigenic Sites of HIV-1

The HIV-1 envelope glycoprotein (Env) consists of approximately 15 spikes per virion which are embedded in the viral membrane. Each spike is composed of trimers of three gp120 surface glycoproteins non-covalently attached to three gp41 transmembrane glycoproteins. In addition to functional glycoprotein trimers, the viral membrane may also display gp120-gp41 monomers, single gp41 molecules and other non-functional fragments (Poignard et al. 2003; Moore et al. 2006) which can elicit virus-binding but non-neutralizing antibodies (Wyatt et al. 1998). Cryo-electron tomography of trimer spikes bound to Mabs as well as in their unliganded form has partly elucidated the structure of Env spikes, although differences in glycoprotein conformation have been reported by different groups (Zhu et al. 2008b; Liu et al. 2008).

There are about 30 N-linked potential glycosylation sites in each gp120-gp41 glycoprotein heterodimer but the number of actual glycans present varies with the mammalian cell type used for virus expression (Go et al. 2008). Computer models of trimers with different numbers of glycans show that they can impede antibody access to protein epitopes that lie underneath (Schief et al. 2009).

The majority of gp120 epitopes are discontinuous and conformation-dependent (Thali et al. 1991; Moore and Ho 1993) and many of them are quaternary structure-dependent neotopes present only on intact and functional trimeric glycoprotein spikes.

HIV-1 exists in numerous genetic subtypes as well as circulating recombinant forms and the Env protein sequence from distinct subtypes can differ by more than 35%, leading to considerable antigenic variation. Despite this high sequence variability, some regions of Env are conserved which preserves the overall 3D structure of the proteins as well as the capacity of the virus to infect host cells. These conserved regions in gp120 and gp41 are important for vaccine development since they are likely to harbour epitopes that could be the target of broadly neutralizing antibodies elicited by different HIV-1 subtypes.

## The CD4-Binding Site

The CD4-binding site (CD4bs) at the surface of gp120 is a highly conserved, conformation dependent region which allows the virus to attach to its primary CD4 receptor on the surface of target cells (Kwong et al. 1998). The conserved CD4bs is also an antigenic site which elicits a wide range of broadly neutralizing antibodies in HIV-1-infected individuals and it has, therefore, been considered a promising target for vaccine development (Zhou et al. 2007c; Wu et al. 2009). The CD4bs is not buried by glycans and remains accessible to CD4 although glycans on the V2 and V3 loops of gp120 are able to restrict antibody access to the CD4bs.

One of the best characterized CD4bs epitopes is the one recognized by Mab b12 (Zwick et al. 2003a). Modifications in individual glycan groups and amino acid residues of gp120 affect the neutralizing capacity of b12 (Duenas-Decamp et al. 2008) and various steric constraints were shown to restrict the ability of b12 to access its cognate epitope by altering the quaternary conformation of the Env spikes. Despite the detailed knowledge that is now available on the binding characteristics of b12, the exhaustive paratope map of this neutralizing Mabs did not facilitate or guide the design of immunogens capable of eliciting b12-like neutralizing antibodies.

## The CD4-Induced Antigenic Site

When gp120 interacts with CD4, it undergoes a conformational change that exposes a conserved previously cryptic binding site and allows the virus to bind to one of its chemokine co-receptors, CCR5 (R5 viruses) or CXCR4 (X4 viruses) (Labrijn et al. 2003; Decker et al. 2005). Since this co-receptor binding site is formed only after gp120 binds to CD4, it harbours the so-called CD4-induced (CD4i) epitopes.

## The V3 Antigenic Site

The semiconserved and flexible loop V3, which also plays a role in co-receptor binding (Huang et al. 2005), is an immunodominant antigenic site which has been called the principal neutralization domain of HIV-1 (Javaherian et al. 1989; Spear et al. 1994). The V3 loop is able to induce neutralizing antibodies (Zolla-Pazner 2004) but it was for many years considered too variable to serve as a target for vaccine-induced antibodies. It was initially believed that V3 antibodies only neutralized T-cell line-adapted (TCLA) strains of HIV-1 and not primary HIV-1 isolates (Van Cott et al. 1995a) because it was not possible to remove antibodies that neutralized primary isolates from anti-HIV-1 human antisera by absorption with linear V3 peptides. This led to the erroneous conclusion that these neutralizing antibodies were not directed against the V3 loop, because it was not appreciated that they actually recognized a specific V3 conformation (Zolla-Pazner 2004). When the human anti-V3 antibodies were allowed to react with a V3 fusion protein that retained the V3 loop conformation present in the virus (Kayman et al. 1994), it was found that they reacted more strongly with the constrained loop than with linear V3 peptides. Such antibodies, which were able to neutralize TCLA strains as well as primary isolates from several HIV-1 clades, did recognize a V3 conformation present in the virus, in the V3 fusion protein and in a disulfide-bonded V3 loop but not in linear V3 peptides (Gorny et al. 2002). These results illustrate the potential pitfalls of using linear peptides in such investigations since these peptides are likely to be inadequate antigenic mimics of constrained cyclized loops. Mabs raised against a consensus cyclized V3 loop reacted more strongly with the cyclized than with the linear V3 form and they cross-reacted more extensively with different HIV-1 variants than antibodies raised to a single V3 sequence (Richalet-Secordel et al. 1994).

It has been found that $N$-glycans can reduce the accessibility of antibodies to epitopes in the V3 and CD4bs regions (Binley et al. 2010) and that their removal can enhance the neutralizing capacity of certain Mabs (Li et al. 2007).

## The MPER Antigenic Site

The binding of gp120 to CD4 and subsequently to the chemokine receptors, CCR5 or CXCR4 triggers major conformational changes in the viral glycoproteins. As a result, gp120 dissociates from gp41 and the *N*-terminal fusion peptide of gp41 (residues 512–527) translocates and inserts into the target cell membrane, ultimately resulting in the fusion of viral and cellular membranes (Wyatt and Sodroski 1998). This process involves the formation of a transient prehairpin intermediate state in which the membrane proximate external region (MPER) of gp41 is present in an extended conformation and is briefly accessible to a number of neutralizing antibodies. When such antibodies bind, they inhibit the subsequent structural rearrangements of gp41 required for the fusion of viral and target cell membranes, thereby preventing viral entry into the target cell and the initiation of virus infection (Sattentau et al. 1993; Frey et al. 2008). The epitopes recognized by such antibodies have been called transitional or transient epitopes because they are exposed to antibodies for only brief periods of time (Zolla-Pazner 2004).

Three linear peptide epitopes have been identified in the highly conserved 25-residue-long MPER of gp41. Mab 2F5 was found to bind to residues 662–668 (EDKWAS) of MPER (Muster et al. 1993; Zwick 2005; Julien et al. 2008) while Mabs 4E10 and Z13 recognized two overlapping epitopes situated around residues 671–676 (NWEDIT) of MPER (Zwick et al. 2001c; Nelson et al. 2007) These three continuous epitopes are usually designated as 'core epitopes'. The three Mabs hardly react with virus particles because their epitopes are poorly accessible, but they react strongly in vitro with their peptide epitopes especially when these are embedded in peptide-liposome conjugates (Ofek et al. 2004; Cardoso et al. 2005; Alam et al. 2007) The membrane activity of MPER-derived peptides in phospholipid vesicles can be assessed with fluorescent probes (Nieva et al. 2011).

The peptides bound to 4E10 and Z13 have very different conformations and this has been interpreted in terms of a single conformational switch (Pejchal et al. 2009). Another explanation is that each antibody is able to select one conformation among the many ones available (Schief et al. 2009) in the highly unstructured MPER region. It has been shown that residues situated at the apex of particularly long CDR H3 regions of 2F5 and 4E10 interact with the viral membrane and are involved in antigen recognition (Zwick et al. 2004; Scherer et al. 2010; Ofek et al. 2010b). This has led some authors to conclude that these Mabs are polyreactive for separate peptide and lipid moieties rather than being specific for a single chemically heterogeneous epitope. Complex models of stepwise interactions with subsites of the 2F5 paratope have been proposed (Alam et al. 2009; Xu et al. 2010) and it has been suggested that the CDR H3 loop interacts with lipids through 'non-specific' hydrophobic interactions while the core peptide epitopes interact through specific interactions (Ofek et al. 2010b). Such a dichotomy in binding modes seems unwarranted since hydrophobic interactions are known to contribute strongly to specific epitope-paratope binding (Van Oss 1995).

It has been shown that when the 4E10 epitope is buried in the viral membrane, it can be extracted from the lipid bilayer when it interacts with the 4E10 Mab (Sun et al. 2008). It is well-established that the 2F5 CDR H3 apex residues are essential for neutralization but that they are not involved in binding to the core epitope residues (Julien et al. 2010). There is little doubt that the epitope recognized by 2F5 is larger than the few residues identified by peptide mapping and that its exact structure will be revealed only by analysing complexes of 2F5 bound to some of the conformational states of gp41.

## *The Glycan Antigenic Site*

One antigenic site of HIV-1 is composed exclusively of a cluster of high-mannose glycans present on the outer face of gp120. This site has been studied mainly with the neutralizing Mab 2G12 but may be of limited relevance for vaccine development.

## Studies with HIV-1 Mabs in the Context of Vaccine Development

### *V3 Epitopes*

As discussed above, all antigenic sites of HIV-1 harbour many individual epitopes recognized by specific complementary paratopes. Despite its designation as the third variable domain of gp120, the V3 loop has many conserved features, including a fixed size of about 35 residues, a conserved type II turn at its tip, a disulphide bond at its base and a net positive charge (Zolla-Pazner 2004). These features allow V3 to play a role in the recognition of chemokine receptors by the virus and also explain the observed structural homology between the V3 loop and β-hairpin structures in the CC and CXC chemokine ligands recognized by the viral coreceptors (Sharon et al. 2003). As a result, anti-V3 antibodies are able to block virus infectivity by interfering with co-receptor recognition (Gorny et al. 2002). The V3 loop is very flexible and its conformation in intact gp120 molecules has not been established. However, the structure of many complexes of V3 peptides with neutralizing Mabs has been determined. Many anti-V3 Mabs produced by infected individuals have been selected using linear V3 peptides and most of them were isolate-specific and did not neutralize primary HIV-1 isolates (Gorny et al. 1997). In spite of the extensive sequence variation of V3, some conformation-sensitive anti-V3-Mabs selected with constrained V3 peptides were able to neutralize many primary isolates from various HIV-1 clades (Gorny et al. 2002) The broadly neutralizing Mab 447–52 D for instance was able to bind to the V3 loop and neutralized viruses in clades A, B, C, D and F. However, it is not possible to generalize about the

**Fig. 1** NMR structures of
V3 peptides bound to HIV-1
neutralizing antibodies and
the hydrogen bond network
in the V3 β-hairpins. The
backbone conformation and
hydrogen bonds
(represented by dashed
lines) network of (**a**) V3IIIB
bound to 447–52D Fv
(blue), (**b**) V3IIIB bound to
0.5b Fv (yellow) and (**c**)
V3MN bound to 447–52D
Fv (red) are shown (Rosen
et al. 2005)

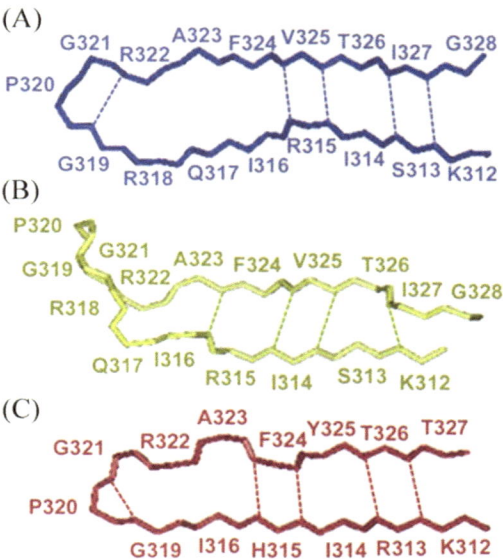

accessibility of V3 to different antibodies since steric masking by glycans or the
V1/V2 loops may prevent certain Mabs to bind to the V3 loop and neutralize the
virus (Krachmarov et al. 2005; Pinter et al. 2004; Pantophlet et al. 2008). Mab
447–52 D makes contacts with residues in the V3 crown via their side chains and
with residues in the V3 stem via main-chain interactions; this explains why this
antibody is able to tolerate some sequence variability in the stem without loss of
neutralizing activity (Stanfield et al. 2004).

Other Mabs which bind to the V3 crown are also able to neutralize a number of
clade B, C and D viruses (Pantophlet et al. 2007) In recent years, many additional
neutralizing anti-V3 Mabs have been described (Gorny et al. 2006; Hioe et al. 2010)
which confirms the potential role of the V3 loop as a candidate vaccine immunogen.
Many investigators have tried to establish which conformation of the V3 loop is
likely to be effective in a V3 synthetic peptide vaccine. NMR studies of V3 peptides
of HIV-1IIIB and HIV-1MN isolates complexed with the two Mab 447–52D and
Mab 0,5b which recognize the V3 loop in intact virions demonstrated that the
conformation of the antibody-bound V3 peptides was dictated by a process of
induced fit to each Mab (Rosen et al. 2005). As shown in Fig. 1, alternative
β-hairpin conformations can be induced in the same peptide (V3 IIIB) depending
on whether it binds to Mab 447–52D that neutralizes both R5 and X4 viruses or to
Mab 5β which neutralizes only X4 viruses. These findings indicate that the epitope
conformation observed in Mab-peptide complexes can be induced by the binding
process and thus that the observed structure of a bound peptide does not necessarily
indicate which immunogen conformation is likely to be able to elicit a particular type
of neutralizing antibody. Even cyclic V3 peptides retain significant flexibility (Mor
et al. 2009) and although it is possible to chemically constrain V3 loops in particular

conformations, there is no evidence yet that better vaccine candidates are obtained by increasing the rigidity of an antigenic site that is naturally flexible and, therefore, more capable of being recognized by a variety of B cell receptors (Garcia et al. 2006; Mester et al. 2009). The belief that a conformationally constrained peptide will probably be a more successful vaccine immunogen is based on crystallographic evidence showing that peptides bound to neutralizing Mabs adopt a particular conformation (Cardoso et al. 2005; Rosen et al. 2005). Such findings, however, do not exclude the possibility that the immunogens that induced the neutralizing antibodies were fairly unstructured and that the conformations observed in the epitope-paratope complexes resulted from processes of mutual adaptation, induced fit or conformational selection (Wilson and Stanfield 1994; Weliky et al. 1999; Berger et al. 1999; Bosshard 2001). There is evidence that many epitopes in proteins are located in surface regions that possess a high segmental mobility (Westhof et al. 1984; Tainer et al. 1985; Uversky et al. 2005) and that highly flexible loops in viruses can elicit a major portion of the neutralizing antibody response (Uversky et al. 2005; Parry et al. 1990). The flexibility of CDR loops also greatly increases the ability of antibodies to adapt to a variety of epitopes (Ramsland et al. 1997; Ma et al. 2002; Leder et al. 1995; James et al. 2003; Jimenez et al. 2003).

## *MPER Epitopes*

Once it had been established that the broadly neutralizing Mabs 2F5 and 4E10 were recognizing two short neighbouring regions in the highly conserved MPER of gp41 (Stiegler et al. 2001), many investigators regarded these two continuous epitopes as promising vaccine candidates (for a review see Montero et al. 2008; Zwick 2005). Various linear peptide constructs incorporating the two hexapeptide epitopes were synthesized using additional flanking residues and constraining the constructs in α-helical or β-turn conformation (McGaughey et al. 2003; Brunel et al. 2006). Some of these constructs had a higher affinity for the Mabs than the free, unconstrained peptides but none of them were able, when used as immunogens, to induce antibodies with significant neutralizing capacity (McGaughey et al. 2003; Neurath et al. 1995; Joyce et al. 2002; Ho et al. 2005). Inserting the core epitopes into carrier proteins to improve their immunogenicity also failed to induce neutralizing antibodies (Muster et al. 1994; Coëffier et al. 2001; Phogat et al. 2008) and several crystallographic studies of MPER peptide–antibody complexes did not lead to the discovery of effective vaccine immunogens (Zwick 2005; Ofek et al. 2004; Pejchal et al. 2009).

The inability of the short peptides recognized by neutralizing Mabs 2F5 and 4E10 to induce antibodies with the same neutralizing capacity is not surprising since these peptides do not represent the full immunogenic epitopes that triggered the appearance of the neutralizing antibodies during immunization with HIV-1. As is usually done with all continuous epitopes, the core epitopes of 2F5, 4E10 and Z13 were mapped either with phage-displayed hexapeptides (Muster et al. 1993) or with synthetic peptides (Zwick et al. 2001c; Stiegler et al. 2001), i.e. using procedures that cannot reveal complete epitope structures. It has been shown, for instance, that MPER residues outside of the core epitope of 4E10 also contribute to the epitope (Cardoso et al. 2005). Furthermore, the fact that the core peptide epitopes in MPER are only briefly exposed during the viral fusion process and are only very poorly accessible in native gp41 has made it difficult to determine the exact antigenic structures that are recognized. The available evidence indicates that the epitopes of 2F5, 4E10 and Z13 are discontinuous (Saphire et al. 2007; Alam et al. 2007; Zwick et al. 2004) and it is, therefore, not surprising that the many attempts made to turn these short peptides into effective vaccine immunogens did not succeed. Research carried out over the last two decades has unambiguously demonstrated that the continuous epitopes of infectious agents are not suitable candidates for developing structure-based vaccines (Van Regenmortel and Muller 1999; Hans et al. 2006).

## Effective Vaccine Immunogens Cannot Be Predicted from the Structure of Epitope-Paratope Complexes

A common assumption underlying most of the structural analyses of HIV-1 epitopes described in this review is that the knowledge gained from such studies is bound to facilitate and guide the design of effective HIV-1 vaccines. This belief stems from the hypothesis that it is possible to extrapolate from antigenicity to immunogenicity and that suitable vaccine immunogens can be deduced by determining the structure of viral epitopes bound to neutralizing Mabs. Antigen structures observed in crystallographic complexes result from processes of mutual adaptation, induced fit and conformation selection that occur during the binding reaction and they do not necessarily correspond to the immunogenic structure that was recognized by B cell receptors during the immunization process. Although it is sometimes acknowledged that the antigenicity of an epitope is not predictive of its ability to elicit neutralizing antibodies, the strategy often advocated for developing an effective vaccine immunogen relies on optimizing the length and conformation of a peptide complexed with a neutralizing Mab (Brunel et al. 2006).

The importance given to the particular structure of an epitope when it is bound to a paratope ignores the extensive cross-reactive potential as well as the multi-specificity and polyreactivity of all antibody molecules (Notkins 2004). Since the binding site of an immunoglobulin molecule comprises 50–70 hypervariable residues distributed over six CDR loops, a single immunoglobulin will always harbour a considerable number of overlapping or non-overlapping subsites of 10–20 residues which will be identified as separate paratopes when a complementary epitope is found that binds to them (Van Regenmortel 1998). Such cross-reactive epitopes may be structurally similar to each other or may be very different as in the case of epitopes known as mimotopes which often show no sequence similarity with the protein antigen used to induce the antibody (Van Regenmortel 2002a, 2009a, b). It is also well-known that many of the residues in a bound epitope can be replaced by any other amino acid without impairing its binding capacity to the same antibody (Getzoff et al. 1988). It is also important to bear in mind that the specificity of a Mab is never absolute (Van Regenmortel 1998) and that it possesses so-called 'true cross-reactivity' (Berzofsky and Schechter 1981) which allows it to react with numerous structurally related epitopes. Even when a Mab recognizes, for instance, a neotope present in a trimeric form of gp41 (Yuan et al. 2009), it may still be able to cross-react with other, less conformation-sensitive epitopes, usually with less affinity (Van Regenmortel 1992a). In the same way, a Mab that reacts strongly with a discontinuous epitope may be able to react more weakly with several continuous epitopes. The multispecificity of Mabs makes it, thus, hazardous to conclude that all the antigens they recognize necessarily possess an identical conformation (Yuan et al. 2009). The structure of a peptide epitope complexed with a paratope differs from the structure present before the conformational changes that occur when the two binding partners interact (Wilson and Stanfield 1994; Bosshard 2001; Goh et al. 2004) and the isolated epitope is usually a poor conformational mimic of the corresponding region in the cognate, native protein of the infectious agent (Chen et al. 2009).

In view of the multispecific cross-reactive potential of antibodies, the structure of an epitope after binding to an antibody is an unreliable guide for identifying the precise epitope structure that was recognized by B cell receptors during the immunization process and should be present in a vaccine (Van Regenmortel 2009a). It must also be emphasized that the crystallographic mapping of a paratope of a neutralizing Mab does not in any way tell us which immunogen must be used to obtain neutralizing antibodies with similar structural characteristics and neutralizing capacity (Van Regenmortel 2002a).

## Does Structure-Based HIV-1 Vaccine Design Amount to Reverse Vaccinology?

The term reverse vaccinology has been used to describe the strategy of developing vaccine immunogens by studying the interaction of human monoclonal antibodies with viral antigens (Burton 2002). It has been claimed that this approach corresponds to antibody-based HIV immunogen design because it aims at 're-eliciting' the immune response that gave rise to broadly neutralizing antibodies during natural infection (Schief et al. 2009). Instead of the classical approach of trying to generate protective antibodies by immunization with a vaccine, reverse vaccinology is supposed to be able to generate vaccines from the known structure of neutralizing Mabs (Burton 2002). Such a project is actually unlikely to succeed because it assumes that it must be possible to derive an effective HIV-1 vaccine immunogen from the structure of an HIV-1 epitope when it is bound to a neutralizing Mab. The underlying hypothesis is that it is justified to extrapolate from antigenicity to immunogenicity, because it is believed that the purely chemical ability of an antigen to recognize an antibody should allow one to predict the capacity of that antigen to give rise to a protective immune response when it is administrated to a competent host (Van Regenmortel 2010). It is well-known, however, that when a peptide fragment of a protein is able to bind to antibodies raised against the protein, this does not at all guarantee that the peptide will necessarily elicit antibodies that react with the native protein (Van Regenmortel 2006). The epitope structure observed in a complex with a neutralizing Mab is only one of the many epitopes that could be accommodated by that antibody and it does not necessarily correspond to the immunogenic epitope which was initially recognized by B cell receptors during the immunization process that induced the production of the Mab (Van Regenmortel 2009a). It must also be pointed out that the term 'reverse vaccinology' is more generally used to describe the analysis and exploitation of the genome sequences of bacterial pathogens in order to discover which proteins should be expressed by recombinant technology and used as candidate vaccine immunogens (Rappuoli 2001; Serruto et al. 2009). One limitation of genome-based reverse vaccinology is that it tends to emphasize linear continuous epitopes rather than the tertiary and quaternary structure-dependent epitopes that are more relevant for viral vaccines (Amela et al. 2007).

## Antigenicity Versus Immunogenicity

The antigenicity of an epitope is a purely chemical property that describes the structural and chemical complementarity with a paratope which can be analysed in terms of specific molecular interactions (Van Oss 1995). In contrast, the immunogenicity of an epitope is a biological property that has a meaning only in the context of a particular vertebrate host. It refers to the ability of an antigen to give rise to an immune response in a competent host and depends on extrinsic factors such as the host immunoglobulin repertoire, the presence of appropriate B cell receptors, self tolerance, the production of cytokines as well as various cellular and regulatory mechanisms meaningful only in a given biological context (Van Regenmortel 2010; Sercarz and Berzofsky 1987). The failure to distinguish antigenicity from immunogenicity is related to the reductionist fallacy that it is possible to reduce biology to chemistry (Van Regenmortel 2002a, 2004b). Although most epitopes or fragments of a viral antigen are immunogenic in the sense that they readily induce antibodies that react with the immunogen, this type of immunogenicity is irrelevant for the purpose of vaccination. What is required is the induction of antibodies which recognize the cognate, native viral antigen (a property known as cross-reactive immunogenicity) and are also able to neutralize the infectivity of the virus (so-called cross-protective immunogenicity (Van Regenmortel 2006). Unfortunately, only some of the antibodies induced by viral antigens possess neutralizing activity and the particular features of the immunogen and immunized host that are responsible for their appearance are unknown. When it is claimed that an effective vaccine immunogen can be developed by rational design, this implies that our understanding of the complexity of the immune system and the induction of cross-protective immunogenicity is sufficient to allow us to predict the outcome of vaccination attempts. Design terminology implies that one has the intention and knowledge necessary to achieve a predetermined goal, in this case protection against disease, but our ignorance of why only certain epitopes are able to elicit a neutralizing and protective immune response in fact prevents us from developing a vaccine by design. It is also impossible to control the process of somatic mutation that occurs during antibody maturation and which gives rise to antibodies endowed with adequate neutralizing capacity.

## Alternatives to the Rational Design of Vaccines

It is often claimed that the rational design of structure-based vaccine immunogens is the best available option for developing an HIV-1 vaccine with sufficient neutralization breath. It has been stated, for instance, that rational design represents the only approach that can 'elevate vaccine research from an empirical exercise to a scientific discipline' (Karlsson-Hedestam et al. 2008). Such a statement which glorifies rational design and denigrates the empirical approach in science is highly misleading since modern science actually blossomed after the seventeenth century when empirical observations replaced the earlier reliance on scholastic and rational analysis for studying natural phenomena. The subsequent immense accumulation of scientific discoveries in the experimental sciences did not arise from deductive thinking and purposeful design but from the unpredictable outcomes of controlled experimental observations that mostly followed a trial and error approach (Van Regenmortel 1999b).

As documented in this review, the extensive data that have been collected on paratope structures of anti-HIV-1 neutralizing antibodies and on the epitopes that bind to them have not helped us to design vaccine immunogens able to elicit neutralizing and protective antibodies (Montero et al. 2008; Joyce et al. 2002). Attempts to predict potential neutralization epitopes for vaccination purposes from the sequence of viral proteins have also failed (Ponomarenko and Van Regenmortel 2009). Empirical data useful for meaningful benchmarking of vaccine predictions are those obtained by probing antipeptide antibodies for cross-reactivity with proteins but such data are rarely available. It has been suggested that predictions of vaccine immunogens from protein sequence will only become feasible if a new benchmarking approach based on continuous dose-response data on antibody-mediated neutralization is used (Caioli 2010). As has been the case with all successful vaccines developed in the past, it seems likely that an HIV-1 vaccine will only be discovered empirically by testing vaccine candidates experimentally. It is likely that our increased knowledge of HIV-1 antigenic sites may help us to select which immunogens should be investigated and that increased understanding of HIV-1 neutralization mechanisms could facilitate the assessment of the neutralizing efficacy of potential vaccines (Zwick and Burton 2007; Willey and Aasa-Chapman

2008; Montefiori and Mascola 2009). Methods used for measuring the levels of humoral and mucosal immune responses are being improved continuously (Hioe et al. 2010; Selvarajah et al. 2005; McBurney and Ross 2009; Binley 2009) and there is evidence that the levels of neutralizing antibodies required for protection may be less than previously estimated (Hioe et al. 2010).

The impossibility of transforming a structurally defined epitope into a successful vaccine immunogen underlines the need for experimental studies of HIV-1 immunogens and the antibodies they elicit. In recent years, the feasibility of such studies has greatly increased since inactivated HIV-1 virions that have retained their native antigenic conformation can now be obtained more easily. Formaldehyde-treated and heat-treated HIV particles have been shown to be able to induce high-titer neutralizing antibodies responses (Poon et al. 2005b) although these treatments are known to sometimes alter the antigenic properties of virus particles (Rossio et al. 1998).

Various other methods are available to chemically inactivate HIV while retaining the conformational and functional integrity of the gp proteins (Race et al. 1995). Aziridines (Brown 2002) such as binary ethylenimine, have been used successfully to inactivate HIV by alkylation of the viral nucleic acid. HIV-1 can also be inactivated with the compound dithiodipyridine by covalently modifying zinc finger motifs in the viral nucleocapsid protein (Arthur et al. 1998; Lifson et al. 2004). Virions inactivated by this method were found to induce CD4- dependent fusion as efficiently as untreated virus, indicating that the envelope proteins had retained their native conformation (Rossio et al. 1998; Arthur et al. 1998).

Another method uses the inactivating compound $N$-ethylmaleimide which targets the nucleocapsid protein and blocks HIV-1 reverse transcription while preserving the functional activity of the gp proteins (Morcock et al. 2005). HIV-1 can also be inactivated by the photoinduced alkylating compound 1,5 iodonaphthylazide which targets the lipid domain of the viral envelope and preserves the capacity of inactivated virions to react with the 4E10, 2G12 and b12 Mabs (Rossio et al. 1998; Raviv et al. 2005). All these methods make it possible to obtain conformationally intact but non-infectious virus which could be used to study its immunogenic properties in animal studies and even possibly in humans provided the total innocuity of inactivated and genetically modified virus can be established.

Live-attenuated viruses have not been considered safe for vaccination because they could revert to a pathogenic virus or could recombine with a wild type virus in an infected host. Another approach that has been investigated is the use of virus-like particles (VLPs) or pseudovirions. VLPs are self- assembling, non-replicating and non-pathogenic particles that are similar in size and conformation to intact HIV-1. It is possible to remove certain genome elements from VLPs to prevent integration of the packaged genome into the host cell chromosomes (Young et al. 2006). Although many expression systems are available for producing HIV VLPs (Doan et al. 2005), the quantities of VLP material that can be obtained at present are inadequate for developing a practical vaccine. Empirical immunogenicity studies offer the best hope of discovering which immunogens are able to induce neutralizing antibodies and protect again disease. Such studies which must involve both animal and human trials require extensive financial resources and do present considerable challenges.

Although the effectiveness of a vaccine can ultimately only be demonstrated by protection against disease, surrogate assays based on surface plasmon resonance (SPR) biosensors may help to decrease the number of biological assays that need to be performed. It is indeed possible to predict in a fairly reliable manner the neutralizing capacity of antibodies induced by a vaccine immunogen because this capacity is correlated with the kinetic off-rate constant of the antibody when it interacts with its antigen (Van Cott et al. 1994; Rich and Myszka 2003; Van Regenmortel 2004c). SPR biosensors allow the kinetics of antigen–antibody interactions to be measured with considerable precision and they should find many applications for predicting the efficacy of different vaccine regimens as well as for assessing the quality control of different batches of vaccines using suitable panels of Mabs for measuring the 'active' concentration of the vaccine material (Zeder-Lutz et al. 1999; Hearty et al. 2010).

The use of Mabs for analysing potential vaccine immunogens has led investigators to focus on individual single epitopes of a viral antigenic site. As a result, single immunogenic sites that may interact with very few B cell receptors have been emphasized as likely elicitors of protective antibodies and the need to study polyclonal responses to different immunogens with their related synergy benefits has received little attention. This shortcoming may be overcome in the future if extensive immunogenicity studies are more widely undertaken.

In conclusion, the failure of structural analysis of epitope–paratope complexes in guiding the design of an HIV-1 vaccine should now be openly acknowledged. Instead of accumulating ever more structural information on HIV-1 epitopes and their paratope partners, what is needed is a change in the conceptual framework of vaccine developers so that they embark on more ambitious, empirical large scale immunogenicity trials of vaccine candidates.

# Two Meanings of Reverse Vaccinology and the Empirical Nature of Vaccine Science

The remarkable success of molecular and structural biology in elucidating many biological phenomena has had the unfortunate consequence that many scientists are no longer aware of the empirical foundations of all experimental science. This is evident from published statements that claim, for instance, that rational design represents "the only approach that can elevate vaccine research from an empirical exercise to a scientific discipline" (Karlsson-Hedestam et al. 2008).

One such type of rational design has become known as "reverse vaccinology" (RV), a term which has two completely different meanings. In the field of bacterial vaccines, RV refers to the strategy of predicting potential vaccine immunogens, using in silico analyses of entire bacterial genomes, and identifying all the antigens that a bacterial pathogen is able to express (Rappuoli and Bagnoli 2011). Investigators choose in a reverse manner, starting from the genome rather than from the organism, which surface-exposed proteins should be investigated as potential vaccine immunogens. Instead of studying a small number of antigens obtained by fractionating bacterial extracts, they evaluate hundreds of expressed bacterial proteins for their capacity to induce a protective immune response. In an effort to develop a Meningococcus B (Men B) vaccine, 350 open reading frames out of the 2158 present in the Men B genome were successfully cloned in Escherichia coli and purified in sufficient amounts to immunize mice. Analysis of the mouse sera revealed 90 previously unknown proteins of which 28 were able to induce antibodies that killed the bacteria (Rappuoli and Bagnoli 2011). In previous studies, only 12 Men B surface-antigens had been identified of which only 4 possessed bactericidal activity.

In virology, RV refers to the strategy of generating a vaccine from the known crystallographic structure of neutralizing antibodies bound to viral epitopes. Instead

Vaccine, 2011, 29, 7875.
Marc H V Van Regenmortel

© Springer Nature Switzerland AG 2019
Marc H V Van Regenmortel, *HIV/AIDS: Immunochemistry, Reductionism and Vaccine Design*, https://doi.org/10.1007/978-3-030-32459-9_13

of the usual approach of trying to generate protective antibodies by vaccination, RV uses the reverse approach of generating a vaccine from antibodies (Burton 2002).

It has been claimed that by analysing the structure of broadly neutralizing Mabs (bnMabs) bound to epitopes of different HIV-1 isolates, it should be possible to design an HIV-1 vaccine immunogen that will re-elicit bnAbs using a "reverse engineering" strategy (Burton 2010). This assumes that if the viral antigen is modified so that it binds bnMabs better, it will have acquired the immunogenic capacity of eliciting bnAbs similar to the Mabs.

Unfortunately no attempts to use this type of RV have been successful (Burton 2010) showing that it is not possible to extrapolate from an antigenic structure observed in crystallographic complexes to the immunogenic structure needed in a vaccine (Van Regenmortel 2011b). Since even a vaccine that induces high levels of neutralizing antibodies does not necessarily protect against virus infection (Mascola 1999), empirical clinical trials will obviously continue to be required to establish the effectiveness of a vaccine. Even in successful genome-based RV, trial-and-error experimentation is also needed to discover which bacterial proteins must be combined to obtain an effective vaccine (Rappuoli and Bagnoli 2011).

To clarify which strategy is being used, genome-based and structure-based RV should be clearly differentiated and it should not be suggested that they contradict the well-known empirical nature of vaccine science.

# Requirements for Empirical Immunogenicity Trials, Rather than Structure-Based Design, for Developing an Effective HIV Vaccine

## Introduction

Human immunodeficiency virus type 1 (HIV-1) is the virus we know most about (Klatt et al. 2010) and its antigenic structure has been analyzed in considerable detail (Norrby 1993; Wyatt et al. 1998; Zolla-Pazner 2004; Montero et al. 2008; Hoxie 2010). Many studies have been devoted to the HIV-1 envelope glycoprotein (Env) present as spikes embedded in the viral membrane. Each functional spike consists of trimers of three gp120 surface glycoproteins non-covalently attached to three gp41 transmembrane glycoproteins. In addition to functional trimers, the viral membrane may also display gp120-gp41 monomers, single gp41 molecules and other non-functional fragments (Poignard et al. 2003; Moore et al. 2006; Crooks et al. 2007). The antigenicity of the spikes has been studied extensively and numerous Env epitopes have been shown to induce a wide range of neutralizing antibodies in HIV-1 infected individuals (Zolla-Pazner 2004; Javaherian et al. 1989; Muster et al. 1993; Zwick et al. 2001c; Poignard et al. 2001; Gorny et al. 2002; Dhillon et al. 2007; Li et al. 2007; Hioe et al. 2010). As a result, the antigenic sites of the Env protein have been considered promising candidates for developing a preventive HIV vaccine able to provide prophylactic immunity (Zwick 2005; Zwick and Burton 2007; Willey and Aasa-Chapman 2008; Binley 2009; Stamatatos et al. 2009).

The strategy of trying to develop a vaccine by studying the interaction of human Mabs with HIV-1 Env epitopes has been called reverse vaccinology because it attempts to generate a vaccine from the known structure of neutralizing antibodies rather than by following the reverse, more common approach of generating protective

Archives of Virology, 2012, 157, 1–20.
Marc H V Van Regenmortel

© Springer Nature Switzerland AG 2019
Marc H V Van Regenmortel, *HIV/AIDS: Immunochemistry, Reductionism and Vaccine Design*, https://doi.org/10.1007/978-3-030-32459-9_14

antibodies by immunization (Burton 2002, 2010). When crystallographers study the structure of an Env epitope recognized by a neutralizing Mab, they often claim that their data will help the rational design of vaccine immunogens able to elicit antibodies endowed with the same neutralizing capacity as the Mab used to define the epitope (Schief et al. 2009; Saphire et al. 2001; Karlsson-Hedestam et al. 2008).

In vaccine development, the term design usually means selecting a vaccine candidate and finding appropriate vaccine formulations, schedules, adjuvants and routes of administration. The strategy of rational, structure-based design discussed here is more circumscribed since it refers to the development and optimization of a vaccine immunogen within the restricted context of a single Mab-antigen pair. This approach assumes that if a viral antigen can be modified so that it binds better to a neutralizing Mab, it will have acquired the immunogenic capacity of eliciting neutralizing polyclonal antibodies (Van Regenmortel 2011a).

It will be argued here that the claims of reverse vaccinology are based (1) on misconceptions regarding the nature of protein epitopes and of immunological specificity, (2) on the assumption that a discontinuous epitope defined by X-ray crystallography can act as an effective vaccine immunogen on its own without being embedded in a multi-epitopic protein that always induces many different types of antibodies, (3) on the belief that effective vaccine immunogens can be predicted from the antigenic structure of viral antigens bound to neutralizing Mabs.

Research carried out over the last two decades has demonstrated that reverse vaccinology has not been able to guide the design of an HIV vaccine. The present review expands on an earlier analysis of the limitations of reverse vaccinology (Van Regenmortel 2011b) and shows that antibody polyspecificity explains why elucidating the structure of viral epitopes recognized by a neutralizing Mab is of little use for discovering immunogens capable of inducing a protective immune response. It may be advisable, therefore, to shift the current emphasis in HIV vaccine research from a rational design approach based on the structural analysis of Env epitopes to a more empirical approach that investigates possible therapeutic benefits that may result from the administration of a variety of HIV-1 vaccine immunogens (Caputo et al. 2009).

## The Nature of Protein Epitopes

The regions of antigen molecules recognized by antibody molecules are called epitopes while the regions of Antibodies that bind to epitopes are called paratopes. Both regions are identified by solving the 3D structure of antigen-antibody complexes and determining which amino acid residues in the two partners make contact with each other.

# Discontinuous Epitopes

The vast majority of protein epitopes have been found to be discontinuous (Atassi and Smith 1978; Barlow et al. 1986; Sundberg and Mariuzza 2002) and to consist of two to five short stretches of residues that are distant in the protein sequence but are brought together by the folding of the peptide chain. Most HIV-1 epitopes that have been characterized are discontinuous epitopes (Thali et al. 1991; Moore and Ho 1993; Van Cott et al. 1995b; Zwick et al. 2003b). The atomic groups forming a discontinuous epitope are not held together by internal chemical bonds and the epitope is a 3D entity that emerges only because the chain acts as a scaffold. If the scaffold is perturbed, for instance by a change in chain conformation or by denaturation of the protein, the epitope ceases to exist (Laver et al. 1990; Van Regenmortel 2009b). Discontinuous epitopes cannot be extracted and isolated in active form from the protein antigen and they possess binding activity only within the context of the 3D protein structure in which they are embedded (Fig. 1). In spite of claims to the contrary, it is not possible to predict the exact structure of a discontinuous epitope from nucleotide or amino acid sequences (Hopp 1993, 1994; Van Regenmortel and Pellequer 1994; Ponomarenko and Van Regenmortel 2009).

# Continuous Epitopes

A second type of protein epitope known as a continuous epitope is defined as any short peptide fragment of the protein, usually five to eight residues long, that is able to cross-react with antibodies raised against the protein. When linear peptides are given the status of continuous epitopes of a protein, the impression is created that these epitopes actually exist as discrete binding and immunogenic sites in the native protein. In most cases, however, only some of the residues of a continuous epitope are located at the surface of the native protein where they are usually part of a more complex discontinuous epitope (Fig. 1). Although epitope databases list thousands of continuous epitopes presumed to exist in individual proteins (Greenbaum et al. 2007; Zhang et al. 2008), these linear peptides are actually poor mimics of the larger antigenic structures present in native proteins and they mostly cross-react only weakly with antiprotein antibodies (Van Regenmortel 1999a). When the structure of continuous epitopes bound to anti-peptide antibodies is compared to the structure of the corresponding regions in the cognate native protein, the two structures are usually found to differ considerably, with the result that most antibodies raised against the peptides are unable to bind to the parent protein because bumps and clashes prevent antibody binding (Chen et al. 2009).

Continuous epitopes always have ill-defined, fuzzy boundaries because they are only defined by their binding activity rather than by showing that all the residues in the peptide interact with antiprotein antibodies (Van Regenmortel 1999a). Attempts are sometimes made to assess the boundaries of continuous epitopes by measuring

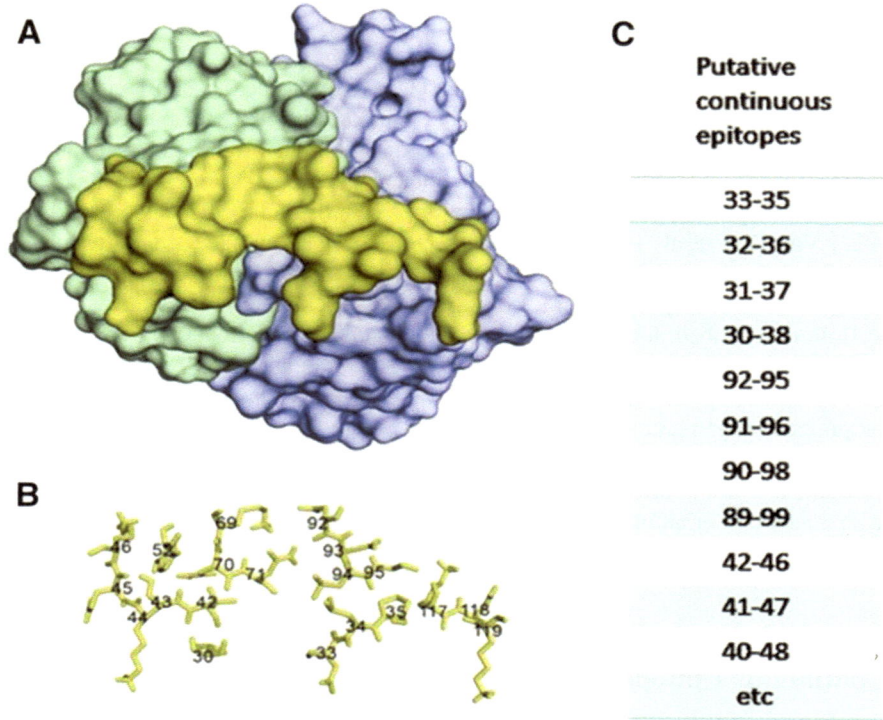

**A**

**B**

**C**

Putative
continuous
epitopes

| |
|---|
| 33-35 |
| 32-36 |
| 31-37 |
| 30-38 |
| 92-95 |
| 91-96 |
| 90-98 |
| 89-99 |
| 42-46 |
| 41-47 |
| 40-48 |
| etc |

**Fig. 1** Discontinuous epitope of the outer surface protein A of the spirochete *Borrelia burdorferi* elucidated by X-ray crystallography from a complex with Mab 184.1. (**a**) Outline of the epitope in yellow, (**b**) Position in space of the residues comprising the epitope. This set of residues cannot be isolated as such from the protein to demonstrate that it possesses binding activity in its own. (**c**) Parts of the discontinuous epitope and other peptide segments of the protein that may be able to bind Mab 184.1, in which case they would be called continuous epitopes (courtesy of Pernille Haste-Andersen, Danish Technical University)

the binding activity of peptides of decreasing size and giving the status of epitope to the smallest peptide that retains detectable activity. However, the results of such analyses depend on the immunoassay used because the peptide conformation usually differs in different types of assay (Muller et al. 1986). In some assays, tripeptides can be shown to possess significant binding activity (Trifilieff et al. 1991). The contribution of individual residues to the epitope activity can also be evaluated by measuring the binding capacity of sets of peptide analogs presenting single residue replacements. When this is done, certain residues are found to be essential for binding while others, called replaceable residues, can be replaced by any of the other 19 amino acids without impairing antigenic reactivity (Getzoff et al. 1988; Geysen et al. 1988). When a residue substitution is found to affect the epitope activity, it could mean that the residue is in contact with the paratope or that the substitution induced a conformational change which affects the reactivity of the peptide.

It has been suggested that most continuous epitopes are not genuine epitopes of native proteins but correspond to unfolded regions in denatured proteins that are recognized by antibodies directed to denatured molecules (Laver et al. 1990). It can, indeed, never be excluded that the antiprotein antiserum used for detecting continuous epitopes contains antibodies specific for denatured protein molecules because some of the protein molecules used for immunization became denatured before or after being injected in the animal (Jemmerson 1987). In the reciprocal situation where antibodies raised to peptides are allowed to react with the cognate protein, it also cannot be excluded that the antibodies recognize denatured protein molecules present in the immunoassay format that was used (Spangler 1991). It is now accepted, for instance, that earlier claims that immunization with peptides always elicits high levels of antibodies that cross-react with the native cognate protein (Lerner 1984) arose because it was not realized at the time that the protein used in solid-phase immunoassays became denatured by adsorption to plastic (Van Regenmortel 1999a; Darst et al. 1988). The classification of epitopes as either continuous or discontinuous may give the impression that the fundamental units of antigenic recognition are amino acid residues, although epitope-paratope interactions involve only a few atoms in each residue which may be side chain or main chain atoms. It should also be appreciated that crystallographers often use different criteria for identifying which contact residues are involved in an interface (Getzoff et al. 1988). For instance, contact residues in the epitope have been defined as those that have any atom less than 0.5 nm away from any atom in the antibody (Hafenstein et al. 2009). Other distance thresholds have been used for identifying different types of contact, for instance 0.35 nm for hydrogen bonds and salt bridges, 0.45 nm for van der Waals interactions and 0.45 nm for $\pi$ cation interactions (Chen et al. 2009). The use of different thresholds may lead to different residues being attributed to the binding interface.

# Neotopes and Cryptotopes

The quaternary structure of the protein subunits assembled in virus particles gives rise to additional types of epitopes known as neotopes and cryptotopes. The term neotope was introduced in 1966 (Van Regenmortel 1966) to refer to epitopes that arise either from the juxtaposition of residues in neighbouring proteins or subunits that are recognized by antibodies as a single epitope, or from conformational changes induced in the proteins by intersubunit interactions (Van Regenmortel 1992a). Neotopes have been found in the capsid and membrane proteins of most viruses (Neurath and Rubin 1971; Van Regenmortel and Neurath 1985) and are also present in HIV-1 (Broder et al. 1994; Cho et al. 2000; Gorny et al. 2005). Since the quaternary structure of virions is dynamic rather than static (Bothner et al. 1998;

Yuan et al. 2006), neotopes tend to be transient epitopes that can assume different conformations (Zolla-Pazner 2004; Binley 2009).

When the glycoproteins in HIV-1 spikes associate to form trimers, a part of their surface is no longer accessible to antibodies. These masked regions harbour epitopes known as cryptotopes (Jerne 1960) that are active only when the subunits are dissociated (Broder et al. 1994).

There is evidence that differences in the quaternary structure of gp120 between primary HIV-1 isolates, i.e. viruses that have been minimally passaged in cell lines, and T-cell line adapted (TCLA) isolates are responsible for their differential sensitivity to neutralization. It seems that a more open gp 120 conformation in TCLA viruses allows neutralizing antibodies to access more easily the binding sites for CD4 and for the coreceptors CCR5 and CXCR4 as well as the V3 loop (Poignard et al. 2001; Bou-Habib et al. 1994; Poignard et al. 1996).

## *Mimotopes*

Another type of linear peptide epitope was recognized in 1986 and given the label mimotope (Geysen et al. 1986). Mimotopes are detected when peptides from combinatorial peptide libraries obtained by chemical synthesis or phage display are tested for their ability to bind antiprotein Mabs that recognize either continuous or discontinuous epitopes (Irving et al. 2001). When the Mabs are directed to discontinuous epitopes, they may bind to peptides showing little or no sequence similarity with the antigen used to elicit the antiprotein antibodies. Antigenic cross-reactions occurring in the absence of sequence similarities may arise from the overriding attraction between hydrophilic and hydrophobic groups that occurs when a pattern of hydropathic complementarity is conserved in dissimilar peptide sequences (Hanin et al. 1997; Van Regenmortel 1998). Such a phenomenon is responsible for the binding of complementary peptides specified by the sense and antisense strands of RNA molecules because their respective codons always code for amino acids of opposite hydropathicity (Blalock 1990; Tropsha et al. 1992).

When Mabs used to detect mimotopes recognize a continuous epitope in a protein, there is sometimes extensive sequence identity between the mimotope and the continuous epitope. In the case of two neutralizing Mabs that recognized the same 7-residue sequence present in the surface glycoprotein of a murine coronavirus and in several mimotopes, it was found that the two Mabs presented little similarity in their heavy chain complementarity determining regions (CDRs), indicating that two paratopes that recognize similar sequences can be very different (Yu et al. 2000). It is also often found that Mabs harbouring unrelated paratopes are able to recognize very similar discontinuous epitopes (Malby et al. 1994; Lescar et al. 1995; Van Regenmortel 2009a), suggesting that it may be difficult to control the type of antibody that will be obtained when very similar epitopes are used as immunogens.

A limited number of mimotopes may also be able to induce antibodies that cross-react with the protein (Delmastro et al. 1997; Puntoriero et al. 1998; Larralde et al.

2007) and show a modest capacity to neutralize viral infectivity (Schellekens et al. 1994; Steward et al. 1995). The antigenic and immunogenic activities of mimotopes often depend on their conformation. For instance, when mimotopes are displayed on a phage, they may better mimic the conformation of a viral epitope than when they are used as free peptides or conjugated to a carrier protein (Irving et al. 2001; Zwick et al. 2001a; Dorgham et al. 2005). Mimotopes displayed on a phage protein may also be better elicitors of antibodies that cross-react with the virus and neutralize its infectivity (Yu et al. 2000). Although several mimotopes of HIV-1 epitopes have been isolated from combinatorial peptide libraries (Zwick et al. 2001a; Dorgham et al. 2005; Keller et al. 1993; Boots et al. 1997; Scala et al. 1999) none of them were able, when used as immunogens, to elicit adequate levels of neutralizing and protective antibodies.

## *What Do Mimotopes Mimic?*

Many peptides that are given the label mimotope because they cross react with an antiprotein Mab do not necessarily bind only to residues of the known antiprotein paratope since they may interact with neighbouring residues that are part of a different paratope recognized by the mimotope. As a result, many so-called mimotopes do not really mimic the epitope recognized by the Mab but simply react with polyspecific immunoglobulin (Ig) molecules that always possess numerous potential binding sites for different epitopes.

   The most commonly used procedure for assessing what a mimotope recognizes is to test whether it is able to inhibit the binding of the Mab to its homologous antigen. If the mimotope is found to compete with the antigen for binding to the Mab, this is usually taken as sufficient evidence that it mimics the epitope recognized by the Mab (Irving et al. 2001; Zwick et al. 2001a). This criterion for establishing mimicry is actually questionable since the binding inhibition could simply be due to the fact that steric hindrance prevents the mimotope from binding to a neighbouring paratope in the Ig binding pocket of the Mab. If an antibody recognizes two unrelated epitopes through separate, non-overlapping paratopes, it would of course make little sense to say that the two epitopes are antigenic mimics of each other. If the mimotope presents some sequence similarity with portions of the epitope, this is taken as evidence that the mimotope binds to the putative paratope of the Mab. However, it is equally plausible that the mimotope binds to its "own" specific site consisting of a neighbouring set of residues. Furthermore, if the mimotope is able to elicit antibodies that recognize the epitope believed to be mimicked, this is taken as confirming evidence that the mimotope exactly mimicks the epitope.

   The antigenic mimicry displayed by a mimotope, however, does not imply that the peptide is a precise structural mimic making identical contacts with the paratope to those made by the epitope that elicited the antibody, since a modest similarity between two antigens may be sufficient to give rise to a cross-reaction involving overlapping but different paratopes. The concept of immunogenic mimicry is even more difficult to define in structural terms than antigenic mimicry, since it involves undefined functional activities in the immunized host as well as unknown characteristics of the immunogen that elicit the appearance of cross-reactive antibodies (Irving et al. 2001).

## *The Specificity of Antibodies Is Limited Rather than Absolute*

The concept of specificity in immunology has been difficult to define (Cohn 1997, 2005; Langman 2000; Cohen et al. 2004). The term specificity is derived from the word species and describes what is characteristic of a species (Van Regenmortel 1998). For centuries, biological species were considered to be separated by clear-cut discontinuities and in the nineteenth century bacteriologists believed that antisera raised against different species of bacteria were completely specific, allowing different species to be identified by serology with absolute certainty (Mazumdar 1995). This view was later questioned by Landsteiner (1962) who showed that serological cross-reactions between different cell types were caused by antibodies that reacted to different degrees with a wide range of cells. He demonstrated that there was no one-to-one relationship between an antigen and its antibody and that an antigen could elicit a wide spectrum of antibodies capable of cross-reacting with many antigens. This led to antigen-antibody interactions being regarded as much less specific than previously thought and is nowadays explained by the widespread occurrence of cross-reactions between different epitopes and paratopes. The work of Landsteiner led Talmage (1959) to suggest that immunological specificity was not the result of a unique antibody molecule recognizing each distinguishable antigen but arose from the combinatorial effect of several different antibodies recognizing separate epitopes on the same antigen. According to that interpretation, a specific antigen recognition or elimination process is always initiated by several binding events involving different antibodies.

Specificity has been defined as "the exact complementary relationship between an agent and something acted on" (Medawar and Medawar 1978), a definition which applies to the stereochemical complementarity observed between antigen and antibody, enzyme and substrate or receptor and ligand. In the case of antibodies, the recognition of a complementary antigen is mediated by the binding site of Ig molecules consisting of 50–70 hypervariable residues distributed over the six CDRs in the variable domains of the heavy of light Ig chains. Each immunoglobulin binding site can accommodate many overlapping or non-overlapping binding subsites of 10–20 residues, corresponding to paratopes able to bind different antigens. An Ig molecule may, for instance, possess two independent binding sites that allow it to bind simultaneously to phosphorylcholine and to $a$-D-galactopyranoside (Bhattacharjee and Glaudemans 1978). A more common situation is that different paratope subsites partly overlap, in which case binding to one epitope would prevent a second related or unrelated epitope from being accommodated at a nearby location in the Ig binding site. All antibody molecules are therefore polyspecific since they always harbor numerous subsites corresponding to paratopes specific for different epitopes. The flexibility of CDR loops also contributes to the ability of Igs to adapt to a variety of epitopes (Ramsland et al. 1997; James et al. 2003; Jimenez et al. 2003). It is important to distinguish between an actual paratope and the virtual functional site of an Ig molecule consisting of all the residues of the six CDRs which, however, do not form a binding site for any particular epitope.

## Degeneracy of the Immune System and Fuzzy Recognition Sites

Many terms such as cross-reactivity, molecular mimicry, plasticity, promiscuity, degeneracy and polyspecificity have been used to describe the finding that a T cell receptor (TCR) and B cell receptor (BCR) can recognize many different ligands (Cohn 2005; Sperling et al. 1983; Parnes 2004). The term polyspecificity is particularly fitting because it captures two essential features of immune receptors, i.e. their ability to recognize multiple peptide/MHC (major histocompatibility complex) ligands or B cell epitopes as well as their capacity to distinguish between closely related ligands that present only very small structural differences (Wucherpfennig et al. 2007).

The term cross-reactivity does not imply that a particular mechanism is responsible for recognition but it suggests that there is a primary ligand for each TCR and BCR and that the binding of all other ligands are cases of cross-reaction. In reality, there is no single intrinsic or "real" epitope for any immune receptor but only a diverse group of potential ligands (Wucherpfennig et al. 2007). Cross-reactivity is also said to occur when an immune response and the resulting secreted antibodies directed at one antigen are able to interact with other antigens that did not stimulate the original reaction (Parnes 2004). Cross-reactivity then refers to a lack of specificity. Degeneracy is the ability of elements that are structurally different to perform the same activity or function. Degeneracy may yield the same or different functions depending on the context in which it is expressed (Edelman and Gally 2001). The genetic code, for instance, is degenerate because there are many more triplet codons than encoded amino acid residues. As a result, an enormous number of structurally distinct mRNA species could be translated to generate the amino acid sequence of any particular protein. Degeneracy is further increased when one includes variations in polynucleotides sequences that result in mutated but functionally equivalent gene products (Edelman and Gally 2001). In immunology, degeneracy refers not only to the ability of structurally different antibodies to perform the same recognition function but also to the ability of a single antibody to perform multiple functions like binding to different antigens. Degeneracy is a property of a system as a whole and not of its single components (Parnes 2004). The degeneracy of T cell recognition refers to the fact that peptide-binding motifs are polyspecific for MHC class molecules and that a single TCR can be activated by a large number of different peptides possessing limited sequence similarity. The degeneracy of the immunoglobulins made by an animal ensures that the animal has the ability to make antibodies which recognize essentially any molecular motif which its immune system may encounter. If each immunological receptor could only recognize a single complementary ligand, the immune system would need one receptor for every possible epitope which would require orders of magnitude more receptors than there are cells in the immune system. It is only because of polyspecificity that a complete immune repertoire can be achieved by a much more restricted number of receptors (Wucherpfennig et al. 2007). Degeneracy is also a prerequisite to allow organisms to adapt over evolutionary time since natural selection can only operate among a population of dissimilar organisms (Edelman and Gally 2001).

The description of protein binding sites as fuzzy recognition sites was introduced by Moodie et al. (1996) to account for the finding that protein-ligand complementarity can be achieved by not just a single arrangement of amino acids but by a large number of alternative arrangements. Epitopes have also been called fuzzy recognition sites because there is no clear-cut minimum difference in binding affinity or in atomic positions at the antigen-antibody interface that can serve as an absolute yardstick for deciding that two epitopes are the same or not (Van Regenmortel 1998, 1999b). The use of fuzziness as a descriptor of reality is useful for handling concepts and entities that do not possess clear-cut boundaries (McNeill and Freiberger 1993). Whereas classical, Aristotelian logic analyses concepts and classes in terms of mutually exclusive dichotomies of the type: a glass is full or is not full and a statement is true or not true, fuzzy logic is a method to handle a continuum with innumerable steps (McNeill and Freiberger 1993; Van Regenmortel 2001b; Sadegh-Zadeh 2001). Linguistic descriptors are often assumed to refer to entities that possess sharp boundaries because they satisfy the propensity of the human mind to want to make clear-cut distinctions. The continuous nature of the spectrum of electromagnetic waves does not prevent us from distinguishing different colors and the continuity of geological rock formations does not stop us from giving different names to individual mountain peaks. In a similar way, the continuous nature of biological variation and of protein evolution does not prevent us from distinguishing and giving separate names to individual biological species, antigenic binding sites and antibodies (Van Regenmortel 1998).

Even the classical tenet of immunological theory which posits the need of the immune system to discriminate between self and nonself (Cohn 2005), has been interpreted in terms of fuzzy recognition sites (Leng and Bentwich 2002) and complex systems (Efroni and Cohen 2002). It used to be believed that T cells in the thymus were either clearly responsive or unresponsive to self and foreign antigens but there is now evidence that many self-reactive T cells which escape from clonal selection as a result of low affinity interactions with self-peptides, eventually become part of the general T cell repertoire (Sandberg et al. 2000). The occurrence of antibodies able to react with foreign antigens results from selective forces operating during immunization which depend on reactant concentration and on the probability of receptor binding and lymphocyte stimulation occurring above a certain triggering threshold (Foote and Eisen 1995; Batista and Neuberger 1998).

## *Epitopes and Paratopes Are Relational Entities*

Once an Ig molecule has been found to bind to an antigen, it becomes possible to solve the structure of the antigen-antibody complex and to identify the paratope and epitope regions. Epitopes and paratopes are actually relational entities defined by their mutual complementarity and they depend on each other to acquire a recognizable identity. An epitope is thus not an intrinsic structural feature of a protein that could be identified in the absence of a particular interaction with a paratope (Van Regenmortel 1999a, 2009b). This relational dependence means that as soon as the epitope is slightly altered and binding to the antibody is affected, the paratope also is no longer the same. As a result, analyzing the antigenicity of a protein amounts to analyzing the size of the immunological repertoire of the host immunized with that protein. The number of epitopes identifiable in a protein is therefore equal to the number of different Mabs that can be raised against it. Using this criterion, the insulin molecule was shown to possess 115 epitopes (Schroer et al. 1983) and the BLysS molecule more than a thousand (Edwards et al. 2003). This analysis differs from the viewpoint advocated by Cohn (2005) which defines a "paratopic clan" as a family of paratopes, distinguishable one from the other, that are all functionally capable to recognize a given single epitope. Cohn further defines a "mimotopic array" as a set of epitopes, distinguishable one from the other, that are all able to interact functionally with a given single paratope.

## *Epitopes and Antigenic Sites*

It is now accepted that the entire accessible surface of a protein harbours a large number of overlapping epitopes that can be detected only if Mabs that bind all these regions are available (Benjamin et al. 1984; Berzofsky 1985). This means that the same residues at the surface of a protein can always be part of different overlapping epitopes. Even if the majority of residues recognized by two different Mabs are the same, the chemical bonding pattern between epitope and paratope may be very different and involve totally unrelated CDR sequences (Van Regenmortel 2009a). When only a small number of Mabs are available for studying the antigenicity of a protein, investigators tend to give undue importance to spurious boundaries assumed to demarcate discrete antigenic regions and they disregard the fact that the protein surface is an antigenic continuum. This leads to an analytical bias that emphasizes immune responses elicited by single epitopes rather than fostering an examination of the beneficial neutralizing synergy that often result from the combined interactions of several independent epitopes and paratopes (Tilley et al. 1992; Laal et al. 1994; Zwick et al. 2001b). Restricting one's attention to single epitopes also negates the insight of Talmage (1959) that specific interactions result from the combined effect of several independent recognition events.

Although it is impossible to draw distinct boundaries between overlapping epitopes, it is customary to refer to the existence of separate antigenic sites in proteins, each one comprising many overlapping or non-overlapping epitopes. Each antigenic site is identified on the protein surface on the basis of a recognizable spatial location as well as by certain structural and functional properties. Several antigenic sites of HIV-1 Env spikes have been extensively studied, for instance the CD4 binding site (CD4bs) which is the region that allows the virus to attach to its primary CD4 receptor on the surface of target cells (Li et al. 2007; Berzofsky 1985; Zhou et al. 2007c), the V3 antigenic site (Zolla-Pazner 2004; Javaherian et al. 1989; Hioe et al. 2010) and the membrane proximate external region (MPER) of gp41 (Montero et al. 2008; Muster et al. 1993; Zwick 2005). It must be stressed that each HN antigenic site consists of several overlapping epitopes that may give rise to neutralizing synergy, a phenomena that may be overlooked of only one epitope recognized by a Mab is used in the study of an antigenic site.

## Antibody Heterospecificity

Antibody heterospecificity is the ability of an antibody to bind more strongly to a different antigen than the one used for raising the antibody and is frequently observed when an antibody is tested against a series of analogs or mutants of the immunogen (Loor 1971; Al Moudallal et al. 1982; Underwood 1985; Harper et al. 1987). Heterospecificity is due to the fact that the clonal selection of B cells, which subsequently leads to antibody secretion in plasma cells, can be triggered by an immunogen endowed with only moderate affinity for a B cell receptor. Antibodies of low or moderate affinity will then be obtained which may react better with heterologous epitopes endowed with a superior degree of complementarity with the paratope (Van Regenmortel 2009b). In some cases, antibodies may have such a low affinity for the immunogen that they will not react with it at all and will only bind to a mutated antigen presenting a substitution that facilitates heterospecific binding. The technique of intragel cross-absorption of antisera is particularly suitable for detecting this phenomenon (Van Regenmortel 1966). It is well-known that following immunization with antigens, antisera are obtained that contain levels of total immunoglobulins that far exceed the levels of antibodies able to react in immunoassays with the administered antigen (Eisen 2001). A likely explanation of this difference is that B cell responses are degenerate and trigger the secretion not only of Igs that can function as detectable antibodies with the homologous antigen but also of heterospecific antibodies that have an affinity for the immunogen below the threshold required for binding to it although they may be able to bind to other, related antigens.

## Specificity Versus Discrimination Potential

The specificity of antibodies is inversely related to their ability to cross-react with a variety of antigens (Berzofsky and Schechter 1981). Part of the confusion surrounding the concept of specificity arises when no distinction is made between a single epitope and a multiepitopic antigen. If an investigator wishes to differentiate between two antigenically related proteins using a panel of Mabs, an antibody that recognizes the same epitope in both molecules would be called nonspecific whereas an antibody recognizing an epitope present in only one of them would be considered specific (Van Regenmortel 1998). Since an antibody would thus be called specific or non specific, depending on what the investigator is trying to achieve, it is preferable to refer to the discrimination potential of antibodies raised against a multiepitopic antigen. An antibody, of course, cannot be specific for a multiepitopic protein and it will at best only have a limited specificity for an individual epitope. That an antibody is always able to cross-react with many related epitopes is also demonstrated by the ability of large numbers of peptide analogs presenting amino acid replacements at each residue position to react with the same antibody (Getzoff et al. 1988; Geysen et al. 1988; Roberts et al. 1993).

## Antibody Specificity and Affinity

It is often believed that antibody specificity is correlated with high affinity because it is expected that highly specific antibodies will possess a better stereochemical complementarity with their antigens than antibodies of lower affinity. There is, however, no necessary link between affinity and specificity since antibodies of lower affinity may discriminate better between two antigens than do antibodies of higher affinity. The reason for this is that low affinity antibodies tend to detect fewer cross-reactions than antibodies of high affinity since cross-reactions are usually weaker than homologous reactions and will more quickly drop below the level of detection (Van Regenmortel 1998). It must also be emphasized that in the same way that epitopes and paratopes are defined only through their mutual relationship, the affinity of an antibody can only be defined with respect to a binary relationship with a single antigen. In contrast, the specificity of an antibody is a ternary relational property that is meaningful only with respect to a more complex set of relations between one antibody and at least two antigens, the reason being that specificity arises from the ability of an antibody to react differently with two or more antigens and thus to discriminate between them.

## Discontinuous Epitopes Cannot Be Extracted from Their Protein Environment in Order to Be Used as Vaccine Immunogens

The failure of reverse vaccinology in developing an HIV-1 vaccine is partly due to the unwarranted expectation that the structural delineation of discontinuous HIV-1 epitopes bound to neutralizing Mabs should allow these epitopes to be used as successful vaccine immunogens. As discussed above, a discontinuous epitope corresponds to a set of non-contiguous residues which cannot be isolated as a separate entity to show it possesses binding activity outside the protein context. Many unsuccessful attempts have been made to reconstitute the exact structure of discontinuous epitopes by chemical synthesis (Van Regenmortel 2009a and references therein, Timmerman et al. 2007) and it does not seem possible to reproduce their exact immunological activity in the absence of the scaffolding contribution of the peptide chain. As a result, the immunogenic capacity of a discontinuous epitope can only be evaluated by using the native protein in which it is embedded as an immunogen, a procedure that will always stimulate an heterogeneous immune response against the numerous epitopes of the protein. For instance, when antibodies are raised against the CD4bs surface of gp120, most of them do not possess the exceptional neutralizing capacity of Mab b12 (Pantophlet et al. 2003b).

When a multiepitopic protein is used as immunogen, various epitopes will selectively bind to the B cell receptors present in the immunized host for which they possess sufficient affinity. However, the epitopes will not preferentially bind to those rare receptors, which in addition to recognizing the epitope, are responsible for the subsequent secretion by plasmocytes of antibodies that neutralize viral infectivity. The selection of B cell receptors during immunization is blind to the presence of neutralizing anti-viral activity in the antibodies that will be secreted since this biological function becomes effective only during the subsequent infection of a host. Furthermore, even if a single discontinuous epitope could somehow be used as an immunogen, the probability is still extremely low that a selectable B cell receptor would be present in the immunized host that exactly matches the variable immunoglobulin gene and CDR sequences of the Mab used for identifying the discontinuous epitope. Unfortunately, we do not know how to elicit neutralizing rather than non-neutralizing antibodies since it is not clear which features of an immunogen and immunized host are responsible for the cross-protective immunogenic activity that is required from any vaccine candidate (Zwick and Burton 2007; Parren and Burton 2001; Van Regenmortel 2006; Reading and Dimmock 2007).

## The Unsuccessful Search for Immunogens Was Able to Elicit B12-like Antibodies

The difficulties encountered when one attempts to design a vaccine immunogen on the basis of the known structure of a neutralizing antibody recognizing a discontinuous HIV-1 epitope are clearly demonstrated in the case of the b12 antibody. This Mab which was obtained from a phage-displayed Fab library derived from an HIV-1 patient (Burton et al. 1994) recognizes a conserved discontinuous epitope overlapping the CD4bs of gp120. It neutralizes a wide range of primary and TCLA HIV-1 isolates by inhibiting the binding of gp120 to the CD4 receptor. Antibody b12 has an 18-residue long CDR H3 region, a feature present in many anti-CD4bs human Mabs, allowing them to penetrate into a hydrophobic cavity on the gp120 surface. Using a docking model of b12 onto gp120, several contact residues of the epitope were identified and their contribution to binding activity was evaluated by alanine mutagenesis (Saphire et al. 2001). The gp120 structure that was used for docking b12 was the only one available at the time and corresponds to that of a monomeric core complexed with CD4 and with the Fab 17b (Kwong et al. 1998), and it may therefore not correspond to the conformation of the trimeric CD4bs seen by anti-bodies and by the immune system. It is known that when gp120 binds to CD4, it undergoes a large conformational change which allows the virus to bind to one of its chemokine coreceptors (Labrijn et al. 2003; Decker et al. 2005).

In an attempt to identify an HIV-1 vaccine immunogen capable of eliciting b12-like neutralizing antibodies, peptide libraries displayed on filamentous phage have been used to select peptides that bind b12 (Zwick et al. 2001a). A 15-residue peptide labeled B2.1 with the sequence $H_1ERSYMFSDLENRCI_{15}$ was found to bind b12 with considerable affinity when it was present as a disulfide-bridged homodimer involving two $C_{14}$ residues. The residues $S_8$, $D_9$, $C_{14}$ and $I_{15}$ of peptide B2.1 were present in several of the selected mimotope peptides and were considered to be responsible for most of their binding activity (Zwick et al. 2001a). The peptide B2.1 shows significant homology to the D-loop region (residues: $R_{273}SVNFTDNAKTII_{285}$) of gp120 and appeared to be an exact conformational mimic of the D-loop region. When the immunogenicity of B2.1 was tested in mice and rabbits no antibodies were obtained that cross-reacted with gp120, indicating that B2.1 was not an effective candidate vaccine immunogen.

In another study, additional mimotopes reacting with b12 were obtained which possessed the same consensus sequence M/V$h$SD (where $h$ represents an aromatic amino acid) found in the D2.1 peptide (Dorgham et al. 2005). Second and third generation libraries were constructed which gave rise to the improved consensus sequence G/SLL/MVWSDEL/H that reacted even more strongly with b12. Phage-displayed peptides containing this sequence competed with the binding of gp120 to b12 and were therefore considered to mimic the epitope recognized by b12. In contrast to what had been found with the B2.1 peptide, immunization with these more active phage-displayed mimotopes were able to elicit Env-specific antibodies, although immunization with the mimotopes conjugated to a carrier protein did not. Cross-reactive immune sera from immunized mice, however, did not neutralize laboratory-adapted strains of HIV-1 in vitro, showing that these improved mimotopes were also ineffective vaccine immunogens (Dorgham et al. 2005). The conclusion that immunization with peptide mimics of discontinuous epitopes is not a feasible approach for developing epitope-targeted vaccines was corroborated in a recent study of mimotopes that mimicked discontinuous epitopes of three well-characterized proteins (Irving et al. 2010).

The inability of mimotopes reacting with the b12 antibody to elicit b12-like antibodies is due to the fact that these linear peptides do not reproduce the complex structural features of the discontinuous epitope recognized by Mab b12 and which presumably induced its formation. This was clearly demonstrated when the crystal structure of a complex between Fab of b12 and the B2.1 peptide was determined at 1.8 Å (Saphire et al. 2007). Only three of the critical binding residues of the B2.1 peptide (residues $F_7$, $S_8$ and $D_9$) were found to form side-chain contacts with the b12 paratope while other residues important for the binding activity ($Y_5$, $C_{14}$, $I_{15}$) only helped to stabilize the peptide structure. The B2.1 peptide is able to bind b12 not because it mimics the D-loop of gp120 but because it reacts with a different paratope subsite in b12 from the one that recognizes the complex discontinuous epitope determined by docking and mutagenesis studies (Zwick et al. 2001a, 2003a; Pantophlet et al. 2003b). Since the B2.1 peptide mimicks at best a very small portion of the discontinuous epitope recognized by b12, it is not surprising that this peptide is unable to elicit antibodies that cross-react with gp120 (Saphire et al. 2007).

In an attempt to understand why Mab b12 possesses potent neutralizing properties whereas the majority of anti CDRbs Mabs do not, more than 60 residues of the CD4bs were mutated to alanine to determine whether these substitutions affected the ability of gp120 to bind to CD4 and to neutralizing and non-neutralizing Mabs (Pantophlet et al. 2003b). Many substitutions were found to affect the binding of neutralizing and non-neutralizing Mabs in the same way, indicating a high degree of overlap in the epitopes recognized by these anti-CD4bs antibodies. Some of the mutations enhanced Mab binding but such an effect was never observed for CD4 binding, suggesting that the virus has evolved toward an optimal gp120-CD4 interaction. Although there were some individual differences between the effects of substitutions on the binding of different Mabs, no clear picture emerged that explained why b12 possessed a unique neutralizing capacity. Since certain residues in the V1/V2 and V3 loop of gp120 were able to affect the binding of b12, it was suggested that non-neutralizing anti-CD4bs Mabs might interact with the gp120 surface using a different orientation from the one used by b12 (Pantophlet et al. 2003b). It is known that all anti-CD4bs Mabs bind monomeric gp120 similarly, independently of their neutralizing capacity, whereas only b12 is able to bind trimeric, functional gp120 presumably through neotopes present only on gp120 oligomers. This may be the reason why only native gp120 trimers are able to elicit an effective neutralizing antibody response (Parren and Burton 2001; Sattentau and Moore 1995). All attempts to find immunogens able to elicit b12-like antibodies were based on the premise that such immunogens must reproduce exactly the structure of the epitope recognized by b12 because it was assumed that the antigenic activity of an epitope is always accompanied by a related immunogenic activity. This assumption implies that antigenicity which is a chemical property arising from a molecular recognition process involving complementary antibodies is equivalent to immunogenicity which is a biological property leading to antibody secretion that occurs only in the context of an immunized host (Van Regenmortel 2010). Immunogenicity depends on intrinsic properties of the immunogen but also on extrinsic factors controlled by constituents and properties of the host immune system such as the host Ig gene repertoire, appropriate BCRs, cytokines, etc. (Berek 1992; Harwood and Batista 2010). All these extrinsic factors did contribute to the occurrence of the neutralizing antibody initially used to identify the putative vaccine epitope but they cannot be expected to be present in identical form in the host that will be vaccinated with the same antigen. Another reason that makes it impossible to extrapolate from antigenic to immunogenic structures is that the structure of an epitope determined in a crystallographic complex is also influenced by processes of induced fit, conformational selection and mutual adaptation of the two binding partners (Rini et al. 1992; Wilson and Stanfield 1994; Berger et al. 1999; Bosshard 2001; Goh et al. 2004). As a result, the particular antigenic structure revealed in an antigen-antibody complex does not necessarily correspond to the immunogenic structure that was recognized by B cell receptors during the immunization process and should be present in the vaccine immunogen (Van Regenmortel 2011b). When tested in an immunoassay, a peptide may bind to an anti-protein neutralizing Mab by an induced fit mechanism driven by the pre-existing structure of the Mab. However, the same induced fit will not occur when the peptide is used as immunogen and is recognized by B cell receptors that have no special affinity for the cognate intact protein.

The assumption that immunogenic potential can be deduced from antigenic structures has led several groups to modify gp120 molecules in order to improve their vaccine potential. For instance, recombinant gp120s were generated that no longer were able to bind non-neutralizing anti CD4bs Mabs while retaining their capacity to bind b12, an approach called immunofocusing (Pantophlet and Burton 2003). It was hoped that such a re-engineered gp120 would acquire the immunogenic capacity of eliciting only neutralizing antibodies, somehow mirroring the altered antigenic reactivity (Pantophlet et al. 2003b). As was to be expected, there have been no subsequent reports indicating that the attempt to induce an immunogenic functionality in gp120 that mirrored the antigenic properties of the engineered constructs was successful. Recently a stabilized gp120 antigen, better able to bind neutralizing antibodies, has been developed by rational design (Wu et al. 2010). The CD4bs of Env was re-engineered by deleting the V1 and V3 loops of the full gp120 molecule and introducing cross-links between different regions of the core protein. One of the re-engineered stabilized surfaces which reacted strongly with Mab b12 but not at all with four non-neutralizing CD4bs Mabs was used to select CD4bs-directed Mabs from memory B cells derived from a clade B-infected donor. A small number of CD4bs-specific memory B cells was obtained from about 25 millions peripheral blood mononuclear cells of the donor. One selected Mab (VRC01) neutralized about 90% of nearly 200 HIV isolates from multiple clades and had a much broader neutralizing capacity than b12 (Wu et al. 2010). The authors suggested that their results do provide an important insight for future HIV-1 vaccine design, although they presented no evidence that their rationally designed antigen is also a superior immunogen able to induce VRC01-like polyclonal antibodies in humans.

## Continuous Epitopes of Viral Proteins Are Not Effective Vaccine Immunogens

Although many linear peptide sequences in proteins cross-react with antibodies raised against the protein, they mostly are unable to elicit antibodies that cross-react with the native protein. The lack of cross-reactive immunogenicity of peptides

(Van Regenmortel 2006) can be explained as follows. When an antiprotein antibody is tested for it ability to react with a peptide in an immunoassay, it may select one of the many available peptide conformations or it may bring it about by induced fit, the result in both cases being a detectable cross-reaction. In contrast, when the same peptide is confronted with a variety of B cell receptors during immunization, different conformations of the peptide may be recognized by separate receptors, leading to the subsequent appearance of a variety of antipeptide antibodies. However, there is no reason why the peptide would bind preferentially to those rare receptors which in addition also cross-react with a more complex epitope present only in the native protein. Since a peptide corresponding to a continuous epitope is nearly always a poor mimic of a protein discontinuous epitope, it is unlikely that a significant proportion of the elicited antipeptide antibodies would react with the cognate protein. Antibodies which do not recognize viral proteins will, of course, possess no neutralizing activity. It is sometimes possible to improve the cross-reactive immunogenicity of a peptide by increasing the conformational similarity between peptide and intact protein, for instance by cyclization of the peptide, but this approach rarely succeeds in producing immunogens able to elicit neutralizing antibodies (Van Regenmortel 2009a, and references therein; Joyce et al. 2002; Ho et al. 2005). As mentioned earlier in the case of discontinuous epitopes, linear peptides corresponding to continuous epitopes, when used as immunogens, will not preferentially select and bind to the rare B cell receptors which in addition to cross-reacting with the native protein also trigger a B cell differentiation process giving rise to antibodies endowed with a functional protective activity. Although a few mimotopes of several continuous epitopes of different viruses, including HIV-1, have been found to be able to elicit antibodies that cross-react with the virus and possess some neutralizing activity (Yu et al. 2000; Steward et al. 1995; Scala et al. 1999), no successful protective peptide-based antiviral vaccine has been developed and marketed for human use (Hans et al. 2006). Two 6-residue long continuous epitopes located in the conserved region of MPER of gp41 have been regarded as promising HIV-1 vaccine candidates (Zwick 2005) and many groups have attempted to turn them into effective vaccine immunogens (Montero et al. 2008). It seems highly unlikely, however, that these short linear peptides which are part of larger complex discontinuous epitopes eliciting potent neutralizing Mabs, will be able to induce an effective protective immune response (Van Regenmortel 2011b).

## The Special Case of Terminal Regions in Proteins

The terminal regions of proteins frequently harbour continuous epitopes because they tend to be surface-oriented and are more hydrophilic and mobile than internal regions (Thornton and Sibanda 1983; Westhof et al. 1984; Tainer et al. 1985; Pellequer et al. 1994). A major purpose of epitope prediction is to replace intact antigen molecules by linear synthetic peptides for use as diagnostic reagents for detecting viral antibodies (Leinikki et al. 1993) or as immunogens for raising antipeptide antibodies able to cross-react with a parent protein (Walter 1986; Muller 1999b). The most effective strategy in this case is to select 10-residue long synthetic peptides corresponding to the N- and C-terminal regions of the protein since these have the highest probability of cross-reacting antigenically with the protein. Methods for predicting which continuous epitopes in proteins are likely to be effective vaccine immunogens have been notoriously unsuccessful (Ponomarenko and Van Regenmortel 2009; Greenbaum et al. 2007; Blythe and Flower 2005; Caioli 2010) and only very limited success has been obtained when certain terminal peptides were used as vaccine immunogens. Vaccination with the N-terminal 15-residue peptide of the VP2 protein of canine parvovirus, for instance, was able to protect dogs against virus infection (Langeveld et al. 1994). Another example is the C-terminal region of the VP1 protein (residues 200–213) of foot-and-mouth disease virus, which is able to elicit high levels of neutralizing antibodies (Bittle et al. 1982). Residues 200–213 of one VP1 subunit are located next to the immunodominant disordered loop (residues 141–160) of an adjacent subunit (Rowlands et al. 1983), and both regions contribute to a single discontinuous neotope (Parry et al. 1989). The ability of the terminal regions of proteins to induce antibodies that cross-react with the native protein is partly due to the fact that these regions possess an intrinsic disorder which makes them very similar to the unstructured peptides (Uversky et al. 2005). Although some disordered regions in proteins may be recognized preferentially by antipeptide antibodies, the majority of viral epitopes capable of inducing a neutralizing antibody response are actually complex, discontinuous topographic sites rather than terminal linear peptides.

## *The HIV-1 V3 Antigenic Site*

Loops and turns in proteins are often immunodominant antigenic regions and the V3 loop in HIV-1 has for many years been considered the principal neutralizing domain of the virus (Javaherian et al. 1989; Spear et al. 1994). Although it is designated as the third variable domain of gp120, the V3 loop has many conserved features such as a constant size of about 35 residues, a conserved type II turn at its tip and a disulfide bond at its base (Jiang et al. 2010). The semiconserved nature of the V3 loop allows it to play an important role in the recognition of viral co-receptors, but its sequence variability has led investigators to question its value as a target for vaccine-induced antibodies. Another reason that led investigators to doubt the value of the V3 loop as a vaccine candidate is that they looked for the presence of anti V3 antibodies in human anti-HIV-1 antisera by absorbing the sera with linear V3 peptides (Van Cott et al. 1995a). Since this absorption step did not remove the antibodies that neutralized primary HIV-1 isolates, it was concluded that the neutralizing antibodies were not directed against the V3 loop. In fact, human sera that neutralize both TCLA isolates and primary isolates from several HIV-1 clades contain antibodies that recognize a cyclic V3 conformation present in the virus and in disulfide-bonded V3 loops but absent in linear V3 peptides (Zolla-Pazner 2004; Gorny et al. 2002;

Hioe et al. 2010; Kayman et al. 1994; Gorny et al. 2006). The V3 loop is very flexible and is able to adopt different conformations when it binds to different Mabs (Stanfield et al. 1999). It was found that alternative $\beta$-hairpin conformations can be induced in the same V3 peptide, depending on whether it binds to Mab 447-52D that neutralizes both R5 and X4 viruses or to Mab 5$b$ which neutralizes only X4 viruses (Rosen et al. 2005). This shows that the epitope conformation observed in a Mab-epitope complex can result from an induced fit process and is therefore not a reliable guide for deciding which immunogen conformation should be used for eliciting a certain type of neutralizing antibody. It is not yet clear whether intrinsically flexible V3 loops should be constrained in a particular conformation to increase their potential value as vaccine candidates (Mester et al. 2009). In the case of the immunodominant GH loop (residues 141–160) of the VP1 protein of foot-and-mouth disease virus, a stabilized all-D retro loop synthesized with D-amino acids was found to induce higher levels of neutralizing antibodies than the corresponding L-peptide (Briand et al. 1997; Carver et al. 1997; Nargi et al. 1999).

## One of the Many Hurdles to Be Overcome by an HIV-1 Vaccine Is the Antigenic Variability of the Virus

There are several reasons why the strategies that were used successfully in the past to develop effective vaccines against many viral diseases are not applicable to HIV-1 (Hilleman 1992; Kusters and Almond 2010; Virgin and Walker 2010): (1) Since the natural immune response in HIV-1 infected individuals does not clear the infection, there is no natural immunological mechanism that the vaccine could imitate. An HIV-1 vaccine would thus have to surpass the normal immune response induced by infection and would for instance have to generate a far stronger response against conserved epitopes than what occurs during infection; (2) HIV-1 integrates into the host genome and establishes a latent pool of infected cells that conceal the virus from immune recognition; (3) The virus progressively destroys the immune system; (4) HIV-1 isolates exhibit an enormous antigenic variability arising from the error-prone viral reverse transcription which can produce as much as 35% sequence diversity in gp120 between viral subtypes and 10% diversity in gp120 in a single infected individual.

Although an increasing number of broadly neutralizing Mabs have recently been isolated from infected human donors (Stamatatos et al. 2009; Scheid et al. 2009; Walker et al. 2009; Corti et al. 2010; Mascola and Montefiori 2010), such antibodies are rarely induced by experimental immunization. It seems that during HIV-1 infection, the immune response is mostly directed against highly variable and accessible Env regions such as the V1, V2 and V3 loops rather than against functionally important but less accessible conserved domains such as the receptor and co-receptor binding sites. The propensity of the immune system to respond to immunodominant, strain-specific highly variable antigenic sites instead of to subdominant conserved epitopes that are able in principle to elicit a more broadly protective immunity, has been called deceptive imprinting because it allows the virus to escape immune pressure (Nara and Garrity 1998). Attempts to overcome deceptive imprinting by redirecting the immune response to less strain-specific epitopes have been made using a strategy called immune refocusing. By introducing several new glycosylation sites into the V1 and V3 loops, it was possible for instance to obtain immunogens that induced a broader neutralization response that was effective against heterologous strains (Tobin et al. 2008). Some of the reported immunization results, however, are difficult to interpret because linear instead of cyclized loop peptides were used in inhibition experiments aimed at determining the specificity of the induced antibodies. Attempts have also been made to redirect the immune response to conserved neutralization epitopes by deleting some or most of the variable loops from Env, but the results were not satisfactory (Lu et al. 1998; Kim et al. 2003). Other approaches to overcome the problem of extensive antigenic variability include using as immunogen either artificial consensus sequences based on the most common amino acid in each position in an alignment (Gaschen et al. 2002; McBurney and Ross 2009), or alternatively combinatorial convergent peptide libraries consisting of thousands of V3 peptides known as mixotopes that are recognized as a single entity by immune receptors (Gras-Masse et al. 1999). Other strategies to deal with the existence of many antigenic variants by means of polyvalent vaccine cocktails have also been proposed (Korber and Gnanakaran 2009). Recent results have shown that the V3 loop may, in fact, be able to induce a neutralizing response against many different HIV-1 clades, indicating that a polyclonal response to V3 may have a much greater neutralizing breadth than that displayed by any single Mab or even a cocktail of Mabs (Hioe et al. 2010; Jiang et al. 2010; Pantophlet et al. 2007).

## *Rational Design Versus Empirical Immunogenicity Trials in HIV-1 Vaccine Development*

A pervading theme in current vaccine research is the belief that rational design offers the best prospects for the much needed development of effective vaccines against HIV-AIDS, tuberculosis and malaria (Bramwell and Perrie 2005; D'Argenio and Wilson 2010; McElrath and Haynes 2010; Sette and Rappuoli 2010). In the case of HIV-1, this viewpoint has led to the claim that "rational design represents the only approach that can elevate vaccine research from an empirical exercise to a scientific discipline" (Karlsson-Hedestam et al. 2008). It is argued here that such a view which denigrates the time-honoured empirical, trial-and-error approach that gave us all our present day effective vaccines is highly misleading and could even possibly jeopardize future efforts at developing an HIV-1 vaccine (Van Regenmortel 1999b, 2001b). It is not always clear what scientists mean when they refer to rational design and it may be helpful to briefly analyze these terms. The label rational tends to be used to describe any approach that relies on the probable outcome of an intervention as predicted from available scientific knowledge. The term rational is also used to describe common sense decisions such as focusing on known antigenic regions of a viral protein as primary targets for a vaccine (Obeid et al. 1995). Since all scientific work is based on reason, logic and currently accepted scientific theories, it makes little sense to oppose rational and empirical approaches since all scientific knowledge is derived from empirical observations made in the course of experiments that are necessarily planned and analyzed in a rational manner. It is therefore misleading to imply that trial-and-error experimentation is not rational since all novel findings and discoveries arise from the unpredictable outcome of controlled experiments rather than from purely rational, deductive thinking (Van Regenmortel 1999b). The term design can be defined as the deliberate and intelligent conceiving of an artificial object or process. The designer's task is to pose and solve an inverse problem, namely to imagine on the basis of existing knowledge what would bring about a desired outcome (Bunge 2003). Plausible solutions can only be tested by trial and error until the preset goal is attained (Van Regenmortel 2007). Since scientists need to secure funding for their research, it is tempting for them to suggest that their current understanding of the complexity of the immune system is sufficient to allow them to design immunogens and to predict the type of antibodies they will elicit.

Rational design terminology implies that there is a causal relationship between the structure of an antibody and its function, giving rise to the assertion: "structure determines function". Although it is true that a biological activity always depends on an underlying structure, the structure does not possess causal efficacy on its own in bringing about a certain activity. Causal relations are relations between successive events and not between two material objects nor between a structure and an event (Van Regenmortel 2002a). There is no unique causal relation between the structure of an antibody and its neutralizing capacity since the occurrence of neutralization is influenced by a large number of immunological and pathophysiological factors arising from complex interactions between antibody, pathogen, and host. The isolation of the potent neutralizing Mab VRC01 (Wu et al. 2010) and the elucidation of its structure (Zhou et al. 2010), as well as the isolation of many additional broadly neutralizing antibodies from the serum of infected donors (Walker et al. 2009; Corti et al. 2010; Mascola and Montefiori 2010; Binley et al. 2008; Gray et al. 2009), has been interpreted as providing valuable molecular information that should facilitate rational vaccine design (Burton and Weiss 2010). The search for additional broadly neutralizing Mabs has been stimulated by the finding that it is possible to provide sterilizing protection in non-human primates by administering neutralizing Mabs prior to viral challenge. However, this type of successful immunotherapy using neutralizing antibodies derived from infected individuals does not in any way tell us how such antibodies can be induced by vaccination.

It is precisely our lack of knowledge regarding which immunogens are able to induce such antibodies in certain infected individuals which prevents us from developing an HIV-1 vaccine by design. Furthermore, the polyspecificity of antibody molecules and the degeneracy of the immune system imply that it is futile to search for the single "intrinsic" epitope of a neutralizing Mab and for the putative unique immunogenic structure capable of inducing protective antibodies. It has been claimed (Virgin and Walker 2010) that our failure to develop an HIV-1 vaccine is due to the fact that we do not understand what type of immune response is needed for protection. We can of course surmise that effective CD4, CD8 and B-cell responses are needed but this is of little help since we do not know how to elicit the required responses by immunization. One could also argue that studying the mechanism of a preventive or therapeutic HIV-1 vaccine will become possible only once an effective vaccination protocol has been discovered empirically. Figure 2 lists many of the questions that we need to answer if we want to be able to rationally design an HIV-1 vaccine (Van Regenmortel 2001b; Virgin and Walker 2010; Ebensen and Guzman 2008). It has been suggested that rational design may not be the best strategy to follow (Tobin et al. 2000) since it is not possible, for instance, to predict which combination of several substitutions in a protein may lead to new desired properties that are not manifested when the protein presents only one substitution. This has led to the conclusion that an improved protein function is more likely to be obtained by successive rounds of random mutation, recombination and selection than by rational design (Van Regenmortel 2002a and references therein). What is of course possible is to rationally design an antigen to fit a single Mab (Wu et al. 2010; Villen et al. 2006) or to improve by molecular design the paratope binding efficacy of a Mab intended for passive immunotherapy. What is not possible is to rationally design an

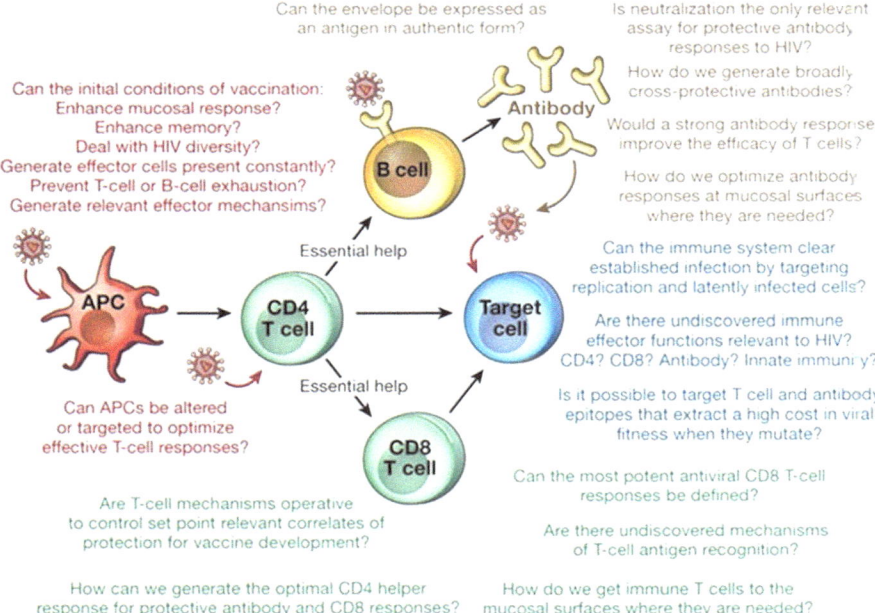

**Fig. 2** Unanswered questions regarding the factors and mechanisms believed to be important for inducing an effective HIV-1 vaccine immune response (Virgin and Walker 2010)

immunogenic molecule that will elicit a protective polyclonal antibody response of predetermined efficacy.

The existence among HIV-1 infected individuals of long-term non-progressors (LTNPs) and elite controllers (ECs) has given rise to the hope that the elucidation of the mechanisms responsible for their occurrence may help HIV-1 vaccine design (Poropatich and Sullivan 2011). LTNPs which make up between 2 and 5% of all HIV-1 seropositive individuals (Okulicz et al. 2009) maintain low plasma levels of HIV-RNA and elevated CD4+ T cell levels whereas the even rarer EC individuals have undetectable levels of viral RNA. A combination of viral, genetic and immunological factors has been implicated in the ability of these slow progressors to control HIV-1 infection. The principal viral factor believed to slow disease progression is HIV-1 strain attenuation arising from mutations in viral accessory genes that cause a reduction in virus infectivity and replication (Poropatich and Sullivan 2011). If control of disease progression in these individuals results from an infection with attenuated, poorly replicating viruses, it is not clear how this could be made use of in vaccine development. The same applies to host genetic and immunological factors that promote control of infection such as chemokine receptor polymorphisms or the presence of certain protective HLA class I alleles in LTNP and EC individuals. Although some slow progressors have been shown to be able to induce broadly neutralizing antibodies against multiple HIV-1 strains (Villen et al. 2006; Simek et al. 2009), the relevance of this anti-Env immune response to vaccine design is also not obvious since we do not know which immunogens were able to induce such as a response. It is, indeed, generally recognized that the major bottleneck in designing an HIV-1 vaccine lies in our ignorance of which immunogens are capable of inducing a broadly neutralizing immune response (Stamatatos et al. 2009; Mascola and Montefiori 2010; McElrath and Haynes 2010). As discussed in this review, a successful design strategy requires that we should know which immunogens and immunological mechanisms are likely to lead to the appearance of protective immunity following vaccination. Unfortunately even a vaccine that induces high levels of neutralizing antibodies will not necessarily protect against virus infection (Mascola 1999). Many investigators have recently stressed the need for embarking on large scale research programs in basic and preclinical immunology which they believe will give us the understanding required for designing vaccines (Virgin and Walker 2010; McElrath and Haynes 2010). Although nobody would quarrel with the need for more basic research in immunology, the track record in vaccinology shows that effective vaccines in the past were always discovered by trial-and-error immunization trials rather than by rational design based on available structural and immunological knowledge. The widespread current expectation that discoveries in biomedicine and vaccinology are likely to result from structure-based molecular studies, "omics" type investigations (Sette and Rappuoli 2010) or systems biology approaches (Pulendran et al. 2010) rather than from empirical trials may well turn out to be counterproductive if it inhibits investigators from undertaking exploratory trial-and-error experiments. Empirical immunogenicity trials based on a less than perfect understanding of the complexities of the immune response and of the nature of dynamic and transitory HIV-1 epitopes may still be the key to discovering a promising HIV-1 vaccine candidate. Although epitope mapping with neutralizing Mabs and structural analyses may help investigators to identify which regions cf the viral surface should be targeted by a vaccine, it must be emphasized that antigen mapping on its own will not lead to the discovery of vaccine immunogens that can only be revealed following experimental immunogenicity trials. Identifying the many cross-reactive epitopes that a neutralizing Mab can recognize is an entirely different task from discovering empirically which vaccine immunogens are able to induce a protective immune response.

Once an effective immunogen has been identified empirically, it may still take many years of basic research to establish its mechanism of action. As is the case with most present day successful vaccines, it is even conceivable that we may never fully understand why an empirically discovered vaccine is actually effective. This is a salutary reminder that discovering an effective vaccine empirically should have a higher priority than elucidating the mode of action of an ideal HIV-1 vaccine. Basic immunological research, therefore, should not stand in the way of rational immunogenicity trials nor should it restrict the financial resources that must be made available to implement them.

## *Conclusions*

Creative imaginative thinking is required to choose credible candidate vaccines and to invent plausible hypotheses to help select judicious vaccine regimens, dosages, adjuvants, delivery routes and effective T cell help. Instead of misrepresenting experimental hypothesis testing as a case of rational design, it would seem preferable to openly acknowledge that the empirical testing of immunogens, which initially should take the form of small human trials, still offers the best hope of discovering an effective HIV-1 vaccine. There is increasing doubt that developing a preventive vaccine able to provide prophylactic immunity to stop HIV-1 infection from occurring altogether is a realistic enterprise. There is indeed no evidence that it is possible to induce sufficiently high levels of broadly neutralizing antibodies at mucosal surfaces to achieve that aim. In recent years, much research has been directed instead to the development of therapeutic HIV-1 vaccines for treating people whc are already infected with HIV-1. It has been shown, for instance, that immunization with the conserved HIV regulatory proteins, Tat, Rev. and Nef, which are the first proteins expressed after HIV infection, is able to stimulate the immune system of patients receiving anti-retroviral therapy, enabling them to partially control virus replication (Caputo et al. 2009). Other therapeutic vaccines are currently being investigated (García et al. 2011; Li et al. 2011) and it may become possible to assess their value more rapidly in the future by using so-called «adaptive trial designs» aimed at screening out poor vaccines while extending the evaluation of efficacious ones (Corey et al. 2011). In addition, since many methods are now available to chemically inactivate HIV-1 while retaining the conformational and functional integrity of the gp proteins (Van Regenmortel 2011b and references therein), it may be justified to re-examine the possible value of a killed HIV-1 vaccine, provided the total innocuity of inactivated and genetically modified virus can be established.

The unexpected promising results of the Thailand "RV144" trial, initially condemned as ill-conceived and unjustified (Burton et al. 2004b; McNeil et al. 2004; Belshe et al. 2004; Jefferys and Harrington 2004) but which turned out to be the first human vaccine trial showing modest protection against HIV infection (Rerks-Ngarm et al. 2009; Letvin 2009), demonstrate that empirical trials will always be needed since their outcomes are inherently unpredictable.

**Acknowledgments** The author is grateful to Pernille Haste-Anderson of the Danish Technically University, Denmark for preparing Fig. 1 (**a**, **b**) and to Florence Diemer for efficient secretarial assistance.

# Paradigm Changes and the Future of HIV Vaccine Research: A Summary of a Workshop Held in Baltimore on 20 November 2013

## Introduction

The concept of scientific paradigm was made popular by Thomas Kuhn in his influential book *"The Structure of Scientific Revolutions"* published in 1962 (Kuhn 1962). He argued that scientists in their research are always guided by assumptions and theoretical presuppositions that determine the lines of investigation they pursue. These underlying assumptions which often are not explicitly stated represent the prevailing paradigms present in every field of scientific enquiry at any particular time. The field of HIV vaccine research is no exception and several paradigms have had a major influence on the type of research undertaken and funded in the last 25 years (Esparza 2013b). Since investigators are not always fully aware of the underlying paradigms that influence their choice of research program and experimental approach, they may not question the validity of a particular paradigm when they regularly obtain results that are not consistent with the hypothesis or theory that gave rise to the paradigm. When this happens, they may continue to pursue unfruitful lines of investigation that impede scientific progress. This pitfall can only be avoided if researchers remain constantly aware of the paradigms that make them pursue a particular research program.

In HIV vaccine research, there is evidence that several prevalent paradigms have not helped the development of the field, and this may partly explain why after

Journal of AIDS & Clinical Research, 2014, 5, 281.

Van Regenmortel, M.H.V., Andrieu, J.M., Dimitrov, D.S., Ensoli, B., Hioe, C.E., Moog, C., Ruprecht, R.M.

© Springer Nature Switzerland AG 2019
Marc H V Van Regenmortel, *HIV/AIDS: Immunochemistry, Reductionism and Vaccine Design*, https://doi.org/10.1007/978-3-030-32459-9_15

25 years of intensive research efforts, it has not been possible to develop an effective vaccine using classical strategies (Van Regenmortel 2011b; Burton et al. 2012; Esparza 2013a). One such paradigm is the assumption that HIV-1 epitopes identified by crystallography of complexes of HIV Env bound to affinity -matured neutralizing (n) Mabs are likely to be effective vaccine immunogens able to induce a protective immune response. Another misleading paradigm is the assumption that anti-HIV-1 Mabs are monospecific for a single viral epitope instead of always being polyspecific and able to bind numerous epitopes different from the one identified when the structure of the Mab-HIV-1 complex was solved. A third one is the expectation that after isolating from HIV-1 infected individuals increasing numbers of nAbs that may be useful for *passive* immunotherapy, this will necessarily facilitate the development of immunogens suitable for *active* immunization (Van Regenmortel 2012a).

## A New Paradigm Advocating Basic Research

Paradigms come in different forms and some correspond to general conceptual frameworks and theoretical assumptions that permeate an entire field at a particular time. In recent years many HIV investigators have become convinced that our current knowledge of the human immune system and of HIV-1 immunopathology is far too limited to permit the development of an HIV-1 vaccine in the near future. This has given rise to a new popular paradigm which assumes that this stumbling block to vaccine development can only be overcome by embarking on large scale basic research programs in immunology (McElrath and Haynes 2010; Virgin and Walker 2010). This paradigm espouses the classic thesis of Vannevar Bush, prevalent since the end of the Second World War (Bush 1945), which claims that all technological innovations are derived from applied research programs that always find their origin in curiosity-driven basic research performed without any consideration of potential practical use. This thesis is no longer universally accepted since the separation between basic and applied research is nowadays often perceived as somewhat of a false dichotomy. A more realistic account of scientific and technological progress proposed by Stokes (1997) follows the so-called quadrant model of scientific research which accepts that research can be driven simultaneously by a quest for fundamental understanding and by considerations of use for solving a practical problem. According to this model, a commitment to try to understand HIV-1 immunopathology need not exclude a commitment to try to control HIV-1 infection by vaccination (Esparza 2005). The remaining difficulty, of course, lies in the impossibility to predict which items of new knowledge derived from basic or applied research will subsequently allow the development of a useful technological product such as a vaccine.

# From Basic and Applied Research to Technological Innovation

Two types of human knowledge are usually distinguished (Mokyr 2002a, b) corresponding to:

1. Knowing *what* is the case, for instance the existence of a natural biological phenomenon. This is usually called *propositional* knowledge which comprises *inter alia* all the scientific facts which can be expressed as true statements or *propositions* that such and such is the case.
2. Knowing *how* to do something while achieving something of practical utility, using what is called procedural or *prescriptive* knowledge.

Any addition to propositional knowledge is a *discovery* that unearths something that existed all along but was unknown to anybody. Any addition to prescriptive knowledge, on the other hand, is an *invention* derived from basic and applied research that makes it possible to do something that was previously unfeasible and that achieves a desirable, practical or societal goal.

An example of such a technological innovation would be an effective preventive HIV-1 vaccine that allows nature to be manipulated successfully by providing protection against viral infection. Unfortunately it is impossible to predict which item of propositional knowledge is required to lead to the prescriptive knowledge needed to manipulate and control a natural phenomenon such as a viral infection. As pointed out by Ian Hacking in his book *Representing and Intervening* (Hacking 1983), it is only by intervening in a material system that we learn to manipulate and control it, thereby deriving the desired prescriptive knowledge. For instance, it is only by comparing various ways of manipulating the human immune system, using trial-and-error empirical experimentation, that we may eventually learn to control it and achieve protective immunity by vaccination (Van Regenmortel 2012b). Whereas elements of propositional knowledge (for instance the 3D structure of an antibody) may be "right" or "wrong", this is not the case with the ability to manipulate the immune system using prescriptive knowledge which is either "successful" or "unsuccessful", the outcome being adjudicated empirically. This means that we achieve useful prescriptive knowledge of the immune system only by a prior successful empirical intervention that would lead for instance to protective immunity. This conclusion is at odds with the currently fashionable paradigm that increasing our knowledge of basic immunology on its own would be sufficient to improve our ability to develop an effective HIV-1 vaccine.

A workshop entitled: *"What type of HIV vaccine research should be promoted"* took place during a virology conference held in Baltimore in November 2013 (www.omicsgroup.com/conferences/virology-2013/). The purpose of this workshop was to discuss new paradigms that better fit our increased knowledge of HIV-1 immune responses and which could therefore be more helpful in guiding future vaccine research than did past unsuccessful paradigms.

In order to structure the discussion, panelists were asked to respond to four questions. Some of their responses are summarized below.

**Question 1: Which new ideas, hypotheses and paradigms should be introduced in the HIV-1 vaccine field?**

1. Since neutralizing antibodies and cytotoxic T cells do not prevent HIV infections nor control viral replication in humans, many virologists believe that classical vaccination approaches will not succeed in the case of HIV-1. In other words, they accept that they need to achieve something that the human immune system is normally not capable of doing when it encounters the virus. Earlier attempts using the approach known as reverse vaccinology (Burton 2002; Van Regenmortel 2011a) failed and the putative nAb germline predecessors were found to bind poorly or not at all to the HIV-1 epitopes that are recognized by their hypermutated mature descendants isolated from HIV-1 infected individuals (Van Regenmortel 2012b). The resulting very poor immunogenicity of HIV-1 epitopes usually prevents the human immune system to initiate and sustain a response that leads to the elicitation of broadly neutralizing (bn)Abs. Even if an immune response is initiated, a complex and lengthy antibody maturation pathway that differs in individual vaccinees is needed for obtaining bnAbs (Dimitrov 2010). New strategies are therefore needed to identify candidate vaccine immunogens that bind germline predecessors of bnAbs or intermediates in the maturation process of these Abs. It is not clear at present whether the unravelling of individual antibody maturation pathways will succeed in identifying HIV-1 immunogens capable of inducing suitable predecessors of known bnAbs in populations of naïve individuals.

2. New strategies should be introduced to identify the immunogens responsible for Ab responses associated with vaccine protection. One such approach called protection-linked biopanning, using recombinant phages encoding random peptide libraries, was found to be able to identify viral epitopes that bind to antibodies present only in vaccinated protected individuals (Bachler et al. 2013). Earlier paradigms unfortunately led investigators to concentrate on epitopes that only *bind* bnMabs (i.e. on viral antigens) rather than on immunogens containing immunogenic epitopes capable of *eliciting* such antibodies (Van Regenmortel 2012b).

3. Since most HIV-1 infections occur through sexual contact, more studies should be devoted to mechanisms of mucosal immunity. Little is known about how HIV-1 disseminates through the mucosa. If dendritic cells are involved, Fc-mediated inhibition could interfere with the infection process. If the first target is a CD4 T lymphocyte, preventing infection of CD4 T cells should be a priority. Since cells at mucosal surfaces express various Fc receptors, immune complex trapping through these receptors may lead either to virus degradation by phagocytosis or to enhancement of virus replication in the case of neonatal receptors. Studies aimed at developing immune responses at mucosal surfaces using specific adjuvants or involving certain bacterial flora also deserve further investigation.

4. Although IgA represents the most abundant immunoglobulin made by the human body and is an important component of mucosal secretions, the role of the various IgA forms in either preventing or enhancing HIV transmission is unclear. Only one study thus far has examined the difference between IgA1 and IgA2 present mucosally as dimers (Watkins et al. 2013). In this study, dimeric IgA1 was significantly more protective than the dimeric IgA2 version of a neutralizing anti-HIV mAb with the same epitope specificity. In contrast, serum IgA responses against the HIV envelope have been linked to an increased risk of virus acquisition in the RV144 trial. The fact that IgA, in different forms and from different anatomical compartments, has been associated with either protection or increased risk of virus acquisition, indicates that the role of serum and/or mucosal IgA in preventing or facilitating HIV transmission should be further investigated. The role of mucus and natural mucosal fluids, in addition to mucosal antibodies, in enhancing or decreasing virion capture should also be examined.

5. Although replicating vectors have scored the best at inducing long-term memory, these vectors are mostly attenuated pathogens and since their long-term safety may be a serious issue, alternative strategies using for instance intradermal injections and intranasal administration (Bomsel et al. 2011) should be studied.

6. Currently available animal models are not satisfactory to allow large numbers of candidate vaccines to be evaluated. In order to test more vaccine candidates and strategies, additional cost-effective high-throughput animal models that are able to be transposed to generate appropriate human immune responses should be developed. The development of novel virus chimeras using Env from different virus clades, similar to SHIV in macaques, should also be encouraged.

7. Since SIV in macaques and HIV-1 in humans can only replicate efficiently in vivo if mucosal CD4 T cells are activated, it has been hypothesized that it might be possible to suppress viral replication by interfering with CD4 T cell activation. This was tested by stimulating mucosal dendritic cells in macaques with a mucosal vaccine containing inactivated SIV associated with the Calmette-Guerin bacillus, in an attempt to induce in vivo the terminal differentiation of SIV-specific CD8 T cells. This mucosal vaccination was found to be extremely effective since after intrarectal challenges with large amounts of SIV, most macaques remained sterilely protected. The protection was found to be induced by CD8 T cells that possessed strong SIV suppressive activity, and surprisingly it was not associated with SIV-specific antibodies or CTLs (unpublished results by Andrieu JM and Lu W). This led to the further hypothesis that this type of vaccine may have acted through a phenomenon of mucosal/oral tolerance. Since probiotic bacteria, for instance lactobacilli, have been suspected for a long time to be inducers of immune tolerance, an oral vaccine made with the same killed SIV immunogen in association with large amounts of *Lactobacillus plantarum* was tested in macaques. This vaccine was found to induce a previously unrecognized class of non-cytolytic MHC 1b/E restricted CD8 regulatory T cells (Tregs) which specifically suppressed the activation of SIV positive CD4 T lymphocytes. This suppression which prevented the initial burst of virus replication in vivo, permanently protected 15 out of 16 macaques from infection (Lu et al. 2012). This

strong protective effect induced by CD8+ Tregs was obtained in experiments that were based on an "out of the box" innovative paradigm, confirming the assumption that SIV requires activated immune cells for its replication. Since CD4+ T cell activation drives both the initial SIV and HIV-1 replication in macaques and humans respectively, this approach should be tested in humans. If successful it could offer an exciting prospect for a preventive or therapeutic HIV-1 vaccine.

8. In recent years, considerable and growing attention has been given to the development of therapeutic HIV vaccines for treating people already infected with HIV-1. Immunization with the conserved HIV regulatory protein, Tat, has been shown to stimulate the immune system of patients receiving antiretroviral therapy, leading to a further and stable CD4 T cell increase and immune restoration (Ensoli et al. 2010). Extracellular HIV-1 Tat can form a molecular complex with trimeric Env, shielding it from anti-Env nAbs and redirecting virus entry to RGD-binding integrins (Monini et al. 2012). Anti-Tat Abs, which are infrequently produced upon natural infection, are able to restore and increase HIV neutralization that would normally be impaired by extracellular Tat (Monini et al. 2012). Since vaccination with Tat decreases the proviral DNA load (Ensoli et al. submitted), slows down the progression to AIDS and can lead to complete or partial protection from infection (Bachler et al. 2013; Monini et al. 2012; Cafaro et al. 1999, 2010), the accumulated evidence supports a novel paradigm that views HIV-1 Tat as an important vaccine candidate either on its own or as part of a multicomponent vaccine (Bachler et al. 2013; Monini et al. 2012; Cafaro et al. 1999, 2010).

**Question 2: How should innovative research on HIV vaccines be funded, especially risky projects that are unlikely to be favored by selection committees because of insufficient confirmed data?**

9. The current prevalent funding structure focusing predominantly on large networks of investigators under strong leadership of a principal investigator has considerable built-in inertia which makes it difficult to rapidly adapt to new paradigms. Generous funding of smaller groups and networks would allow the pursuit of more flexible lines of research based on original paradigms and would foster greater intellectual flexibility. Currently, at most 10% of available funding is devoted to supporting high-risk projects based on innovative paradigms. Funding redistribution should be encouraged and a significantly larger funding percentage (perhaps up to 50%) should be used for smaller scale innovative projects. Such a change would diversify the vaccine approaches that can be investigated and would increase the probability of finding an effective vaccine.

10. It could be argued that the reigning paradigms followed in the past by well-funded large networks of investigators were usually nothing more than hypothetical assumptions for which very limited confirmed HIV data existed. Since it is impossible to predict which proposed experimental approaches will later be found to be empirically successful, funding agencies should utilize selection committees comprising accomplished open-minded experts who themselves have made original contributions outside the framework of conservative,

conventional wisdom. The validity of human judgments cannot easily be quantified and it is questionable whether the majority vote of a large selection committee will necessarily back the most promising and original proposal rather than projects that follow the accepted scientific consensus.

11. High-risk projects based on novel paradigms should be examined by selection committees different from those in charge of traditional 4 year grants requiring extensive supporting data and a solid track record of publications. Many traditional grants support fundamental immunological research that is considered to be of high quality because it produces excellent scientific papers published in high impact journals such as *Science, Nature, Cell, Journal of Virology* etc. These papers tend to include statements claiming that the newly acquired scientific information is likely to help the future development of an HIV-1 vaccine, although they rarely present new prescriptive knowledge relevant to the goal of producing a technological innovation in the form of an effective vaccine. Projects considered to be risky because they follow innovative, unorthodox paradigms that lack widespread support could be funded by two-phase grants, first to establish that they do produce prescriptive knowledge of potential utility, and then subsequently transferring the pilot study to a larger study. However, the emphasis should be on useful prescriptive knowledge relevant to vaccines rather than to curiosity-driven basic research that only increases our immunological knowledge without providing the new descriptive knowledge necessary for developing an invention.

12. Innovative projects could also be funded by a procedure similar to that used by the MacArthur Foundation (http://www.macfound.org). Funding would be allocated for a period of several years and a comprehensive report would only have to be produced at the end of the contract and not annually. Applicants should demonstrate high levels of past creativity and accomplishments in vaccinology and related fields and the allocated funding should suffice for carrying out small phase I or phase II clinical trials.

13. The surprising extensive protection against SIV achieved in macaques with an oral tolerogenic vaccine that did not elicit SIV-specific antibodies nor CTLs was obtained by an investigator-driven research project that was not funded by a governmental or large scale international research organization (Lu et al. 2012). It was entirely sponsored by a private benefactor who agreed to fund a risky project based on an unorthodox paradigm for which minimal preliminary data existed. This illustrates the benefits that sometimes accrue from moving away from reigning, fashionable paradigms.

**Question 3: What should be the balance between empirical vaccine research (for instance testing various hypotheses on how to induce protective immunity) and so-called "rational" vaccine design based on structural knowledge?**

14. Empirical and rational approaches to vaccine development are sometimes opposed as if they were incompatible and it has for instance been claimed that "Rational design represents the only approach that can elevate vaccine research

from an empirical exercise to a scientific discipline" (Karlsson-Hedestam et al. 2008). However, all existing vaccines have been developed using trial-and-error immunization trials that are entirely rational and there is no example of an effective vaccine ever having been obtained solely by rational design based on structural analysis (Van Regenmortel 2012a). Claims to the contrary arise because the structure-based rational design of a viral epitope engineered to better fit a single bnMab is masquerading as vaccine immunogen design, a confusion which is due to the erroneous assumption that antigen binding reactivity necessarily entails an immunogenic capacity to induce protective Abs (Van Regenmortel 2012b). As discussed above, structural knowledge of viral epitopes and antibody paratopes corresponds to propositional knowledge which does not tell us how to successfully manipulate the immune system so that immunization with an appropriate immunogen will elicit a protective anti-viral response. Only after an effective immunogen has been identified empirically does it become possible to investigate its possible mode of action. This should involve a study not only of possible neutralizing antibodies but should include the study of barrier functions at mucosal surfaces, virion trapping properties of mucosal secretions and cellular as well as innate immunity phenomena that may prevent cell-to-cell virus transmission in primary and chronic infections. Developing an effective vaccine empirically is therefore the first priority and should precede attempts at elucidating the theoretical mode of action of an ideal HIV-1 vaccine on the basis of our knowledge of the immune system.

15. The lack of success of the reverse vaccinology/engineering paradigm should be acknowledged as such and should lead to a paradigm shift emphasizing the relevance of the bnAb germline ancestors and their maturation. In 2013, Bruce Alberts, the Editor of *Science* wrote the following: "Scientists need to develop a value system where simply moving on from one's mistakes without publicly acknowledging them, severely damages, rather than protects, a scientific reputation." Knowing what leads researchers astray is as important to science as knowing which hypothesis is corroborated by experimental observations.

**Question 4: If you had the authority to do it, what type of "out of the box" vaccine concepts would you support for funding?**

16. In view of the high level of somatic hypermutation observed in all anti-HIV-1 bnAbs, any studies addressing the maturation pathways of such Abs should be encouraged. There is, however, no guarantee that studying stochastic maturation processes in individual immune systems would provide the prescriptive knowledge required for reproducing similar maturation phenomena in large populations of vaccinees.

17. Methods for inducing strong mucosal antibody responses involving dimeric or multimeric IgA1 should be investigated. New immunogens and adjuvants should be tested and the role of various Ab inhibitory functions in protection should be analyzed.

18. More studies of HIV-1 immunopathogenesis should be undertaken in an attempt to identify potential key targets for intervention as well as new biomarkers of disease progression and of infection control. Other markers than viral load and CD4 T cells counts in blood are needed. It is known that it is the level of CCR5+ CR4+ T celles in the gut that is the relevant parameter and should be considered rather than the CR4+ T cell level in the blood that is easier to measure.
19. In view of the totally unexpected "out of the box" results observed in macaques orally vaccinated with killed SIV plus *Lactobacillus plantarum*, a duplication trial in the Chinese macaque model and subsequently a phase one clinical study in humans should be performed as soon as possible. If the hypothesis of Andrieu and Lu et al. (2012) is confirmed in such simian and human experiments, considerable funding should be made available to advance its possible clinical development.

# Editorial: Paradigm Changes Are Required in HIV Vaccine Research

In his influential book "The Structure of Scientific Revolutions" published in 1962, Thomas Kuhn argued that researchers in every field of scientific enquiry are always guided by theoretical assumptions, presuppositions, and hypotheses that constitute the prevailing scientific paradigm under which they operate at any given time (Kuhn 1962). When they operate within a shared paradigm, a scientific community working in a particular field will be committed to using certain strategies and experimental approaches that are accepted as being essential for trying to solve the problem at hand.

In the field of HIV vaccine research, there is in fact evidence that in recent years a number of paradigms based on invalid assumptions had such a detrimental effect (Van Regenmortel 2014b). This drawback can only be avoided if scientists keep in mind that if the paradigms they have adopted are based on erroneous assumptions, this could lead them to select inappropriate research strategies that are unlikely to succeed. Questioning the validity of paradigms is thus an important safeguard since it may reveal which invalid assumptions have led investigators astray in the past. Only when past mistakes are acknowledged can novel paradigms be introduced, which better fit our improved knowledge of HIV-1 immune responses, and are therefore more likely to help future vaccine development (Esparza 2013b; Van Regenmortel et al. 2014).

The following unwarranted assumptions underlying some popular paradigms in HIV vaccine research may have contributed to our inability during the past 25 years to develop an effective HIV-1 vaccine.

Frontiers in Immunology, 2015, 6, 326.
Marc H V Van Regenmortel

© Springer Nature Switzerland AG 2019
Marc H V Van Regenmortel, *HIV/AIDS: Immunochemistry, Reductionism and Vaccine Design*, https://doi.org/10.1007/978-3-030-32459-9_16

## Assumption No. 1: Vaccine Immunogenicity Can Be Predicted from Viral Antigenicity

Most fragments of a viral antigen are immunogenic and are able to induce antibodies that will react with the fragment. However, this type of immunogenicity is irrelevant for vaccination since these antibodies rarely recognize the cognate, intact antigen, and even more rarely neutralize the infectivity of the viral pathogen that harbors the antigen (Van Regenmortel 2006). A confusion between biological immunogenicity (the ability to induce antibodies in a host) and chemical antigenicity (the capacity of an antigen to bind antibodies) lies at the heart of the reverse vaccinology paradigm that has been pursued vigorously for more than 10 years. According to this paradigm, HIV-1 epitopes identified by X-ray crystallographic analysis of complexes of HIV Env bound to affinity-matured neutralizing monoclonal antibodies (nMabs) were expected to be also effective vaccine immunogens able to induce a protective immune response. However, an epitope that binds to a nMab will not necessarily be able to induce the same type of neutralizing antibody in an immunized host. As discussed at length elsewhere (Van Regenmortel 2015a), investigators who claim they are designing a vaccine immunogen are only improving the binding reactivity (i.e., the antigenicity) of a single epitope–paratope pair and are not actually designing a vaccine immunogen able to generate protective antibodies. Immunogenicity does depend on numerous factors that exist only in the context of the host immune system, and these are independent of the binding properties of the viral antigen used for immunization (Van Regenmortel 2012a).

## Assumption No. 2: There Is a Primary and Intrinsic Epitope Specific for Each B Cell Receptor and Its Corresponding Antibody

In reality, there is no single intrinsic epitope for any antibody molecule but only a diverse group of potential ligands able to bind to it with various degrees of fit. Vaccinologists have been slow to accept that antibodies are not monospecific for a single epitope and that the degeneracy of the immune system always makes antibodies polyspecific for numerous related or unrelated epitopes present in different antigens (Sperling et al. 1983; Notkins 2004; Eisen and Chakraborty 2010; Dimitrov et al. 2010). Once it is accepted that the epitope identified by X-ray crystallography of an HIV-1 Env-nMab complex is only one of the many epitopes that could be accommodated by that antibody, there is no justification for assuming that this particular HIV-1 epitope must correspond to the immunogen that elicited the nMab and should therefore be considered as a possible candidate vaccine.

## Assumption No. 3: HIV-1 Epitopes Recognized by Mature nMabs Isolated from HIV-1 Infected Individuals After a Lengthy Process of Antibody Affinity Maturation Will Be Able to Induce a Protective Immune Response in Naive Individuals

It has been established that the initial HIV-1 immunogen that triggers the affinity maturation process leading to mature neutralizing antibodies usually recognizes a BCR germline version that differs considerably from the BCRs corresponding to mature antibodies (Dimitrov 2010). The very extensive affinity maturation required to obtain neutralizing HIV-1 antibodies is one of the main reasons why the structure-based reverse vaccinology approach did not succeed in developing an effective HIV-1 vaccine (Van Regenmortel 2014b). A new paradigm based on the germline/maturation hypothesis was subsequently developed, which assumed that it may be possible to discover effective HIV-1 vaccine immunogens by analyzing putative germline antibody intermediates of known HIV-1 nMabs (Prabakaran et al. 2014). It is not clear at present whether the unraveling of large numbers of different antibody maturation pathways will allow the identification of HIV-1 vaccine immunogens suitable for vaccinating large human populations.

## Assumption No. 4: The So-Called Rational Design of HIV-1 Immunogens Is More Effective than the Classical Trial-and-Error Screening of Immunogens Used in the Past for Developing Successful Vaccines

Many authors claim that so-called "rational design" offers the best prospects for developing an HIV-1 vaccine. Design means the deliberate conceiving of a novel object or process by an intelligent being, while the term "rational" is mostly used to indicate that the designer makes use of available molecular data (Van Regenmortel 2012a, b). When they discuss the rational design of HIV-1 vaccines, these authors refer only to studies that try to improve the structural complementarity between one epitope and one particular Mab, which means they actually discuss antigen design and not immunogen design (Douek et al. 2006; Nabel et al. 2011). Designing vaccine immunogenicity actually means modifying an antigen, so that it becomes an immunogen capable of inducing a protective immune response and this requires investigating empirically the numerous factors, which in an immunized host determine the formation of neutralizing antibodies. Unfortunately, we know very little about the immunological mechanisms in the host that produce neutralizing rather than non-neutralizing antibodies and our ignorance is what prevents us from deliberately "designing" vaccine immunogens of predetermined, known efficacy. Rationally designing an antigen so that it better fits a single Mab is certainly possible but improving an immunogen so that it is better able to elicit protective antibodies can only be achieved by trial-and-error experimentation with various immunogens and not by design (Van Regenmortel 2007, 2012a). It is currently fashionable to denigrate empirical approaches in vaccine research (Karlsson-Hedestam et al. 2008), as if time-honored empiricism and trial-and-error experimentation were not an entirely rational enterprise compatible with high quality science (Van Regenmortel 2012b, 2014b, 2015a). Advocating rational design instead of stressing the need for more empirical research may even be counterproductive since it is likely to make it more difficult for investigators to undertake the small exploratory trial-and-error trials that are needed to move the field forward (Van Regenmortel 2014b). Although epitope mapping with nMabs may indicate which regions of a virus surface should be targeted by a vaccine, such mapping on its own will not make it possible to discover effective vaccine immunogens if no immunogenicity trials are performed.

## Assumption No. 5: Reactions of Viral Antigens with Mabs Are More Specific than the Combined Reactivity of Polyclonal Antibodies Found in an Antiserum and Are Able to Inform Vaccine Design

The introduction of Mabs revolutionized our ability to dissect immune responses to proteins but it also introduced a bias in the analysis of antigens by encouraging investigators to focus on artificial boundaries between overlapping epitopes and to concentrate on single, discrete epitopes as potential vaccine immunogens. The emphasis on single epitopes recognized by Mabs for developing vaccines tends to obscure the beneficial effect of the neutralizing synergy achievable with polyclonal responses (Van Regenmortel 2011b).

The use of Mabs for characterizing epitopes also introduced another bias that occurs because the apparent specificity of a Mab very much depends on the selection process that was used to obtain it. When a nMab, for instance, binds to a short peptide region of the membrane proximate external region (MPER) of HIV-1 gp41, this may simply reflect the fact that the Mab was selected for its ability to bind to that peptide. Since all antibodies are polyspecific, the Mab may actually recognize better a more complex or transient epitope of gp41 that might have been the immunogen that elicited the antibody. However, if it is assumed that the Mab was induced by the linear MPER peptide region because it reacts with it, searching for an effective vaccine immunogen in gp41 may be compromised.

It is often believed that a Mab that binds to a given antigen is a more specific reagent than a polyclonal antiserum raised against that antigen. In reality, an antiserum has a greater collective specificity for a multiepitopic antigen due to the additive specificity effect that arises from the presence in the antiserum of antibodies directed to several different epitopes of the antigen (Van Regenmortel 2014a).

It is sometimes suggested that the isolation of additional broadly neutralizing Mabs will facilitate the future rational design of an HIV-1 vaccine. Why this should be the case is not made clear since these Mabs only have the potential to perhaps 1 day become useful reagents for *passive* immunotherapy. However, their isolation does not improve our ability to discover which vaccine immunogens will elicit neutralizing antibodies by means of *active* immunization (Van Regenmortel 2012b). Once again, confounding antigenicity and immunogenicity is giving rise to unrealistic expectations.

The aim of this Research Topic is to evaluate the shortcomings of some of the paradigms that guided HIV vaccine research in the past as well as to stimulate the search for novel paradigms that better fit our current understanding of immunological specificity and could be more helpful in guiding the future search for an effective HIV-1 vaccine.

# An Outdated Notion of Antibody Specificity Is One of the Major Detrimental Assumptions of the Structure-Based Reverse Vaccinology Paradigm, Which Prevented It from Helping to Develop an Effective HIV-1 Vaccine

*It makes no sense to do the same thing over and over again and expect a different result.*

Albert Einstein

## Introduction

The development of science is promoted when scientists adhere to so-called paradigms that correspond to theoretical presuppositions and assumptions that guide the lines of investigation they pursue. When trying to solve a particular scientific puzzle, for instance how to develop an effective HIV-1 vaccine, the paradigm will also commit scientists to using particular experimental approaches and tools believed to be essential for finding a solution to the problem.

The importance of paradigms in guiding scientific research was proposed by Thomas Kuhn (1962) in his influential book *"The Structure of Scientific Revolutions"* published in 1962. Unfortunately, he never clearly defined the concept of paradigm and he used the term in various ways, for instance, to refer to a collection of procedures and ideas that instruct scientists what to believe and how to work (Horgan 1996). Kuhn later conceded that the term paradigm had become hopelessly overused because it came to signify virtually any dominant idea that binds a scientific community together. He subsequently recommended that *paradigm* be replaced by *exemplar* to mean exemplary instances of successful puzzle-solutions for scientific problems. An exemplar captures the way in which a theory or model is believed to solve a problem while at the same time defining, which new problems

Frontiers in Immunology, 2014, 5, 593.
Marc H V Van Regenmortel

could be addressed in a similar way. However, the ill-defined term paradigm was never abandoned.

Kuhn argued that when scientists in the course of their work obtain results that contradict the theory or hypothesis that gave rise to a paradigm, they do not conclude that the paradigm has been refuted and must be abandoned. Scientists, therefore, do not follow the injunction of Karl Popper that their aim should be to try to disprove or falsify their theories rather than prove them. Popper maintained that observations are never able to prove a theory but can only sometimes logically refute a mistaken theory (Popper 1959). He argued that when scientists obtain reproducible results that are at odds with their working hypothesis, they are logically obliged to accept that the hypothesis has been falsified and they should therefore abandon it (Horgan 2006). Kuhn disagreed and claimed that this is not the way scientists behave because their main commitment is not to test or seek to confirm the implicit theories and hypotheses that underlie the paradigms they adhere to. Scientists in fact tend to ignore anomalous results and will devise new ad hoc hypotheses in an effort to explain away apparent contradictions between theory and experimental observations. Kuhn further claimed that science can make progress only if scientific communities remain committed to their shared theoretical beliefs and experimental techniques and do not abandon a paradigm or hypothesis as soon as incompatible results are obtained (Bird 2018). Only if troublesome anomalies keep accumulating over many years may scientists eventually start questioning their presuppositions and lose their confidence in a given paradigm. This could then usher a scientific revolution that occurs when a paradigm is superseded by a new one and gives rise to a paradigm shift. Periods of so-called normal science are then replaced by a short period of revolutionary science (Kuhn 1962).

In HIV vaccine research, there is evidence that several prevalent paradigms have not helped the development of an effective vaccine (Esparza 2005, 2013b; Van Regenmortel et al. 2014). One such paradigm, which gave rise to the strategy known as structure-based reverse vaccinology (RV) (Burton 2002) was pursued vigorously for more than a decade although it did not lead to the development of an effective HIV-1 vaccine. The theoretical underpinnings of this paradigm have been discussed previously because they illustrate the need for investigators to question the implicit underlying assumptions that make them pursue unfruitful lines of investigation (Van Regenmortel 2012a, b). Only when the presuppositions or hypotheses that gave rise to unsuccessful paradigms are shown to be invalid will investigators become aware that a paradigm shift is required in a particular scientific field (Van Regenmortel et al. 2014).

## Structure-Based RV Paradigm in HIV-1 Vaccine Research

The approach known as RV was introduced in the field of bacterial vaccines by Rino Rappuoli (Rappuoli 2001; Rappuoli and Bagnoli 2011) and refers to the strategy of predicting potential vaccine immunogens using bioinformatics analyses of entire bacterial genomes in order to identify all the surface-exposed proteins that a bacterial pathogen is able to express. The strategy is called RV because investigators operate in a reverse manner, i.e., starting from the genome rather than from the organism, to discover, which bacterial proteins should be studied as potential vaccine immunogens. This allows hundreds of bacterial proteins to be identified as candidate immunogens even when bacteria cannot be cultivated and bacterial extracts cannot therefore be fractionated to establish empirically which proteins are able to induce a protective immune response.

In virology, RV has a different meaning and refers to a strategy which attempts to generate a vaccine from a knowledge of protective antibodies (Abs) rather than from the usual reverse task of generating such Abs by immunization with a vaccine (Burton 2002; Walker and Burton 2010). It was suggested that effective vaccine immunogens might be discovered by exploring the interaction of anti-HIV-1 neutralizing (n)Abs with HIV-1 envelope (Env) spikes, using X-ray crystallography of Env–Ab complexes. This expectation was based on the assumption that once the 3D structure of a broadly neutralizing monoclonal antibody (bnMab) bound to an Env epitope had been elucidated, it would be possible to use the bnMab as a template to reconstruct its complementary epitope outside the context of the natural Env antigen, using structure-based design. It was further hypothesized that this reconstructed epitope designed to fit the bnMab would possess the immunogenic capacity of inducing a polyclonal Ab response with the same neutralizing capacity as the bnAb used as template. The assumption was that if an HIV-1 epitope is able to bind an Ab, it will also possess the immunogenic capacity to elicit the same type of Ab in an immunized host. That this is not necessarily the case, however, is demonstrated by the common observation that when a peptide fragment of a protein is able to bind Abs raised against the protein, the peptide will frequently be unable to elicit Abs that react with the native protein (Van Regenmortel 2009a).

Since the RV approach used to develop bacterial vaccines has been highly successful (Rappuoli and Bagnoli 2011) whereas, the RV approach used in the HIV-1 vaccine field has failed so far (Van Regenmortel 2012a; Burton et al. 2012; Corti and Lanzavecchia 2013) it has been suggested that the two approaches should be clearly differentiated and could be called genome-based RV and structure-based RV, respectively (Van Regenmortel 2011a).

## Unwarranted Assumptions of the Structure-Based RV Paradigm

The structure-based RV approach suffers from several unwarranted expectations that jeopardized the ability of developing an effective HIV-1 vaccine. It did not clearly distinguishing between antigenicity and immunogenicity (Rappuoli 2001; Kong and Sattentau 2012) and analyzed epitopes in antigens and paratopes in Abs solely in terms of which amino acids in the two partners made contact with each other

(Sundberg and Mariuzza 2002). As a result, little attention was given to the fact that the binding activity of an Ab often depends on structural features distant from the paratope itself (Schildbach et al. 1993; Chatellier et al. 1996) and that residues in the antigen that are not in contact with the paratope may nevertheless be able to affect the binding activity and the immunogenic activity of epitopes (Greenspan and Di Cera 1999; Moudgil et al. 1998). Such findings confirm an insight reached years ago that paratopes and epitopes defined in terms of contact residues often differ from the binding sites defined in functional assays (Cunningham and Wells 1993). The structures visualized in Ab-antigen complexes also tend to differ from the structures of the binding sites in the free molecules, before they have been altered by the mutual adaptation and induced fit that occur when the two partners interact (Wilson and Stanfield 1994; Bosshard 2001). This means that the particular antigenic structure revealed in an Ab–antigen complex does not necessarily correspond to the immunogenic structure that was recognized by B-cell receptors (BCRs) during the immunization process and which therefore is often presumed to be needed in the vaccine immunogen.

In an earlier review (Van Regenmortel 2012b), more than 50 original science publications were mentioned, which attempted to reconstruct active HIV-1 epitopes using conformational constraints, protein scaffolds, and other structure-based engineering approaches. Although some of the engineered epitopes possessed increased antigenicity and reacted better with bnMabs, none of them were found to be effective vaccine immunogens, illustrating the shortcomings of the structure-based RV approach (Joyce et al. 2002; Ho et al. 2005; Burton 2010; Pejchal and Wilson 2010; Van Regenmortel 2011b).

## Overlooking the Need for Antibody Affinity Maturation to Obtain Effective Anti HIV-1 Neutralizing Abs

Another unjustified hypothesis of the structure-based RV approach was that the HIV-1 epitopes recognized by the matured bnMabs that are present in HIV-1 infected individuals after a lengthy process of Ab affinity maturation will be able to trigger a protective immune response in naïve individuals. However, studies involving the deep sequencing of all the HIV-1 Abs present in the serum of infected individuals demonstrated that the initial immunogen triggering the Ab affinity maturation process that leads to neutralizing Abs usually recognizes a germline version of BCRs that differs considerably from the BCRs corresponding to mature bnAbs (Xiao et al. 2009; Dimitrov 2010). This means that the epitopes engineered by structure-based RV to mimic HIV-1 epitopes recognized by mature bnMabs are unlikely to be effective vaccine immunogens because they are mostly unable to recognize the germline BCRs present in naïve individuals (Zhu et al. 2011; Chen et al. 2012; Klein et al. 2013; Jardine et al. 2013).

The extent of affinity maturation observed in HIV-1 bnAbs is much more extensive than the 5–10% mutation frequency in the Ab hypervariable regions observed with Abs directed to other viruses (Klein et al. 2013; Jardine et al. 2013; Zhu et al. 2008a) and this characteristic feature of HIV-1 bnAbs together with the enormous antigenic variability of the virus explains why the structure-based RV approach in the case of HIV-1 presented unsurmountable challenges compared to other viral vaccines (Chen et al. 2012; Burton et al. 2005; McLellan et al. 2013). HIV-1 elicits bnAbs only in a minority of infected individuals after several years of infection and attempts are currently made to unravel the mutational pathway that leads from germline BCR ancestors to mature Abs (Liao et al. 2013a). The goal is to identify various immunogens capable of stimulating successive B-cell responses through multiple rounds of antibody maturation processes (Gray et al. 2011; Euler et al. 2012). This is clearly no mean task since large numbers of maturation pathways are likely to exist (Wu et al. 2011b; Mouquet and Nussenzweig 2013; Haynes et al. 2012a).

Current attempts to modify Env epitopes so that they are able to bind germline BCRs or maturation intermediates depart from the original structure-based RV paradigm because they no longer attempt to directly transform epitopes of known structure recognized by mature bnMabs into immunogens capable of eliciting similar bnAbs (Mouquet and Nussenzweig 2013; Haynes et al. 2012a; Fauci and Marston 2014). This new approach represents a new paradigm based on the germline/maturation hypothesis (Dimitrov 2010), which assumes that it may be possible to discover effective HIV-1 vaccine immunogens that bind putative germline antibody predecessors of known HIV-1 bnAbs although they do not bind the highly somatically mutated bnAbs themselves (Prabakaran et al. 2014).

In recent years, increasing numbers of bnMabs have been isolated from HIV-1 infected individuals (Scheid et al. 2009; Stamatatos et al. 2009; Sattentau and McMichael 2010) and it has been suggested that these Mabs provide valuable molecular information that could inform HIV-1 vaccine design (Mascola and Montefiori 2010; Burton and Weiss 2010). Such bnMabs may be of value for passive immunotherapy since they could provide sterilizing immunity to humans and non-human primates when they are administered prior to viral challenge (Mascola et al. 1999; Hofmann-Lehmann et al. 2001). However, their usefulness does not at present extend to the vaccine field since we have no idea how Abs that possess the protective capacity of such bnMabs can be induced by vaccination (Van Regenmortel 2012b). There is also considerable evidence that a protective immune response requires the combined neutralizing activities of several Abs that target different non-overlapping HIV-1 Env epitopes (Doria-Rose et al. 2010) as well as Abs that act in synergy (Laal et al. 1994; Mascola 2007; Zwick et al. 2001b). What may be required, therefore, is to find immunogens able to elicit classical types of protective, polyclonal immune responses rather than elusive immunogens that would elicit single Ab specificities endowed with the exceptional neutralizing capacity of individual bnMabs such as the well-studied b12, VRC01, PG9, or PG16 bnMabs (Burton et al. 1994; Walker et al. 2009; Wu et al. 2010).

## Reliance on Reductionist Thinking: Another Pitfall of the Structure-Based RV Paradigm

Another pitfall of the structure-based RV paradigm is its reliance on reductionist thinking (Van Regenmortel 2012b). Reductionism has been prevalent in molecular biology for half a century and is still popular today because it has been very successful for dissecting biological systems into their constituent parts (Van Regenmortel 2002a, 2004b). The reductionist mindset made immunologists accept that the biological activities of Abs could be explained by their 3D structures and that the immunogenic potential of a viral epitope could be deduced from its antigenic properties. Biological immunogenicity was thereby reduced to chemical antigenicity, which is a variation of the claim that biology can be reduced to chemistry (Van Regenmortel and Hull 2002). Such a claim fails to recognize that the protection achieved by vaccination is a biological phenomenon that has meaning only in the context of an entire organism since organs, tissues, or molecules cannot be vaccinated. Protection always results from a complex network of dynamic interactions between pathogen, host, and immune system and it cannot be satisfactorily understood when innumerable, individual molecular interactions are analyzed separately.

The use of Mabs also introduces a reductionist bias in the analysis of protein antigenicity and immunogenicity because it leads investigators to focus on artificial boundaries between overlapping epitopes and to ignore the fact that the apparent immunological specificity of a Mab very much depends on the selection process that was used to obtain it. Since a Mab is always polyspecific, the fact that it binds, for instance, to one particular peptide of the membrane proximate external region (MPER) of HIV-1 gp41 simply reflects the fact that the Mab was selected for its ability to bind to that peptide. Such a Mab may, however, recognize better a more complex epitope of gp41 that could actually have been the immunogen that gave rise to the Mab. If it is assumed that the Mab was elicited by the linear MPER peptide epitope because it binds to it, this may induce investigators to only investigate peptides as possible vaccine immunogens, a choice, which is likely to be self-defeating (Van Regenmortel 2009a; Montero et al. 2008).

## The Structure-Based RV Paradigm Suffers from Another Major Misguided Assumption, Namely an Outdated Notion of Ab Specificity

According to this view, which is usually not explicitly acknowledged, a Mab that would bind for instance to an epitope on the HIV-1 Env surface is believed to be able to recognize only a single defined target area (the so-called complementary epitope of the Mab) corresponding to the surface residues found to be in contact with paratope residues in the Ab–antigen complex. Although the belief that Mabs are monospecific for a single epitope is no longer held and was shown to be invalid already 30 years ago (Sperling et al. 1983; Berek and Milstein 1987; Chen et al. 1998; Parnes 2004; Cohen et al. 2004; Notkins 2004; Eisen and Chakraborty 2010) many vaccinologists have been slow to accept that there is no single intrinsic or specific epitope for any Ab but only a diverse group of potential epitopes able to bind to it with various degrees of fit (Van Regenmortel 2012a, 2014a; Wucherpfennig et al. 2007). It seems that the failure to recognize that Abs are always polyspecific and possess considerable plasticity, promiscuity, degeneracy, and cross-reactive potential has made the structure-based RV approach appear more plausible to many investigators (Van Regenmortel 2012a).

It is now well-established that most Abs derived from Ab germline genes and expressed before any antigenic stimulation are highly polyreactive and react with a wide variety of autoantigens such as DNA, cytoskeleton proteins, nuclear antigens, and carbohydrates as well as numerous bacterial and viral antigens (Chen et al. 1998; Manivel et al. 2002; Yin et al. 2003; Haynes et al. 2005b; Zhou et al. 2007b; Mouquet and Nussenzweig 2012). The various mechanisms that allow a polyspecific Ab to recognize a multiplicity of epitopes and antigens have been elucidated (Mariuzza 2006). Once investigators realize that the epitope structure observed in a bnMab–HIV-1 Env complex is only one of the many epitopes that could be accommodated by what is always a polyspecific Ab, they have no difficulty to accept that any one of these epitopes could correspond to the immunogen that gave rise to the Mab. They will then no longer assume that the one epitope structure observed in the crystallographic analysis of an antigen complex must necessarily reveal, which immunogenic structure should be used to elicit such a bnAb (Van Regenmortel 2012a).

Furthermore, the fact that Abs are not only polyspecific but are also heterospecific, i.e., able to react more strongly with other antigens than with the immunogen used for eliciting the Ab, also helps to clarify why antigenic and immunogenic properties are not always simultaneously present in the same region of a protein (Van Regenmortel 2014a). The immunogenic capacity of an epitope to induce heterospecific Abs that do not react with the protein used for immunization demonstrates that immunogenicity need not be accompanied by an antigenic reactivity that enables the epitope to bind to the induced Abs. Similarly, the antigenic reactivity of a viral protein revealed by its ability to bind a given polyspecific Ab is also not necessarily accompanied by an immunogenic capacity to induce that same Ab in a particular immune system (Van Regenmortel 2014a).

Although the advent of Mabs has completely transformed our ability to dissect immune responses to proteins, their utilization has also introduced a bias in the antigenic analysis of viruses because investigators tend to focus on artificial boundaries between overlapping epitopes and overlook the fact that the surface of a protein is an antigenic and functional continuum (Berzofsky 1985). When an nMab has been isolated, it tends to become associated with a unique, discrete epitope, which leads to the expectation that a vaccine containing this epitope will induce similar nAbs. However, most protective immune responses are polyclonal and involve the collective and synergistic neutralizing activities of Abs directed to different epitopes (Burton et al. 1994). When the therapeutic efficacy of mixtures of HIV-1 neutralizing Mabs was tested in HIV-1 infected humanized mice, it was found that mixtures of five Mabs were more effective than single Mabs or mixtures of three Mabs (Klein et al. 2012). Mixtures of five Abs suppressed viral loads below the level of detection and also failed to select escape viral mutants. However, chronically HIV-1 infected humans from whom bnAbs were isolated do not appear to benefit from such Abs for controlling virus replication (Corti and Lanzavecchia 2013; Doria-Rose 2010; Euler et al. 2010), nor do HIV-1 infected long-term progressors compared to non-progressors (van Gils et al. 2010). The role of Abs in controlling chronic HIV-1 infection therefore remains an issue that should be further investigated.

Many unsuccessful attempts have been made to elicit by immunization bnAbs similar to potent bnMabs isolated from HIV-1 infected individuals (Corti and Lanzavecchia 2013; Euler et al. 2011). The past research emphasis on unraveling the neutralization mechanisms of individual bnMabs did not help because it did not provide any information on which immunogens were capable of eliciting the different types of bnAbs. The poor success rate so far in discovering even one such effective HIV-1 vaccine immunogen by structure-based RV is certainly a cause of concern. As discussed elsewhere (Van Regenmortel et al. 2014), there is a clear need to study new vaccine immunogens and new methods to induce strong mucosal antibody responses using for instance specific adjuvants. Other markers than viral load and CD4+ T cells in blood for assessing vaccine efficacy should also be investigated (Van Regenmortel et al. 2014).

It is nowadays commonly claimed that rational design offers the best prospects for developing an HIV-1 vaccine and that this approach is superior to the empirical screening and trial-and-error strategies used in the past. When it is claimed that "rational design represents the only approach that can elevate vaccine research from an empirical exercise to a scientific discipline" (Karlsson Hedestam et al. 2008), the essential contribution of empirical trials to vaccine development is actually denigrated as if trial-and-error experimentation were not an entirely rational enterprise (Van Regenmortel 2012a, 2012b; Kong and Sattentau 2012). When authors discuss the rational design of an HIV-1 vaccine (Douek et al. 2006; Nabel et al. 2011) they only refer to studies that improve the degree of complementarity in a single epitope–Mab pair and they do not clarify how an improved antigen could actually be "designed" to also become an immunogen capable of inducing protective Abs. Optimizing the binding activity of a viral antigen by structure-based design using a single Mab as a template is certainly feasible but this is not equivalent to immunogen design, which requires the intentional optimization of numerous factors extrinsic to the epitope–paratope recognition such as the various cellular and regulatory mechanisms of the host that exist only in the context of the vaccinated host and control the generation of neutralizing Abs (Van Regenmortel 1999b, 2012b). Antigen design is simply masquerading as immunogen design when it is assumed that an improved viral antigen will also be an effective vaccine immunogen capable of inducing a protective immune response. The so-called rational design of vaccine immunogens by mimicking the rational approach used in drug design (Van Regenmortel 2007, 2012a; Kuntz 1992) is not feasible without extensive empirical clinical trials of vaccine candidates and to suggest otherwise contradicts the well-established empirical nature of vaccine science. The concern that too much funding may be diverted to empirical clinical trials at the expense of basic structure-based HIV vaccine research (Schiffner et al. 2013) seems unwarranted since the lion share of current funding is devoted to structural studies whereas, the required small scale clinical trials based on innovative paradigm concepts remain poorly funded (Van Regenmortel et al. 2014).

## Causes, Explanations, and Understanding in Vaccinology

The selection of a paradigm to guide the empirical search for an HIV-1 vaccine always depends on framing hypotheses about causal mechanisms, which could provide a plausible explanation for a possible successful vaccine. Most scientific explanations take the form of a causal mechanistic explanation, which means that causes, explanations and understanding are usually intimately linked (Salmon 1998; Lipton 2004). It is relevant therefore to analyze how these three terms are used in vaccinology.

In the physical sciences, explanations are usually presented as logical deductions derived from one or a few relevant laws of nature, together with certain initial conditions. However, this is not feasible in the biological sciences because of the absence of universal biological laws (Dupré 1993; Van Regenmortel 2012b). Since the probability of a biological event is always affected by a very large number of causal factors, causal links become diluted and it is usually not possible to provide an explanation in vaccinology in terms of a single cause.

The reductionist dissection of the immune system into its components severs the dynamic connections that link the parts of a biological complex system in a functionally integrated manner. This allows any level in the resulting biological complexity to be the starting point of a causal analysis, provided a certain state of affairs is considered to be in need of an explanation. As a result, explanations are usually framed in terms of a complex type of probabilistic causality that attempts to take into account the numerous factors that together contribute to an effect in a given biological context. This means that the classical notion of causality is of limited value for providing explanations in immunology (Van Regenmortel 2012b; Salmon 1998). The contribution of one causal factor in a complex multicausal biological system can actually only be investigated by altering that factor experimentally and assessing whether the observed effect is then no longer the same in a given context (Van Regenmortel 2007). This has led to the suggestion that many biological phenomena may be too complex to be comprehended or explained by human intelligence (Gannon 2007).

Attempts to achieve understanding in immunology often start with information about observed effects and by an awareness that certain phenomena share underlying similarities, which then leads one to propose a theory or hypothetical model to explain them. The structure-based RV paradigm is one such model, which assumes that bnMabs that are isolated from HIV-1 infected individuals and recognize particular Env epitopes are likely to have been elicited by these epitopes. When the model was tested by assessing the immunogenicity of the epitopes recognized by different bnMabs, it was found that the results did not fit the model and that no bnAbs were elicited (Ho et al. 2005; Burton 2010; Diskin et al. 2011). This should remind us that a proposed explanatory model does not predict the data to which it is fitted since the model is actually chosen to fit the data and it makes no sense to fit what one wants to explain (Gunawardena 2014). Understanding consists of knowledge about relations of dependence and should make it possible to derive inferences about the consequences of our interventions and give us an ability to predict and control phenomena. Unfortunately, it seems that scientists tend to overestimate the detail and depth of their understanding, which often consists of nothing more than an informed guess about the future prospects of their scientific work. This leads to a frame of mind that has been called an "illusion of depth of understanding in science" (Ylikoski 2009). In the absence of experimental data that back their explanatory model or paradigm, scientists do not choose which theory to accept but choose which theory they are actually going to work with. Scientists are then driven by the promise of future understanding rather than by past convincing explanatory evidence and such an expectation is easily influenced by wishful thinking about their pet theories (Ylikoski 2009).

Since understanding a phenomenon is ultimately displayed by: (1) making right predictions, (2) successfully intervening in a system, and (3) answering explanation-seeking questions about it, the inability to do any of these things is a clear indication that alternative explanatory hypotheses and paradigms are required. As far as HIV-1 vaccines are concerned, our ignorance of why all Env immunogens investigated so far have been unable to induce adequate levels of potent protective Abs is a clear indication that we do not understand the complex mechanisms that are involved in achieving protective immunity against HIV-1.

Although the complexity of the immune system may prevent us from identifying all its internal regulatory mechanisms, it is by comparing the various ways of manipulating the system using empirical experimentation that we may eventually control it and achieve protective immunity by vaccination (Van Regenmortel 2012b).

Empirical, scientific knowledge is based on experimental and observational facts and on the rule that "nothing trumps experience." However, it has been argued that empirical evidence only allows scientists to draw plausible but tentative conclusions when the obtained results make it possible to successfully control and manipulate the experimental system under investigation (Van Fraassen 2002; Hacking 1983). If this does not happen, empiricism dictates that we should investigate additional constituents of the complex biological systems we study until we improve our ability to predict the results of our experimental interventions. It is indeed the ability of investigators to successfully intervene in a material system that gives them the knowledge needed to manipulate and control it.

Since the aim of biologists is both to explain and control biological phenomena, explaining goes hand-in-hand with intervening and it has been suggested that "explanations in biology are always obtained through direct intervention on models of the phenomenon to be explained" (Leonelli 2009). It is important to recognize that such a view is at odds with the widespread expectation that we will succeed in developing an effective HIV-1 vaccine only when we have significantly increased our general knowledge of basic immunology and of HIV-1 antigenic structure (Van Regenmortel 2012b). This means that we need to interfere with the material world in order to obtain empirical knowledge about it and that our scientific understanding increases when we are able to successfully manipulate the system we investigate (Kelley and Scott 2008).

## Conclusion

There is evidence that the popular structure-based RV paradigm used in HIV-1 vaccine research has not been helpful for developing an effective HIV-1 vaccine. Some of the reasons for this lack of success are summarized, one of them being the failure to recognize that all Abs as well as bnMabs isolated from HIV-1 infected individuals are always polyspecific and able to bind to a variety of related and unrelated epitopes. Since the epitope structure identified by X-ray crystallography of a bnMab–HIV-1 Env complex is only one of several epitopes that could be accommodated by the Mab, there is no reason to assume that this epitope of known structure must correspond to the immunogen that elicited the antibody.

Recently, one of the major groups committed to the structure-based RV paradigm reported a study of bnMabs that recognize a high-mannose epitope patch centered on the N 332 residue on HIV Env. They demonstrated that these Abs actually did not have a single defined target point at N 332 but were in fact polyspecific and able to bind various glycan patches as well as a glycan site located at N334 (Sok et al. 2014). For the first time, these authors admitted that antibody polyspecificity was a relevant concept in HIV-1 vaccine research and they claimed that their results represent an extension of the concept of antibody promiscuity and degeneracy that has been widely accepted in immunology for many years (Van Regenmortel 2012a; Sperling et al. 1983; Notkins 2004; Eisen and Chakraborty 2010; Wucherpfennig et al. 2007). They also concluded that polyclonal Abs are more effective for neutralizing viruses than individual Mabs and that the polyspecificity of vaccine-induced Abs should receive increased attention. The new insights arrived at by these authors might in time alter the expectations of the proponents of the structure-based RV paradigm and diminish their reliance on structural data derived from the study of individual bnMabs.

The structure-based RV paradigm has been followed by several large networks of investigators who operate under the strong leadership of principal investigators (Alberts et al. 2014). This situation leads to considerable built-in inertia and does not encourage funding agencies to back large numbers of high-risk projects based on alternative innovative paradigms that could diversify the vaccine strategies that are investigated (Van Regenmortel et al. 2014; Alberts et al. 2014). These issues are currently receiving increased attention and it is hoped that this will in due course lead to increased funding for new original science based on novel paradigms and that it will stimulate a rigorous evaluation of existing HIV vaccine programs (Alberts et al. 2014).

# More Surprises in the Development of an HIV Vaccine (General Commentary)

In the current issue of Frontiers in Immunology, Jean-Marie Andrieu and collaborators, report results from non-human primate experiments designed to explore a new vaccine concept aimed at inducing tolerance to the simian immunodeficiency virus (SIV) (Andrieu et al. 2014). This approach, which is significantly different from other vaccine concepts tested to date, resulted in a surprisingly high level of protection. If the results are confirmed and extended to the human immunodeficiency virus (HIV), this approach may represent a game changing strategy, which should be welcomed by a field that has been marred by mostly disappointing results. When HIV was discovered and established as the cause of the Acquired Immune Deficiency Syndrome (AIDS) in 1983–1984, there was an expectation that a preventive vaccine would be rapidly developed (Esparza 2013a). Vaccines against several major human viral diseases (polio, measles, mumps, rubella, etc.) were successfully developed during the preceding two or three decades, mostly using live-attenuated viruses, and designed to induce the same type of protective immune responses that develop after natural infection. Moreover, recent advances in molecular biology and recombinant DNA technologies were offering exciting new opportunities for vaccine development, first achieved with the licensure in 1986 of a recombinant vaccine against hepatitis B (Valenzuela et al. 1982; AcAleer et al. 1984). Since the use of whole-inactivated or of live-attenuated vaccines was

Frontiers in Immunology, 2014, 5, 329.
Esparza J., Van Regenmortel M.H.V.
To view the original article, visit https://www.frontiersin.org/articles/10.3389/fimmu.2014.00329/full
A commentary on Mucosal SIV vaccines comprising inactivated virus particles and bacterial adjuvants induce CD8+ T-regulatory cells that suppress SIV positive CD4+ T-cell activation and prevent SIV infection in the macaque model *by Andrieu JM, Chen S, Lai C, Guo W, Lu W. Front Immunol (2014) 5:297. doi: 10.3389/fimmu.2014.00297.*

© Springer Nature Switzerland AG 2019                                                    217
Marc H V Van Regenmortel, *HIV/AIDS: Immunochemistry, Reductionism and Vaccine Design*, https://doi.org/10.1007/978-3-030-32459-9_18

considered too risky for a pathogen such as HIV, the molecular approach was the one selected by early HIV vaccine developers. That decision was also based on the confidence that new knowledge on the structure and function of the virus, as well as of the pathogenesis of the disease, will provide the information needed for the rational development of a much needed HIV vaccine (Hilleman 1998). In that environment of optimism, the first phase I clinical trials of HIV vaccines started in the United States in 1988. Since then, more than 200 clinical trials have been conducted globally, the majority of them phase I and II trials, to assess the safety and immunogenicity of different vaccine candidates. Those candidate vaccines were developed and tested according to prevailing paradigms that sequentially explored the role of neutralizing antibodies, cell-mediated immunity (CMI) and, more recently, other potential mechanisms of immune protection (Esparza 2013a; Excler et al. 2014). Although much has been learned from those small-scale clinical trials, the results from phase IIb/III efficacy trials are the ones that have driven major changes on how HIV vaccine research is advanced. Those trials have also given us a few surprises. Fortunately, the field has been able to learn from those lessons and steadily move forward. Perhaps the first major surprise was when in 1994 we learned that field isolates of HIV were more difficult to neutralize in vitro than laboratory-adapted strains, and that proposed existing candidate vaccines could not induce the appropriate type of neutralizing antibodies, a problem that we are still struggling to solve. Nevertheless, in the early 2000s, 2 gp120 candidate vaccines from VaxGen were tested in efficacy trials and, as many predicted, they failed to protect. That failure shifted the field to CMI vaccines and to the suggestion that perhaps the best that an HIV vaccine could do is to decrease virus load in vaccinated individuals who became infected (Johnston and Fauci 2007). Unfortunately, the STEP study, which tested the CMI concept using an adenovirus 5 vectored, and which was a favorite approach of the HIV vaccine community, was halted in 2007 because of lack of efficacy (Buchbinder et al. 2008). That was a major surprise that led to calls to slow down clinical trials and to go back to basic science (Kaiser 2008). The next major surprise came in 2009, when the results from the Thai RV144 were announced. The trial, which evaluated a canary pox prime followed by a gp120 boost, was strongly opposed by some of the leading HIV vaccine scientists (Burton et al. 2004a). Unexpectedly, the trial showed for the first time that prevention of HIV infection was achievable by an HIV vaccine (Rerks-Ngarm et al. 2009). In a commentary authored by the late Norman Letvin (2009), who himself expressed concerns about the conduct of the RV144 trial, he indicated that the findings were not expected based on preclinical studies and human immunogenicity data, concluding with the lapidary remark that "*we have learned to expect the unexpected in our efforts to generate an effective HIV vaccine.*" Although the observed protection in RV144 was modest (31.2%), those results not only brought new optimism to the field, but also triggered a major collaborative effort to try to identify immune correlates of protection (Haynes et al. 2012b). In this regard, novel and more promising vaccines are being developed that may result in higher levels of protective efficacy, including the use of vectors based on adenovirus 26 (Ad26) and cytomegalovirus (CMV) (Baden et al. 2013; Hansen et al. 2013). Another surprise came when a careful statistical analysis of the step study confirmed that vaccination in fact enhanced HIV acquisition among a subset of the volunteers (Duerr et al. 2012), an observation that was also made in the Phambili study conducted in South Africa using the same vaccine as in the step study (Gray et al. 2014). The most likely explanation of the observed enhancement is a specific immune activation induced by the adenovirus 5 vectored vaccines. Although the mechanism is poorly understood, it does not seem to be present with another adenovirus 5 vectored HIV vaccine (Hammer et al. 2013), and it is not clear how relevant it could be to other vaccine approaches (Fauci et al. 2014). Nevertheless, it is well- known that activation of CD4+ cells is important for HIV replication, which creates a dilemma for vaccinologists, who have to thread a compromise between the desire to induce strong vaccine responses and, at the same time, avoid the immune activation that may enhanced HIV acquisition. In this and other regards, HIV/AIDS is different from other viral diseases for which vaccines have been developed, because it forces vaccine developers to explore mechanisms that nature has not developed, especially when dealing with chronic infections (Johnston and Fauci 2011). It is in this context that Jean-Marie Andrieu and collaborators report in this journal (Andrieu et al. 2014) additional results from an approach that they first reported in 2012 (Lu et al. 2012; Van Regenmortel 2013). The investigators used Chinese macaques to explore the concept that the induction of immune tolerance to SIV could induce protective immunity in the absence of usual humoral or cellular immune responses. To achieve that goal, inactivated SIV was intragastrically administered together with living bacterial adjuvants (BCG, *Lactobacillus plantarum*, or *Lactobacillus rhamnosus*) with the goal of inducing tolerance to the SIV antigens. In a series of experiments, the investigators showed that their approach protected the experimental animals from mucosal and parenteral challenges. Vaccination neither elicited SIV-specific antibodies nor cytotoxic T-lymphocytes but induced a previously unrecognized population of non-cytolytic MHCIb/E-restricted CD8C T regulatory cells that suppressed the activation of SIV positive CD4C T-lymphocytes. Although the number of monkeys is relatively small, the levels of protection are impressive, with 23 out of 24 animals protected in one of the experiments, when animals were challenged 48 months after vaccination. The 2012 publication from this group (Lu et al. 2012) had very little impact in the field, perhaps because it was received with a degree of skepticism. After all, 30 years of intense vaccine research had not resulted in a practical effective vaccine, although an HIV vaccine is sorely needed to bring the HIV epidemic under control. No stone should remain unturned in its search, and the approach reported in this journal should not be dismissed a priori. Instead, it should be carefully considered by other scientists and appropriately confirmed or refuted by additional research. In order to accelerate the development of an HIV vaccine, one of us has proposed a number of actions, including the suggestion to establish a program of truly innovative research with protected funding to explore out-of-the paradigm approaches, perhaps allocating to this program not <10% of the total HIV vaccine investment (Esparza 2013b). Out-of-the- paradigm approaches, such as the one proposed by Andrieu et al., should be further explored (Van Regenmortel et al. 2014). Paraphrasing Dean K. Simonton (2013), the University of California psychologist who has dedicated his professional life to the study of creativity: good science contributes ideas that are original and useful, and we have plenty of those in the HIV vaccine field. However, the solution to the HIV vaccine challenge will require genius which, according to Simonton, is characterized not only by originality and usefulness, but also by surprising results.

# Why Does the Molecular Structure of Broadly Neutralizing Monoclonal Antibodies Isolated from Individuals Infected with HIV-1 Not Inform the Rational Design of an HIV-1 Vaccine?

## Introduction

It has been suggested that our inability over the past 25 years to develop an effective HIV vaccine is partly due to the fact that investigators adhered to several unwarranted assumptions and paradigms that made them pursue unfruitful research strategies (Esparza 2013b; Van Regenmortel et al. 2014). One such misleading assumption central to the structure-based reverse vaccinology approach (Burton 2002) was the belief that when an HIV-1 Env epitope is found to bind to a broadly neutralizing monoclonal antibody (bnMab), this epitope should also be able to induce similar neutralizing antibodies when used as an immunogen (Van Regenmortel 2014b). A related assumption was that HIV-1 Env epitopes, targeted by hypermutated bnMabs that are produced in HIV-1 infected individuals after a lengthy process of antibody affinity maturation, would be able to trigger a protective immune response in naive individuals (Dimitrov 2010; Van Regenmortel 2012b).

The present review will discuss another detrimental assumption that impeded progress in the HIV vaccine field, namely the belief that a Mab that binds to the HIV-1 Env protein is a more appropriate and specific reagent for studying HIV immunology and vaccine immunogenicity than a polyclonal anti-HIV antiserum.

Such a belief arises when antibodies are perceived to be monospecific for a single epitope rather than polyspecific for a number of related or unrelated epitopes. In an antiserum containing antibodies directed to different epitopes of a multi-epitopic viral antigen, each individual antibody will also cross-react with numerous epitopes

AIMS Public Health, 2015, 2, 183–193.
Marc H V Van Regenmortel

present in other antigens. However, since these cross-reactive epitopes will be different for each type of antibody found in the antiserum, the cross-reactions will be diluted out in the antiserum and may not be apparent. In contrast, the cross-reactions of a single Mab will not be masked in this manner, and the Mab may therefore appear to be less specific than the antiserum. A polyclonal antiserum will thus have a greater collective specificity for a multi-epitopic viral antigen than a Mab since it contains many antibodies, directed to several different viral epitopes, that give rise to an additive specificity effect (Richards et al. 1975; Van Regenmortel 2014a). The presence of such antibodies in the antiserum often also produces a beneficial, protective neutralization synergy. Most protective immune responses against pathogens are polyclonal and involve the collective neutralizing activities of antibodies directed to separate epitopes. When one antibody present in an anti-HIV antiserum binds to the Env glycoprotein, it may induce a conformational change in the protein and this could then allow another antibody in the antiserum to bind to a newly exposed epitope in Env which could lead to neutralization synergy (Vijh-Warrier et al. 1996; Li et al. 1998). When an antiserum contains certain combinations of antibodies directed to separate epitopes of a virus, antibody synergy may achieve a higher degree of neutralization than would arise from the simple additive effect of any two randomly chosen neutralizing antibodies (Mascola et al. 1997; Li et al. 1997a). It is now generally accepted that the most effective vaccine-induced antibody response will always be polyclonal and directed to multiple neutralizing epitopes (Mascola and Haynes 2013).

Early attempts at developing an HIV vaccine were based either on humoral immunity, on cell-mediated immunity or a combination of both (Esparza 2013b). More recently, it became accepted that a vaccine eliciting neutralizing antibodies was likely to be adequate to confer protection against HIV infection and the present review will only discuss that particular vaccine strategy. Most successful viral vaccines act by mimicking the protective antibody immune response that is usually generated when a virus causes a natural infection in a host. However, during HIV infection, most individuals do not develop protective antibodies able to control viral replication. This means that a successful HIV vaccine must be able to achieve something that the human immune system is normally not able to do when it encounters the virus. This is probably one of the reasons why attempts to develop a preventive HIV vaccine have failed so far.

It has been claimed (Nabel et al. 2011) that if we understood why the human immune system is mostly unable to elicit antibodies that protect against HIV infection, this would facilitate the ultimate development of an effective vaccine. The rationale for such optimism is not obvious since evidence for the absence of protection in one case may throw little light on the absence of evidence for protection in other cases. There is, in fact, no ground for claiming that if we understood why most anti-HIV Mabs are not protective, this would necessarily improve our ability to develop a vaccine able to elicit neutralizing antibodies. A related unwarranted claim is the assertion that if we study the epitope targets of large numbers of bnMabs, this will give us useful knowledge that will "inform" rational vaccine design (Nabel et al. 2011). The aim of this review is to analyse the validity of such claims since their acceptance may lead vaccinologists to pursue unfruitful lines of investigation.

# The Antibody Response to the HIV-1 Envelope Glycoprotein Complex (Env)

The HIV-1 Env glycoprotein consists of about 15 spikes per virion which are embedded in the viral membrane (Schiller and Chackerian 2014). It has been suggested that this low number of spikes makes it difficult to achieve bivalent antibody binding to two adjacent spikes and that this could be responsible for the high degree of polyreactivity and self reactivity of anti-HIV-1 bnMabs and for the fact that these antibodies require a lengthy process of affinity maturation to acquire neutralizing activity (Schiller and Chackerian 2014; Wardemann et al. 2003; Verkoczy and Diaz 2014). Each spike is made up of trimers of three gp120 surface glycoproteins that are non-covalently attached to three gp41 transmembrane glyco-proteins. Besides functional glycoprotein trimers, the viral membrane also displays gp120-gp41 monomers and single gp41 molecules as well as other non-functional fragments (Poignard et al. 2003; Moore et al. 2006). The structure of the spikes has been analyzed by cryo-electron microscopy using Env bound to Mabs as well as unliganded Env (Liu et al. 2008; Schief et al. 2009).

Most gp120 epitopes are structurally discontinuous and many of them are so-called neotopes (Van Regenmortel 1966) that arise from the native quaternary structure that is present only in intact, functional spikes (Pancera et al. 2010). Neotopes arise either from the juxtaposition of residues in neighbouring protein subunits that are recognized by antibodies as a single epitope, or from conforma-tional changes induced in the proteins by intersubunit interactions (Van Regenmortel 1992a).

Many anti-Env bnMabs target conserved protein regions such as the CD4 recep-tor binding site and the transiently exposed membrane—proximal external region (MPER) located at the base of the spike's stem (McElrath and Haynes 2010). Such antibodies inhibit HIV infection by either blocking binding to cellular receptors or by interfering with the fusion of viral and cellular membranes. Both these activities must be preserved if the virus is to maintain fitness (Burton et al. 2004b; Corti and Lanzavecchia 2013).

It is commonly accepted that the most effective way of dissecting the antibody response to HIV infection consists in using neutralizing Mabs derived from infected individuals. In recent years, new antibody screening and selection technologies have made it possible to isolate many additional neutralizing Mabs (Corti and Lanzavecchia 2013; Walker et al. 2009; Zhou et al. 2010) and this has reinforced the expectation that if the structure of these antibodies was elucidated, it would facilitate the design of vaccine immunogens able to elicit a neutralizing immune response (Sattentau and McMichael 2010). There is considerable interest in these Mabs because numerous immunotherapy studies have demonstrated that they are able to provide sterilizing protection in vivo (Mascola et al. 1997; Lewis 2010). However, the isolation of Mabs useful for passive immunotherapy does not improve our ability to discover which vaccine immunogens will elicit such antibodies by active immunization (Van Regenmortel 2012b). The structure of a neutralizing, protective Mab, unfortunately does not in any way tell us which immunogen induced its formation nor which vaccine immunogen would be able to elicit similar neutralizing antibodies in an immunized host.

The difficulties one encounters when trying to make use of the known structure of certain bnMabs for developing effective HIV-1 vaccine immunogens are well illustrated in the case of the many anti-HIV-1 bnMabs found to possess very long CDRH3 regions of 20–34 residues compared to the average length of 16 residues of this region usually found in human B cells (Yu and Guan 2014). These long CDRH3 regions protrude from the paratope surface (Stanfield et al. 2004) and can penetrate the glycan shield of the Env trimer to interact with gp120 loops or reach conserved residues on gp41. It was initially believed that knowledge of the immunological mechanisms responsible for the generation of bnMabs with long CDRH3 sequences would help the design of an effective vaccine but this was found not to be the case. Long CDRH3 sequences are present in the mature naïve B cell repertoire and arise during recombination of VDJ gene segments before the antigen-driven process of antibody affinity maturation (Briney et al. 2012). Long CDRH3 sequences can also arise from VH replacement that occurs when a short stretch of nucleotides from previously rearranged VH genes are left within newly formed CDRH3s. However, knowledge of these mechanisms did not make it possible to increase the generation of bnAbs endowed with long CDRH3s (Yu and Guan 2014).

It is a reductionist fallacy to believe that the biological, neutralizing activity of an antibody which involves a context-dependent ternary interaction between antibody, virus and host can be reduced to and explained by the physico-chemical structure of the participating antibody molecules (Van Regenmortel 2012b). Chemical antigenicity is confused with biological immunogenicity when it is assumed that the immunogenic, protective potential of the Env glycoprotein can be deduced and predicted from its antigenic properties.

During the immune response to HIV-1, neutralizing antibodies are not preferentially selected compared with non-neutralizing antibodies (Corti and Lanzavecchia 2013). B cell receptors are actually selected for their capacity to capture viral antigens and this selection process is blind to the presence of virus neutralizing activity in the antibodies that will subsequently be secreted by plasmocytes. In fact, it seems that individuals from whom nAbs are isolated do not benefit from the presence of these antibodies for controlling viral replication, possibly because only a small fraction of virus-specific memory B cells will be quickly reactivated and will differentiate into plasmablasts (Corti and Lanzavecchia 2013). Furthermore, since bnAb responses typically appear only 2–3 years after infection, the kinetics of the evolving B cell response also lag behind the rapidly diversifying virus, preventing the antibodies from controlling the infection (Ackerman and Alter 2013). The Env-directed antibody response is believed to reflect the continual stimulation by evolving antigenic variants of Env rather than the continued production of antibodies elicited by the Env protein of the originally infecting virus (Klasse et al. 2012) When the extent of change in bnAb susceptibility of HIV-1 within individual progressors was studied during the course of infection, it was found that viral variants resistant to one or more bnAbs could develop in most individuals (Bunnik et al. 2009). This led to the suggestion that since viral resistance against bnAb- mediated neutralization generally developed when autologous serum neutralization had faded, these changes were unlikely to have been driven by escape from autologous humoral immunity (Bunnik et al. 2009). The relevance of these findings to vaccine development is unclear since protection against HIV infection requires high titers of nAbs at mucosal sites that are seldom obtained during chronic infection or by Env immunization (Klasse et al. 2012).

## Dissecting the Antigenic Structure of HIV-1 with Mabs Has Unforeseen Consequences

In spite of their usefulness for dissecting immune responses to proteins, the use of Mabs introduced a distorted view of the antigenic surface of proteins because it encouraged investigators to concentrate on small numbers of single, discrete epitopes recognized by individual Mabs that were considered to be potential vaccine immunogens. As a result, researchers focused on artificial boundaries between the numerous overlapping epitopes present on the surface of viral proteins and disregarded the fact that a protein surface is an antigenic and functional continuum consisting of a large number of epitopes (Berzofsky 1985; Kunik and Ofran 2013; Sela-Culang et al. 2013).

The use of Mabs introduced another bias because the apparent specificity of a Mab very much depends on the selection process that was used to obtain it (Van Regenmortel 2011b). For instance, Mabs are often selected for their ability to bind to individual linear peptides in a peptide library that are believed to mimic a viral epitope. It should then come as no surprise that a Mab selected in this way will tend to bind more strongly to a peptide than to the virions used for eliciting the Mabs. When a bnMab binds to a short region of a virus protein, this could simply reflect the fact that the Mab was selected for its ability to bind to that peptide in an immunoassay. When short peptide regions of the HIV-1 gp41 MPER are found to bind to bnMabs, there is thus no reason to believe that these Mabs were induced by the peptides they react with and not by chemically heterogeneous, more complex or transient epitopes present in gp41 (Van Regenmortel 2012a, b).

When a bnMab is found by X-ray crystallography of a Mab- Env complex to bind to a discontinuous epitope, it is not possible to isolate the epitope in its active form by extracting it from the native protein in order to show that it possesses binding activity on its own (Van Regenmortel 2011b). Since a discontinuous epitope possesses binding and immunogenic activity only when it is embedded in a protein, its immunogenicity can only be tested by using the native protein as immunogen. This inevitably produces a heterogeneous response against the numerous epitopes present in the protein and makes it impossible to derive any useful information regarding the separate immunogenicity of the discontinuous epitope of known crystallographic structure (Van Regenmortel 2012a). Neutralizing Mabs are thus useful for analyzing the antigenicity of HIV-1 proteins but not for studying the immunogenicity of the discontinuous epitopes they recognize.

Another limitation inherent in the use of Mabs for helping to develop an HIV-1 vaccine arises from the polyspecificity of antibody molecules (Van Regenmortel 2014a, b; Dimitrov et al. 2013). In view of the degenerate nature of the immune system, there is no single intrinsically specific epitope for any Ab but only a diverse group of potential epitopes able to bind to it with various degrees of fit (Notkins 2004; Wucherpfennig et al. 2007; Eisen and Chakraborty 2010; Mariuzza 2006). Since the epitope structure determined by the crystallographic analysis of a bnMab-HIV Env complex corresponds to only one of the many epitopes that could be accommodated by that Mab, there is no reason why the epitope identified in the complex should indicate which immunogen induced the Mab, thereby making it an hypothetical vaccine candidate.

## What Does It Mean to Rationally Design an HIV-1 Vaccine?

Proponents of the rational design approach to HIV-1 vaccine development maintain that when they have determined the structure of a bnMab and its HIV-1 Env epitope, this information should allow them to design a vaccine immunogen that will elicit antibodies able to neutralize the virus. The rational design of an HIV-1 vaccine has been advocated for many years (Nabel et al. 2011; Bramwell and Perrie 2005; Douek et al. 2006; D'Argenio and Wilson 2010; Walker and Burton 2010; Burton 2010) and its adepts do not seem to be disheartened by the persistent failure of that approach to develop an HIV-1 vaccine. It may be instructive, therefore, to analyze the rationale of those who remain committed to such an unsuccessful strategy. This is best done by clarifying what vaccinologists actually mean when they claim that they are engaged in "rationally designing a vaccine".

Doing something by design is doing it intentionally. For instance, adepts of so-called "intelligent design" argue that a mythical, intelligent deity is responsible for designing all living forms on our planet and they pretend to explain the occurrence of life and evolution by the creationist credo: "God did it" (Van Regenmortel 2007). Such a belief is today rejected by most scientists because they no longer accept that living organisms and the biological functions they exhibit were designed by the preconceived plan of a designer rather than being shaped by the filter and pressure of Darwinian natural selection. It has been argued elsewhere (Van Regenmortel 2007) that design terminology is inappropriate both for explaining the appearance of biological organisms on earth and for describing the activities of vaccinologists when they empirically select a vaccine immunogen able to elicit antibodies endowed with the rather complex biological function of HIV-1 neutralization.

The term design refers to the deliberate conceiving of an articial, novel object or process by an intelligent being. In the case of a vaccine immunogen, it describes the structure-based strategy used in rational drug design (Kuntz 1992) when it is applied to the improvement of the binding complementarity between an antigen and an antibody in the narrow context of a single Mab-epitope pair. In this case, it is indeed justified to refer to the rational design of the HIV-1 Env glycoprotein with respect to its ability to bind a single bnMab molecule (Wu et al. 2010) or to the rational design of one HIV antibody molecule with respect to one HIV epitope (Diskin et al. 2011). In both cases, it is one epitope or one paratope that is being designed to achieve improved binding complementarity and the authors in fact only study the antigen binding capacity or antigenicity of a viral protein and not its immunogenicity. The claim that what is being rationally designed is actually an HIV-1 vaccine immunogen (Nabel et al. 2011) is based on the misconception that improving the binding capacity of one viral epitope with respect to a single Mab amounts to designing an immunogen capable of generating protective antibodies. This confusion is due to the assumption that studying the binding capacity of bnMabs can reveal which HIV-1 immunogens are able to elicit neutralizing antibodies. It is ultimately the consequence of confounding antigenicity and immunogenicity and of assuming that if an HIV-1 epitope reacts with a bnMab, it should also be able to induce similar, protective antibodies in an immunized host. In the case of HIV, the enormous antigenic variability of the virus and the requirement for extensive antibody affinity maturation to obtain neutralizing antibodies invalidates such an approach that may however work with other viruses or antigens. On the other hand, the positive results obtained when rational vaccine design was used with respiratory syncytial virus (Correia et al. 2014) may not be a valid proof of concept for what can be expected to happen in the case of HIV-1.

It is actually not feasible to rationally design an HIV-1 vaccine immunogen because the designer would need to know beforehand which features and components of the human immune system are responsible for the induction of a protective immune response. Since such knowledge is not available, it is impossible for a vaccinologist to conceive what type of immunogen is capable of eliciting protective antibodies and how its structure must be intentionally optimized in order to bring about the desired outcome. What is possible, of course, is to investigate the numerous factors, external to epitope-paratope recognition, that originate in the immunized host and control which antibodies will be produced by its immune system This would require studying the host immunoglobulin repertoire, the presence of appropriate B cell receptors and cytokines, the occurrence of antibody affinity maturation and various cellular and regulatory mechanisms present only in the context of a given immune system (Van Regenmortel 2011b). Such studies require trial-and-error experimentation involving characteristics of the host and they fall outside the framework of a rational design strategy taken as a purely molecular engineering project for constructing effective vaccine immunogen molecules solely on the basis of the antigenic properties of the virus.

The task of a designer is to pose and solve an inverse problem, namely to imagine, using available knowledge, what would bring a desired outcome (Bunge 2003). Various possible solutions can then be tested by trial and error experimentation until the preset goal is attained. It is thus misleading to oppose structure-based rational design and empirical approaches in HIV-1 vaccine research as if design is better science than empiricism (Karlsson-Hedestam et al. 2008). Trial-and-error testing is entirely rational and forms an integral part of any attempt to develop an effective HIV-1 vaccine (Van Regenmortel 2011b).

## Conclusions

The structure-based rational design of an HIV-1 vaccine is based on the assumption that knowledge of the structure of vulnerable Env epitopes targeted by bnMabs will lead to the discovery of effective HIV-1 vaccine immunogens (Nabel et al. 2011; Van Regenmortel 2011b, 2012a, 2014b). After many years of unsuccessful experimentation by numerous laboratories, Mascola and Haynes in 2013 concluded that "our best attempts to construct vaccine immunogens that present these key epitopes to the immune system have failed to generate antibodies that neutralize most strains of HIV-1" and they acknowledged that "a structure-based approach in and of itself will likely not solve the HIV-1 vaccine problem" (Mascola and Haynes 2013).

The unsuccessful RV paradigm utilizes the deceptive metaphor of a rational design terminology which is as inappropriate for designing improved vaccine immunogens as it is for attributing Darwinian natural selection to the activity of a designer involved in developing useful biological functions (Van Regenmortel 2007). The RV paradigm has lost its popularity with HIV vaccinologists who no longer believe that HIV-1 epitopes of known structure can be directly transformed into effective vaccine immunogens. A new paradigm has replaced it which stresses that vaccine immunogenicity depends mainly on the biology of the immunized host. This paradigm advocates the study of bnAb lineages in infected individuals and their development from early infection through the process of antibody maturation to eventual potent neutralization activity. The aim is to drive the human immune response towards highly mutated anti-HIV-1 bnAbs by using sequential immunizations with various Env immunogens. Several teams have successfully analyzed the paths along which anti-HIV-1 bnAbs develop (Bunge 2003; Karlsson-Hedestam et al. 2008; Haynes et al. 2012a; Moore et al. 2012; Klein et al. 2013; Liao et al. 2013a; Verkoczy et al. 2014; West et al. 2014; Doria-Rose et al. 2014) but it is not clear yet whether unravelling large numbers of different antibody maturation pathways (Prabakaran et al. 2014) will make it possible to identify a series of effective immunogens suitable for vaccinating large human populations.

# Old and New Concepts and Strategies in HIV Vaccinology: A Report from a Workshop Held in Rome on 17 June 2016

## Introduction

The HIV pandemic has a detrimental impact that is vast and ongoing, despite therapeutic progress. An estimated 37 million people were living with HIV/AIDS in 2014, most of whom did not know their HIV status and an estimated 1.2 million people died of AIDS related illnesses during that year (UNAIDS 2015). The brunt of this impact is borne by vulnerable and disadvantaged groups, such as those found in Sub-Saharan Africa, which accounts for about 70% of those living with HIV/AIDS and represent an important site of new infections.

Potent biomedical interventions are required to stem the more than two million new cases of HIV seen each year, a stark reminder that our current armament is deficient, and that the development of an HIV vaccine remains a priority. In fact, given the scale of this harm, the benefits of prophylactic interventions such as vaccination (and other interventions with therapeutic or prophylactic effects) to reduce the disease are potentially enormous. This is especially so if we consider that benefits to the worse off have greater value than benefits to the better off, on egalitarian or other grounds (Arneson 2013).

However, in spite of the substantial efforts and resources that have been devoted to curbing the pandemic during the last 30 years, it has not yet been possible to develop an effective vaccine. Initially, AIDS was perceived as a disease that destroys

Journal of AIDS & Clinical Research, 2016, 7, 634.
Ensoli, B., Cafaro, A., Amicosante, M., Andrieu, J.M., Boyer, J.D., Garcia, F., Gray, G., King, M.R., Rios, A., Sandstorm, E., Van Regenmortel, M.H.V

© Springer Nature Switzerland AG 2019
Marc H V Van Regenmortel, *HIV/AIDS: Immunochemistry, Reductionism and Vaccine Design*, https://doi.org/10.1007/978-3-030-32459-9_20

the immune system and this seemed to exclude the possibility of devising an effective method of immunological prevention. As we became more familiar with its physiopathology, it was realized that afflicted subjects did not die immediately as a result of HIV infection. Rather, the opportunistic infections and tumors that caused such havoc and terror during the initial years of the pandemic occurred at the end of the natural course of a disease which, in the absence of therapy, actually spanned years of asymptomatic infection. By way of contrast, we only have to compare it to the natural course of an emergent infection such as Ebola, to clearly see that the natural course of HIV disease is one of a chronic disease and that HIV-infected subjects do not die of acute disease but rather from the consequences of chronic HIV disease. As a result of this perception, for many years HIV disease was characterized by its terminal rather than by its initial phase. This characterization led to many studies being devoted to the specific properties of the virus and of the immune response present at the end rather than at the beginning of the infection. Highly mutated viruses that can only be neutralized by rare antibodies possessing unique structural characteristics that make them broadly neutralizing, became the "holy grail" target of massive HIV vaccine efforts. Unfortunately, these searches failed to deliver a successful HIV vaccine. In recent years, considerable information has become available about the early stage of HIV infection. Nonetheless, investigators are still embarked today in the search for antibodies that appear very slowly following intense internal evolutionary pressures in individual hosts that bear little relationship with the situation during the initial infectious episode.

To date, the only HIV vaccine to show modest efficacy was the RV144 trial, conducted in over 16,000 heterosexual adults in Thailand (2003–2009), which indicated that a vaccine approach could reduce HIV acquisition and was a major lift for the field (Rerks-Ngarm et al. 2009). Even though modest efficacy was seen, the RV144 HIV vaccine regimen [a canarypox vector (ALVAC) prime, followed by protein boosts with an alum adjuvant] induced no neutralizing antibody responses, and protection appeared to be correlated with binding antibodies (e.g. IgG antibodies to the V1 V2 region of gp120), polyfunctional CD4 T cell responses as well as other non-neutralizing antiviral activity (Corey et al. 2015). These findings were a surprise, and a reminder of the importance of advancing candidates to further clinical development. In fact, before this vaccine trial was initiated, numerous renowned and influential HIV researchers published an opinion letter in Science calling into question the rationale for this study of combining two vaccines that had both failed in prior human trials to generate immune responses that they were designed to elicit. This letter stated that spending $119 million when "the overall approval process lacked input from independent immunologists and virologists who could have judged whether the trial was scientifically meritorious" was an ill-advised use of precious resources (Burton et al. 2004a). Thus, an apparently scientifically unsound approach with no former evidence of potential efficacy proved effective at reducing transmission and this protection was not conferred by neutralizing antibodies, which most vaccinologists still consider an absolute requirement for preventing HIV acquisition.

But there are also other obstacles to HIV/AIDS vaccine development, as Mike King points out: "It is apparent that HIV vaccine development is freighted with many empirical, scientific, philosophical and applied ethical considerations. Factors that delay scientific progress towards vaccine development risk increasing the total harm of the HIV/AIDS epidemic. This can be thought of as a moral form of opportunity cost. Delays can come from a variety of sources, including lack of political, social, ethical, and financial support, as well as inefficient organisation of global HIV research efforts, pursuit of ineffective research programs, and failure to pursue more effective options for vaccine development. Although competing values may explain or justify some impediments, the immense moral value of an effective vaccine remains. Some impediments impose the moral opportunity cost of foregone counterfactual benefits (greater good that could have been created through other options). As an example of this, it has been argued that structure-based reverse vaccinology (SBRV) was an ineffective research strategy for HIV vaccine development, and that other approaches should have been used instead (Van Regenmortel 2012a, 2016b).

The more those working in HIV vaccinology are able and willing to consider and discuss their work in all these terms, and not restrict their focus to vaccine science narrowly conceived, the more likely the great good that vaccinology is capable of achieving can be realized.

A number of investigators have proposed outlier approaches and have investigated unconventional HIV/AIDS vaccine candidates, some of which were discussed at the Rome Workshop. Participants confronted their opinions regarding which approaches were most likely to succeed and their viewpoints are summarized below.

# Question 1: What New Approaches Should Be Followed in HIV-1 Vaccinology?

### Barbara Ensoli and Aurelio Cafaro (BE and AC)

It is apparent from the course of the infection that HIV-1 is immunogenic and capable of inducing strong cellular and antibody responses. CD8 responses (CTL) are responsible for bringing virus replication down to the viral setpoint, indicating a partial control, which eventually is overcome, and the infected individual progresses to AIDS if not treated with antiretrovirals. Anti-HIV antibodies do not appear to play a clear protective role throughout the entire course of the infection, although broadly cross-neutralizing anti-Env antibodies do develop in a minority of chronically infected patients a few years after infection, too late to be effective, as indicated by the rapid appearance of escape mutants. Natural immunity may also play a role and contribute to the relative resistance to HIV-1 infection, as shown by a low HIV-1 transmission rate. The crucial question, however, is whether it is possible to protect individuals from infection.

For many years, the prevailing view was that it should be possible to develop preventative vaccines capable of inducing the same cellular and antibody immune responses observed in the course of HIV infection. By pre-arming people with anti-HIV immune responses, it was hoped that one could prevent them from either getting infected (mainly due to neutralizing antibodies against HIV Env) or that it would be possible to control and clear the infection soon after exposure, mainly by anti-HIV CTL. An alternative view was that it may be possible to modify HIV antigens and facilitate the recognition of numerous virus subtypes.

Over the years, these approaches have substantially failed, with the notable exception of the RV144 trial. New immunogens (Env trimers) and delivery systems (adenoviruses, adeno-associated viruses, attenuated CMV vectors, just to mention a few) have been investigated, with the goal to improve specificity, strength and durability of responses, and to elicit new type of responses (MHC-E and class-II restricted CTL responses, as in the case of the CMV vector). In this regard, the protection afforded by MHC-E and class-II restricted CTL responses in the preclinical, attenuated CMV vector model (Hansen et al. 2013, 2016) or, even more surprisingly, by the induction of suppressive CD8 T cells (Andrieu et al. 2014) (see also response to Q2 and Q4) indicate that vaccine efficacy may result from the induction of completely novel and unexpected responses.

We took a different approach, briefly summarized herein. To start, we hypothesized that it might be easier, when studying natural infection processes, to learn which responses are associated with a milder infection (i.e., low progression) that could perhaps be countered by vaccination. Thus, we decided to target the HIV Tat protein, which seldom elicits specific antibodies during natural infection although, when anti-Tat antibodies do develop, progression to disease is significantly delayed (Rezza et al. 2005; Bellino et al. 2014). The Tat vaccine was first tested in nonhuman primates, demonstrating protection from overt infection after intravenous challenge with the pathogenic SHIV89.6P (Cafaro et al. 1999). Protection was prolonged (2 years), resisted to activation of macaques' immune system by tetanus toxoid boosting (Maggiorella et al. 2004), a stimulus known to burst virus replication and controlled virus replication to undetectable plasma levels when macaques became overtly infected upon rechallenge with a fivefold higher dose of the same SHIV, while all controls progressed to simian AIDS (Borsetti et al. 2009). These results were further confirmed in subsequent preclinical studies, which also revealed the contribution of the Major Histocompatibility Complex (MHC) and of virus challenge dose to the outcome of the experimental infection in cynomolgus macaques (Cafaro et al. 2010). These findings prompted us to evaluate the Tat vaccine in humans, and preventative and therapeutic phase I studies were conducted, which confirmed the safety and immunogenicity of the vaccine in both healthy individuals and infected subjects (Ensoli et al. 2006, 2008, 2009; Bellino et al. 2009; Longo et al. 2009). Moving onwards, we decided to test whether the Tat protein vaccine, administered to infected individuals undergoing effective cART and who were negative for anti-Tat antibodies was able to elicit anti-Tat immunity and protective anti-Tat antibodies. We reasoned that in such individuals, cART may have restored the immune system sufficiently to lead to a vaccine response. Furthermore, the initial lack of anti-Tat antibodies may have avoided potential interference by a pre-existing immune response, which could have made it easier to observe evidence of efficacy in a therapeutic setting compared to a preventive one, as indeed we found (Ensoli et al. 2010, 2015, 2016a, b) and discuss later (see response to Q4).

### Massimo Amicosante (MA)

Information relative to host-HIV interaction has greatly increased in the past decade and has given us a better understanding of the intimate interaction between HIV proteins and their targets in host cells as well as of the role that highly variable human genes play in controlling HIV replication. In addition, studies focused on HIVspecific immune responses have demonstrated that the innate immune response plays a major role in delaying HIV disease progression.

Novel vaccine approaches should be able to trigger efficient innate immune responses using both new adjuvants and specific targets. This might lead to the production of antibodies that could help identifying HIV immunogens suitable for either preventive or therapeutic vaccines.

### Jean-Marie Andrieu (J-MA)

Since most HIV-1 infections occur across the sexual or rectal mucosal barriers, it is important when designing HIV-1 vaccines to understand how the virus penetrates these vulnerable sites. Recently two new pieces of information have suggested alternative strategies for developing an HIV vaccine. The first piece of information is that as high as 80% of HIV-1 mucosal infections are established by only one infecting virus, called transmitted/founder virus (T/F virus). This T/F virus targets one CD4+ T cell (whereas other types of immune cells, such as dendritic cells and macrophages, are not initially the seat of productive infection). Moreover, the total number of different T/F HIV-1 particles seems to be very limited and each of them bears the same small number of specific amino acid (AA) "signatures" at certain strategic positions on the envelope glycoprotein (Oberle et al. 2016). Without those AA "signatures", the virus of infected "donor" patients cannot penetrate the mucosa of still uninfected "receivers".

The second piece of information (already hypothesized more than 10 years ago) is that an activated state of the CD4+ T cell is a prerequisite for productive HIV infection although in vivo, replication in quiescent CD4+ T cells is essentially nonproductive and generally abortive (Fauci 1996).

These two pieces of knowledge are also valid for the macaque model, i.e. quiescent CD4+ T cells are not easily infectable, whereas activated CD4+ T cells are the first cells to be productively infected by T/F SIVs that possess specific AA signatures on their Env glycoproteins.

### Jean Boyer (JB)

HIV vaccine development is a multidisciplinary task that requires the participation of virologists, immunologists, immunochemists, molecular biologists, geneticists, epidemiologists, statisticians and bioinformatics experts as well as the involvement of the pharmaceutical industry. In addition, it can only succeed if the search for basic immunological knowledge is intimately associated with the search for an effective, vaccine solution to what is the worst pandemic of modern times (Esparza and Bhamarapravati 2000; Esparza and Yamada 2007).

### Felipe Garcia (FG)

Most of the therapeutic vaccines that have previously been tested were previously used as preventive candidates using classical approaches such as whole inactivated virus or recombinant gp120 proteins. The capacity of these early vaccines to increase HIV-specific responses was limited and results were discouraging, as no consistent immunogenicity and no clear impact on viral load could be demonstrated. Subsequently, other approaches using innovative vectors such as DNA, recombinant virus and dendritic cells were investigated in exploratory trials with small numbers of patients. The best results were observed with DC-based vaccines (Lu et al. 2004; García et al. 2013). Several other strategies will be explored in the near future such as mRNA-based vaccines, coformulations with nanoparticles and various combined strategies for targeting virus reservoirs.

### Glenda Gray (GG)

Our knowledge and experience in HIV vaccine development has increased our capacity to design immunogens that induce competent immune responses. HIV vaccines have become a pivotal part of the prevention research agenda and we are optimistic that we can develop both a preventive and a therapeutic HIV vaccine.

Currently, only four vaccine approaches have been evaluated in 6 phase 2b/3 studies. Shortly after the announcement of the RV144 results, a public-private partnership (P5 Partnership) convened to develop the pox-protein regimen for sub-Saharan Africa. HVTN 097 evaluated the RV144 regimen in South Africa, a different setting to Thailand in terms of circulating HIV clade, predominant modes of transmission and average body mass index. Immunologic response in South Africans was comparable to, if not slightly better than, responses induced in Thailand.

The P5 supported the development of the pox-protein regimen for testing in sub-Saharan Africa. In order to better match the "Thai vaccine" to the sub-Saharan population, P5 redesigned the ALVAC vector with a clade C *env* insert and constructed a new bivalent clade C recombinant gp120 and tested this in HVTN 100 in South Africa, a phase 1 study. Based on promising immunogenicity from this study a large scale efficacy study, HVTN 702 is scheduled to commence at the end of 2016 in South Africa, which will evaluate this regimen in 5400 HIV uninfected individuals aged 18–35 years old.

Thus, it appears that a heterologous prime-boosting approach utilising viral vectors carrying carefully selected antigens that include *env* for priming and adjuvanted Env based protein(s) for boosting may be a good approach to induce a durable cellular and humoral immune response. To advance from the modest efficacy observed in the Thai trial, additional strategies should be considered, which may include the passive infusion of neutralizing antibodies or the utilisation of more potent vectors like Ad26, MVA or Ad35. Promising viral vectors receiving attention in preclinical development include CMV, while NHP challenge models provide an argument for advancing to clinical development.

## Mike R King (MK)

Among HIV vaccinology projects, priority should obviously be given to those approaches that seem to provide the most likely means for developing an effective HIV vaccine. Or, minimally, those approaches that have been found to be unsuccessful should be given a very low priority. Such decisions should be made through open, informed, unbiased evaluation of research projects. If new approaches seem to offer a better chance of successful development of a vaccine, they should be the ones that receive funding.

An important consideration is that research aimed at obtaining prescriptive knowledge, i.e., knowledge in the form of a new invention that allows one to do something of practical utility that was previously not feasible, should be prioritized. This is the type of knowledge required for obtaining a new vaccine which is nearly always obtained empirically by successfully manipulating and controlling the immune system (Van Regenmortel 2014a). Unfortunately, most selection committees tend to prioritize research that is likely to yield novel factual scientific knowledge (i.e., knowledge and discoveries in immunology) that reveals something that existed all along but was unknown to anybody), the reason being that they fund projects proposed by highly productive scientists who have excellent publication records and made important discoveries in basic science. However, curiosity-driven basic research, undertaken without a commitment to the aim of inventing something of such enormous human importance as an effective HIV vaccine, may be unable on its own to lead to the required prescriptive knowledge that can only be obtained empirically (Van Regenmortel 2016b). This is not to deny that other good outcomes of research, such as factual, scientific knowledge (whether practically useful or not), are also valuable. However, in a research environment that is finite in its resources, it seems reasonable to prioritise research that is aimed at generating both factual and prescriptive knowledge, which means that the research should not only generate new immunological knowledge but should also strive to obtain the prescriptive knowledge that is only achievable by intervening empirically in the immune system (Van Regenmortel 2012a).

This would require that current and future proposed research programs should be evaluated in a way that assesses their possible outcomes in scientific, but also ethical terms, taking the issue of AIDS seriously as a major public health matter (Esparza 2013a). This requires the involvement of scientists, ethicists, as well as all the stakeholders who stand to potentially benefit most from the research.

## Adan Rios (AR)

With the new understanding of how HIV is transmitted by a single variant in most sexual mucosal transmissions, a strong case can be made that the focus of investigation should be on the initial events in the infection process. It is now obvious that the biology of HIV follows the pattern of an evolutionary process and that the initial transmission of the disease is not a random phenomenon but follows the universal evolutionary principle of strong natural selection. This means that understanding the specific characteristics of the Transmitted/Founder (T/F) virus that initiates the infection is likely to be crucial for developing a preventive HIV vaccine. The antigenic structure of T/F viruses may thus hold the key for identifying which immunogens should be used in a preventive vaccine since it is the neutralization of the viruses that are present initially and can spread easily from cell to cell that should be achieved in order to prevent further viral dissemination.

There is good evidence that the current, improved methods of HIV inactivation offer a unique possibility of developing vaccines that would target specifically the T/F viruses responsible for initiating the infection. Paraphrasing Willie Sutton's reply to the question: "Why do you rob banks?" "Because that's where the money is!" It seems equally relevant to direct a preventive HIV vaccine to where it matters most. The possibility of using safe, chemically inactivated HIV creates a biological scenario that allows for a potential new beginning in the search for a preventive HIV vaccine, either by rational design, notwithstanding its failure thus far or by the use of empirical strategies that have a past track record of success (Sheppard and Dorman 2015; Rios et al. 2015, 2016).

Given that over the last 30 years, there have been only six clinical trials of potential HIV vaccine candidates, all based on rational vaccine design, with only one trial showing a less than modest success due to the inherent difficulties in the approach, it could be argued that there are compelling scientific and moral reasons to explore the potential development of the HIV inactivation approach. This argument was made painfully valid during the last 2016 Durban International AIDS Conference, where it became clear that once again timetables and milestone projections to cure HIV are far from being achieved and are in fact becoming a cruel mirage (Table 1).

**Eric Sandstrom (ES)**

In the search for 'new' approaches, it is often forgotten that when the 'new' gets well documented, it soon is no longer new and may fall by the wayside to be replaced by other 'new' approaches; as a result the initial new approach will not get the benefit of a definitive evaluation. This is the case with the DNA prime MVA boost regimen that we have pursued for a number of years (Cebere et al. 2006; Hanke et al. 2007;

**Table 1** Factors affecting the UNAIDS "ending AIDS" goal (Cohen 2016)

1. Worldwide funding for ARV therapy is becoming increasingly and woefully inadequate
2. South Africa has 6.6 million persons infected, of which 3.4 million are on ARV. With a decline in gross domestic product and currency devaluation, expenses are increasing making more difficult to sustain the current expenses
3. Worldwide new yearly infections have remained at nearly two million for 5 years
4. Eastern European and several Asian countries where new infections jumped 57% between 2000 and 2015 only have 18% of the infected persons currently receiving ARV
5. There is a looming crisis of insufficient supply of ARV as profit margins of generic companies' compromise production capacity
6. Only 30% of HIV infected persons in the United States fully suppress the virus and the number is far lower in many countries, making the UNAIDS's goal of the "90-90-90" by 2020 clearly unrealistic
7. Infected children and adolescents constitute an increasing burden because of limited access to care and poor retention in therapy

Sandstrom et al. 2008). In spite of similar or higher immune responses reminiscent of the correlates of protection found in RV144 (such as impressive CD8 induction and a durable immune response that can be boosted by a single late MVA administration), it has proven impossible to find resources for another efficacy trial. If indeed laboratory assays suggest that there may be protection from infection, then an efficacy trial is needed to corroborate these assays and provide the material for further investigation of possible correlates of protection. If, on the other hand, there is no protection, then an in depth reevaluation of all the criteria for selecting which vaccines should move forward for efficacy trials is mandatory. Not to proceed to an efficacy trial of these vaccine candidates remains a major missed opportunity.

**Marc Van Regenmortel (MVR)**
It is essential that HIV-1 vaccine developers critically examine the validity of the research paradigms and conceptual frameworks they use when trying to develop a preventive HIV-1 vaccine. For more than 10 years the paradigm of structure-based reverse vaccinology (SBRV) proposed by Dennis Burton (2002) was pursued by large teams of investigators, although it was not based on sound immunological theory (Van Regenmortel 2016b). SBRV assumed that it should be possible to generate an HIV-1 vaccine by first determining the crystallographic structure of a complex between an HIV-1 Env epitope and a neutralizing (n) monoclonal antibody (Mab) and then reconstructing the epitope by reverse molecular engineering outside the context of the native Env, to fit the nMab. I was hoped that the reconstructed Env would have acquired the immunogenic capacity to elicit a polyclonal response with the same neutralizing ability as the Mab used as template. The proponents of SBRV called this approach "rational vaccine design" although they were only improving the capacity of an epitope to bind one particular Mab (i.e., its antigenicity) and were actually not designing a hypothetical immunogenic epitope able to elicit Abs endowed with the same neutralizing capacity as the Mab. Antigenicity was simply confounded with immunogenicity (Van Regenmortel 2012b).

When an antigen or epitope is introduced in a host immune system (IS), it becomes known as an immunogen, although it is of course the IS that produces the antibodies, the epitope being only a triggering agent that initiates a chain of reactions in the IS. This is only successful if the given host IS also possesses B cell receptors that can recognize the immunogen, as well as various types of T cells and other regulatory mechanisms. Reverse vaccinologists, however, assume that these required features are always present in the individual IS to whom they administer the engineered epitope, although this is obviously not the case (Xiao et al. 2009; Dimitrov 2010). Since they only considered the specificity of the epitope- Ab interaction and not the relevant biological characteristics of the IS that allows it to produce neutralizing Abs, reverse vaccinologists never succeeded in obtaining an effective HIV vaccine. They also disregarded the fact that the epitope engineered by SBRV is only one of the many epitopes that the polyspecific nMab is able to recognize, which means that their engineered epitope is not necessarily the one that elicited the nmAb and should not be expected to be able to elicit protective antibodies (Van Regenmortel 2016b).

The SRBV approach used in hundreds of studies thus failed to yield a preventive HIV-1 vaccine (Van Regenmortel 2012a; Pejchal and Wilson 2010; Kong and Sattentau 2012) because it did not take into account that Abs are polyspecific and that the antigenicity and immunogenicity of a viral protein are different properties that are not necessarily located in the same regions of the protein (Van Regenmortel 2004a).

In a similar way, the success of the currently popular paradigm in HIV-1 vaccine research which assumes that it may be possible, by sequential immunization with various Env immunogens (Doria-Rose and Joyce 2015), to drive human immune responses towards the production of highly mutated anti-HIV nAbs will depend on whether the stochastic nature of successive mutations in Ab genes can indeed be controlled to achieve the required degree of Ab affinity maturation in large numbers of human vaccinees.

# Question 2: Should HIV-1 Inactivation Be Reconsidered?

**BE and AC**

Recent technical progresses make chemically inactivated HIV a valuable vaccine candidate (Rios et al. 2016). However, just to pre-expose an individual to a nonviable form of the virus may not be sufficient to induce the protective responses that occur when he/she gets exposed to the viable counterpart. Many variables make it difficult to predict the outcome of using the different delivery systems and adjuvants that need to be tested. It must also be kept in mind that a preventive vaccine against HIV/AIDS is intended for the general population, whereas a therapeutic vaccine will be directed to a restricted number of people with altered immune systems.

**MA**

The use of whole virus particles as vaccine presents a number of advantages compared to purified viral components or synthetic HIV peptides. Both the modalities of HIV-inactivation and the type of HIV used (wild type strains or modified strains carrying different viral proteins) may influence viral antigen preservation and determine whether an efficient and protective immune response is obtained. This is particularly critical for preventive vaccines where an efficient protective antibody response is likely to be crucial for blocking virus.

**J-MA**

During the last 5 years, Wei Lu and I developed a strategy, based on inactivated virus, that was tested in the Chinese (Guangxi) macaque model to determine whether it was possible to suppress the activation of SIV-positive CD4+ T lymphocytes in vivo in order to prevent SIV replication and protect the animals from virus challenge (Andrieu et al. 2014; Lu et al. 2004). We developed a new oral vaccine made of inactivated SIVmac239 adjuvanted by *Lactobacillus plantarum*, a commensal bacterium of the digestive tract known to be associated with immune unresponsiveness/immunological tolerance. We also used *Lactobacillus rhamnosus* (a commercially available Lactobacillus prepared under GMP conditions) as well as the bacillus of Calmette Guerin (BCG). In contrast to all known vaccines, these oral vaccines elicited neither SIV-specific antibodies nor cytotoxic T-lymphocytes. Instead, they induced a previously unrecognized population of non-cytolytic MHCIb/Erestricted CD8+ T-regulatory cells that specifically suppressed the activation of SIV positive CD4+ T lymphocytes. The suppression of activation also interfered with SIV reverse transcription in CD4+ T cells, thereby preventing the initial burst of virus replication, which in turn protected the macaques from infection. Of the 24 macaques orally vaccinated in this way and challenged intra-rectally 3–14 months later with a high dose of SIVmac239 or with the heterologous strain SIVB670, 23 remained solidly protected for up to 5 years whereas all 24 control macaques became infected (Andrieu et al. 2014). This approach will be investigated in humans in the near future.

A second strategy based on inactivated virus which is potentially applicable for constructing a prophylactic vaccine (to our knowledge this has not yet been tested) would be to immunize macaques, and later humans, with a vaccine based on a mixture of several killed/inactivated T/F viruses (Rios et al. 2016). Presently available technologies (i.e., sequencing, cloning, gene modification, cell culture and different methods of virus inactivation/killing that do not modify the Env glycoprotein) should allow the preparation of T/F HIV-1-based vaccines. Such vaccines may be the most promising candidates for inducing antibodies able to neutralize the T/F viruses that have entered the mucosa but not yet their target CD4+ T cells.

## FG

It has not yet been demonstrated that vaccination with whole inactivated HIV can be successful although some encouraging new data have been obtained. This might be particularly suitable for DC-based vaccines that have shown promising results. We have recently started a new clinical trial with an intranodal DC-based vaccine pulsed with whole inactivated virus.

## GG

The most important consideration for HIV inactivation or a killed virus approach would be immunogenicity, and whether this approach would yield an immunogenic vaccine. Novel approaches to inactivate the HIV vaccine should continue and be evaluated in a pre-clinical challenge model to demonstrate proof of concept efficacy. Inactivated HIV vaccine approaches have, to date been limited by scientific, technical and sociological issues (Sheppard 2005). Use of physical and chemical elements to inactivate HIV have impacted on the structural integrity of the HIV virus (Seo 2015) and other technologies such as radiation technology or the use of "light" therapy may leap-frog this approach, making for a cheap and effective vaccine, should issues of safety and immunogenicity be resolved.

A live-attenuated, or genetically attenuated HIV vaccine approach, may solve the issue of immunogenicity, but safety issues abound, limiting its evaluation in humans. Future work aimed at ensuring the stability of a genetically attenuated HIV vaccine should be funded, as this approach may lead to high vaccine efficacy that is not attainable using current methods.

**MK**

There is good reason to believe that HIV-1 inactivation may offer a viable approach to HIV vaccine development. This strategy was initially rejected on grounds of safety and the initial scepticism still persists today in spite of the development of greatly improved inactivation techniques that preserve the structure and antigenic properties of the Env glycoprotein (Sheppard and Dorman 2015; Rios et al. 2015, 2016; Sheppard 2005; Esparza 2015). Research funding agencies as well as ethics committees and institutional review boards should thus reconsider their earlier systematic rejection of projects utilizing chemically inactivated HIV. In addition to considering any possibility of harm to trial participants, review boards have an obligation to allow promising research (even if previously considered too risky), provided it is conducted in a scientifically and ethically robust way and offers the possibility of promoting the development of an HIV vaccine.

**ES**

The use of chemically inactivated HIV in a vaccine could be considered if it can be demonstrated that the vaccine is able to induce neutralizing Abs that bind to naturally folded Env glycoproteins, provided the viral spikes are not depleted by the inactivation and purification procedures used for preparing the vaccine.

## Question 3: Can Therapeutic Vaccines Help the Development of Preventative Vaccines? Should Therapeutic Testing Precede Testing in Healthy People?

**BE and AC**

This is certainly the case since it is exactly what has been done in the case of the Tat vaccine. If a vaccine is safe, immunogenic and effective in a person with an immune system that is incompletely restored, it is plausible that it may work in a healthy individual. In addition, if vaccine efficacy is first evaluated in therapeutic trials, this could significantly speed up the development of preventative vaccines, since (1) a smaller number of volunteers may be needed to assess efficacy, (2) trials may be of shorter duration which would allow faster comparisons of multiple vaccine candidates, (3) logistic and infrastructure requirements may be reduced which would make testing more feasible and less costly in developing countries. The adoption of an adaptive design of the trial may further speed up the vaccine development process (Corey et al. 2011).

**MA**

Information obtained during trials of therapeutic vaccines together with other immunotherapeutic treatment of HIV might lead to the identification of novel approaches for blocking virus entry and replication. In view of the different stages of HIV infection, going from persons with a fairly healthy condition to people with advanced disease, it may for safety reasons only be acceptable to evaluate certain products in non-infected HIV subjects. Safe efficacy trials could, however, be designed in HIV-infected persons.

## FG

The therapeutic vaccine pipeline is complementary to the development of preventive vaccines. Since immune responses that prevent infection are different from those that help to control the viral load, the development of both types of vaccines could run in parallel, and findings obtained with the one vaccine may be relevant for the other type of vaccine.

## GG

Therapeutic vaccines can assist the development of preventive vaccines. Understanding the mechanisms which control viral replication may lead to a vaccine approach that may assist in the attenuation of disease post HIV infection by lowering viral load setpoint, viral control and immune preservation. Concomitant evaluation in both HIV infected and uninfected participants will advance the clinical development of these candidates.

## MK

Testing therapeutic vaccines may offer some advantages compared to the testing of preventive vaccines in healthy people. Trial participants who are already HIV-positive may view the potential benefits of a vaccine more favourably than do HIV-negative individuals who expect only the benefits of prophylaxis and this may facilitate the recruitment of participants in therapeutic trials. A draw-back is that such trials usually require an interruption of antiretroviral therapy which may be perceived as entailing a certain risk. This could be mitigated by recruiting individuals for whom ART is less effective because of the development of drug resistance. On the other hand, participants in preventive vaccine trials risk being exposed to HIV infection, which is an additional consideration for being able to evaluate the effectiveness of prophylaxis in comparison to the control group (Guenter et al. 2000). Such a risk is obviously not present in therapeutic vaccine trials with HIV-positive participants.

The perception of risk taking by potential trial participants is influenced by numerous factors, trial design being one of them (Mills et al. 2004; Dubé et al. 2015). Regardless of what type of vaccine is being tested, researchers should of course strive to build trust among communities in which they work. Open, understandable, and consistent communication about the risks and potential benefits of trials, including community and broader societal benefits, is important, and not only when seeking informed consent (Henderson 2015; Dubé et al. 2014).

It should also be borne in mind that it can be rational for individuals who participate in a trial to accept certain risks and that this should not be seen by ethics committees as being unreasonable (Buchak 2016; Wertheimer 2011; Largent 2017). The enormous social benefits of preventing HIV infection can be a value that participants and ethics committees must promote by allowing reasonable decisions to be made concerning both trial participation and granting permission of undertaking certain types of research (Wertheimer 2011).

### AR

Although it cannot be denied that the study of a therapeutic vaccine might shed some light on immunological responses to HIV, it is not clear whether such information would be relevant to immune responses occurring in healthy uninfected subjects and would inform the design of a preventive vaccine. It may therefore be counterproductive to use the results of a therapeutic vaccine trial as a gateway for developing a preventive HIV vaccine for healthy subjects.

In view of the efficacy of current therapies for HIV infection, additional advantages afforded by a therapeutic vaccine for patients undergoing interrupted ARV may perhaps be limited although this will of course only become clear once the trials have been conducted.

### ES

Generally therapeutic trials should not be required for developing a preventive vaccine. There are many difficulties in interpreting the results of therapeutic vaccines since the immune system may already be too altered in order to respond optimally when the burden of virus is too great or the virus has already reached many hidden reservoirs. Thus a potentially effective vaccine might erroneously be ruled out because of the failure a therapeutic vaccine trial. However, if preclinical work has indicated that an established infection can be curbed, moving to a preventive vaccine may be a rational option.

## Question 4: Do You Think That Therapeutic Vaccines Could Lead to a Functional Cure?

### BE and AC

Felipe Garcia was the first to show a significant, although transient, control of viremia by the use of a DC-based personalized therapeutic vaccine (García et al. 2013). This was the first step towards a functional cure, i.e., a condition of long-term remission without need of therapy. We took a different approach by intensifying cART with the Tat vaccine (Ensoli et al. 2010, 2015, 2016a, b). Apart from the amelioration of several immune parameters (most noticeable were the restoration of CD4 T cell and B cell numbers beyond the cART-induced level, and the restoration of central memory CD4 and CD8 T cells accompanied by a concomitant decrease of effector cells) we also observed an increased decay of proviral DNA. DNA decay started late (3 years) after vaccination, but continued to decline 8 years after vaccination, as indicated by the increasing number of volunteers (30 out of 92, i.e., 33%) with undetectable ($<2$ copies/µg) proviral DNA; the proportion that was even higher for vaccinees who had received 30 µg Tat three times (12 out of 26, i.e. 46%), the most effective of the four regimens compared in the trial. To our knowledge this is unprecedented and very promising. It should be stressed that cART was never interrupted and may have contributed to viremia control and DNA decay. This is intriguing, because it would suggest that exit from latency may not be necessary to reduce the virus reservoir and that residual ongoing replication is key to replenish it. Of course, these results need confirmation and it should be emphasized that we only examined the blood, whereas virus reservoirs might be present elsewhere in the body.

Of importance, the immunogenicity and safety of this B-clade Tat vaccine was confirmed in a subsequent phase II study (ISS T-003, ClinTrials.gov NCT01513135) conducted in cART-treated South African adult volunteers, that is in a population with a different genetic background and mainly infected with a C clade virus (Ensoli et al. 2016a). Strikingly, B clade Tat induced anti-Tat antibodies with a kinetics and titers that mirrored very closely those observed in the Italian trial, antibodies that cross-recognized Tat from different clades (A, B, C, D) and cross-neutralized the Tat-mediated entry of both oligomeric B- and C-clade envelope in dendritic cells. This is a novel assay we set up upon the discovery that Tat binds oligomeric Env and enters dendritic cells through a Tat-mediated pathway involving Tat binding to integrins (Monini et al. 2012). Anti-Tat antibody titers correlated positively with neutralization. Tat vaccination increased CD4+ T-cell numbers, particularly when baseline levels were still low after years of therapy, and this had a positive correlation with HIV neutralization. Interestingly, in some patients non-compliant to cART (n = 24), vaccination contained viral load rebound and maintained CD4+ T-cell counts above study entry levels as compared to placebo, suggesting that Tat vaccine intensification of cART may indeed counterbalance and hopefully abrogate, the consequences for reduced adherence to treatment, including selection and transmission of cART resistant strains.

Another issue that we believe is very important is the ample cooperation, which we deemed as necessary and actually build up with the South African public authorities, to make the vaccine development programme a successful one. The agreement, signed in 2007 by the Directorate General for Development Co-operation (DGCS) of the Ministry of Foreign Affairs of Italy and the National Department of Health (NDOH) of SA, included, in fact, several components, all considered important to provide the South African government with instruments for undertaking preventive and therapeutic vaccine programs. In particular, the bilateral program foresaw three components: (i) to support the development and/or the strengthening of a network of clinical sites and laboratories located in the area of intervention capable of providing quality health care, particularly in the HIV/AIDS sector; (ii) to support the development of a GMP line of production to manufacture vaccines in South Africa; and (iii) the conduct of a Phase II therapeutic clinical trial with the Tat vaccine developed at ISS. All the three objectives were met, and both the United Nations Industrial Development Organization (UNIDO) and a Panel of Experts convened by the National Department of Health (NDOH) evaluated as highly relevant the program outcomes, advocating completion of vaccine clinical development and registration. To this end, a Private Public Partnership (PPP) is being established in SA.

Thus, we think it is ethically important not only to evaluate vaccine candidates in the countries that need them the most, but also to undertake all those initiatives that render the countries independent with respect to vaccine production and conduction of vaccination campaigns.

## MA

The complete clearing of HIV from infected subjects has been a "chimera" pursued for the past 30 years. The existence of small numbers of elite controllers and long term non-progressors among HIV–infected individuals gave rise to the hope that it may be possible to uncover which biological features or biomarkers in these subjects allowed them to evade or overcome the development of AIDS. It seems that these individuals were able to improve the immune response sufficiently to control the infection without drug treatment, thereby avoiding the problems linked to prolonged antiretroviral treatment. The limited success of therapeutic vaccine trials obtained in recent years has given rise to the hope that it might become feasible to replace ART treatment with therapeutic vaccines.

## J-MA

When attempting to develop therapeutic vaccines, vaccinologists face the problem that the presence of large number of HIV mutants in every patient is unlikely to allow the immune system to produce the variety of neutralizing antibodies required for completely blocking virus multiplication. This seemed to exclude the possibility that a classical immune response based on neutralizing antibodies could be achieved that would entirely stop viral replication in infected patients. On the other hand, if it proved possible to induce an immune response towards the unmutated HIV epitopes present in T/F viruses, it is conceivable that it might be possible to prevent HIV transmission from one patient to another or to interfere with the spread of the infection in individual subjects.

Among the promising current strategies tested in clinical trials, the Tat vaccine has not yet been shown to be able to completely stop viral replication in the absence of antiviral therapy, although there is evidence that it could increase the number of CD4+ T-cells and could slowly decrease the number of HIV DNA reservoir cells (Ensoli et al. 2010, 2015, 2016a, b). An alternative approach aims at reinforcing the capacity of dendritic cells to activate CTLs specific for Gag epitopes. We were partially successful with this strategy in 2002–2004 (Lu et al. 2003, 2004) and other groups are currently pursuing this approach that could perhaps be improved (García et al. 2013).

Another, alternative strategy, similar to the one we used with Chinese macaques, would be to suppress virus-specific CD4+ T cell activation with the objective of suppressing viral replication (Andrieu et al. 2014; Lu et al. 2012). At the moment, a therapeutic vaccine composed of inactivated HIV-1 associated with *Lactobacillus rhamnosus* has been prepared and is expected to be tested in 2017 with a group of 20 patients undergoing antiviral therapy. Three months after oral vaccination, the antiviral treatment will be interrupted and the ability of the vaccine to suppress viral replication will be determined.

## FG

It seems unlikely that a therapeutic vaccine could be completely successful on its own, and a functional cure may require the combination of several different approaches. Numerous pathogenic issues need to be considered such as the possible inability of the immune environment (because of inflammation and Tregs) to generate an effective immunity, the expansion of pre-existing clones targeting escape variants, ineffective DC antigen presentation and the existence of B cell follicle sanctuaries that would allow persistent, productive virus infection.

## GG

Therapeutic vaccines could lead to a functional cure by impacting on viral replication with an immune response that assists in the containment or eradication of latent viral reservoirs. HIV vaccination in combination with other agents that seek to stimulate the resting/latent virus may assist in functional cure. Immune based therapy may be a promising strategy in this respect.

## ES

The results of Lu et al. (2012) indicate that viremia in SIV-infected macaques that are depleted of HLA E restricted CD8 cells, are controlled when the CD8 cells return. Furthermore, Hansen et al. (2016) have demonstrated that a complete eradication is possible after induction of HLA E restricted CD8 cells.

Vaccines that induce HLA E restricted CD8 cells, therefore, hold great promise both for a functional cure and as part of "shock and kill" protocols aimed at eradication.

# Question 5: If You Had the Authority to Do It, What Vaccine Concepts Would You Support for Testing?

### BE and AC

Despite all our efforts to rationalize vaccine design, it must be accepted that only efficacy trials can demonstrate whether a vaccine is effective. The results of Jean-Marie Andrieu show that a tolerogenic vaccine may possibly be as protective as a sensitizing one, at least in the macaque model, indicating that it is extremely difficult to decide upfront what vaccine to go for. Rather than selecting *a priory* vaccine candidate on the basis of fashionable design, technical appeal, preliminary in vitro results or preclinical testing, it may be best to use a therapeutic setting in order to facilitate and speed up the clinical testing and comparison of multiple candidates. As Maurice Hilleman used to say, "The only correlate of protection is protection".

### MA

Although multiple-epitope based vaccines have not been found to be effective against HIV, it might nevertheless be possible to modulate and manipulate different arms of cellular immune responses (both specific and innate as in the case of NK cells) together with antibody responses, in order to achieve some degree of protection against HIV. This could be coupled with other interventions that may be able to lead to either the complete maturation of the immune response against HIV or could contribute to other critical features that could make it possible to control virus replication over time.

## FG

I think that Tat-based vaccines, personalized vaccines, mRNA and certain coformulations with nanoparticles should be explored. However, such prototypes need to take into account the numerous issues that were discussed in this workshop.

## GG

I would support any concept that has been found to be safe, addresses a novel approach with new immunogens or platforms with a different humoral or cellular immune response or correlates of protection elicited, than seen previously. Nonhuman primate challenge data supporting protection with putative correlates of protection that can be evaluated in the human would be advantageous for advancement into clinical efficacy.

## MK

Based on the points discussed previously, support should be given to concepts that hold the greatest promise for the development of an effective HIV vaccine, while also abiding by considerations dictated by the need for ethical conduct of research. Such an assessment is largely, although not purely, a scientific matter, and requires that the assumptions and hypotheses underlying any proposed research strategy should be critically examined by scientists free of any potential conflict of interest. The continued pursuit of the unsuccessful structure based reverse vaccinology approach that was based on unsound immunological hypotheses (Van Regenmortel 2016b) illustrates the need for critically appraising fashionable strategies even when they are followed by large teams of highly respected scientists.

## AR

For the reasons outlined above (questions 1 and 2), novel methods of inactivation and novel approaches for rationally designing an HIV vaccine, focused on the initial stage of HIV infection, should be investigated. Other strategies such as those developed by Barbara Ensoli and Jean-Marie Andrieu are also worth of further support and development since these concepts are backed by preliminary data that require continued investigation.

## ES

Given my bias stated under section "Introduction", I would rapidly proceed to an efficacy trial of the DNA prime MVA boost concept as the most rapid way to move the traditional HIV vaccine field forward. In parallel, a vigorous effort should be made to evaluate the new finding that HLA E restricted CD8 T cells are present during HIV infection. HLA E restricted CD8 T cells can home in on the most important targets for HV infection, namely T helper cells in germinal follicles to which HLA E and HLA B have poor access. Recently discovered discrepancies in the protective effect observed in Chinese and Indian rhesus macaque subspecies can possibly be accounted for by the relative inbreeding of Indian macaques in US colonies which could have severely restricted immune genetic regulation.

## Concluding Remarks

In the vaccine field in general and in the HIV/AIDS vaccine in particular, it seems that the rational design is gaining an increasing consensus and it is proposed as the only way to get a vaccine. The substantial failure of most efficacy trials and the moderate protection unexpectedly afforded by the ALVAC-HIV/AIDSVAX vaccine in Thailand has forced the scientific community, which had relied mostly on a rational design and preclinical efficacy results in a nonhuman primate model that mimic but does not reproduce entirely what happens in the human, to reconsider various issues of HIV vaccine development.

Here a few investigators in the field convened and discussed freely what they think is the way to go. The results presented in the workshop indicate that empirical approaches, although driven by the current knowledge of virus host interaction, may be as valuable, if not more, in the run for inventing an effective HIV/AIDS vaccine.

All the participants agreed that the natural history of infection is key to vaccine design, from the choice of the target (Tat, T/F viruses, etc.) to the identification of relevant immune responses to control/eradicate the virus and the discovery of relevant effector and regulatory responses in elite controllers. Recent acquisitions have clearly shown that the selection of viral antigens (biologically relevant viruses or proteins) and of the delivery system (dendritic cells, DNA, MVA, Adenoviruses, CMV vectors, bacterial vectors) are also critical although they could generate unexpected immune responses, underscoring the limitations of our current knowledge. This is illustrated, for example, by the still incompletely understood negative influence of pre-existing immunity to Adenovirus serotype 5 as well as by the newly discovered protective role played by both cytotoxic and regulatory MHC E restricted CD8 T cells in nonhuman primates. It is also apparent that the regimen utilized affects the outcome: the moderate success observed in the RV144 trial was obtained with a prime-boost regimen of two vaccine candidates that had both failed when given singly. The preclinical testing in nonhuman primates appears very controversial, and a reappraisal of what can actually be obtained and relied on in these models is needed. Conceivably, a similar reasoning applies to the novel and improved humanized mouse models, although they were not discussed here, possibly because they are still seldom used.

It also appears that vaccine development is a multidisciplinary task, requiring strict collaborations of investigators with expertise in different fields as well as the involvement of the pharmaceutical industry, always keeping in mind that the vaccine to develop has to be a feasible one, that is, easy to produce, store and deliver to the population.

The speed at which the HIV/AIDS vaccine field is moving forward is sometimes perceived negatively, since promising candidates (for example the DNA prime-

MVA boost approach) may be neglected because a promising new vaccine approach has come into the arena. Since the capability of evaluating vaccine efficacy is always limited, a selection must necessarily be made. However, recent data suggest that virus reactivation and reservoir replenishment in the chronic phase occur with modalities similar to those HIV exploits to establish the primary infection. This has led to the proposal that a therapeutic setting could be used to evaluate preventative vaccine candidates, since this would make it possible to compare them at an affordable cost (Cafaro et al. 2015; Cohen 2016). An adaptive design of the trial could further speed up the development process.

Evidence from the therapeutic trials presented at the workshop suggest that a functional cure may be achievable using different strategies by (1) instructing the host to control virus replication without cART, as demonstrated by Felipe Garcia with his DC-based vaccine, (2) suppressing the immune response against HIV as achieved by J-M Andrieu, and (3) attacking the hidden reservoirs of the virus that continuously generate new copies of viral proteins and virions, as shown by the cART intensification observed with the Tat vaccine.

Of all the discussed candidate vaccines, a modified version of the preventative vaccine tested in the Thai trial and a therapeutic Tat vaccine are advancing to phase III trials in South Africa. An intranodal DC-based therapeutic vaccine pulsed with whole inactivated virus will soon enter clinical testing, and a preventative vaccine based on a mixture of T/F viruses that have undergone a chemical inactivation that preserves the Env structure will also be evaluated in nonhuman primates in the near future. We all hope that these novel strategies will lead to an effective vaccine.

# Structure-Based Reverse Vaccinology Failed in the Case of HIV Because It Disregarded Accepted Immunological Theory

## Introduction

The development of bacterial vaccines using the reverse vaccinology (RV) approach was pioneered by Rino Rappuoli more than 15 years ago (Rappuoli 2001). RV refers to the strategy of using bioinformatics analyses of entire bacterial genomes in order to identify all the surface-exposed proteins that the pathogen is able to express. These antigens are then produced by high-throughput technologies and tested for their immunoreactivity with patient sera and for their ability to induce protective antibodies. The strategy was called RV because the investigators operate in a so-called reverse manner, i.e., starting from the genome rather than from the organism, to discover which proteins are potential vaccine immunogens. Compared to the classical approach of fractionating bacterial extracts to identify a small number of potential vaccine antigens, RV makes it possible to evaluate hundreds of expressed bacterial proteins for their capacity to induce a protective immune response. Furthermore, RV allows immunogens to be tested even if the bacteria cannot be cultivated. In an effort to develop a Meningococcus B (Men B) vaccine, 350 open reading frames (ORFs) out the 2158 ones present in the Men B genome were cloned in *Escherichia coli* and the proteins were purified in sufficient amounts to immunize mice. A total of 90 previously unknown antigens were discovered in this manner, of which 28 were able to induce antibodies (Abs) that killed the bacteria (Adu-Bobie et al. 2011, pp. 225–241). In earlier studies using fractionated bacterial extracts, only 12 Men B surface antigens had been identified, of which only four induced Abs with bactericidal activity. By sequencing the entire genome of Men B, the complete

International Journal of Molecular Sciences, 2016, 17, 1591–1622.
Marc H V Van Regenmortel

© Springer Nature Switzerland AG 2019
Marc H V Van Regenmortel, *HIV/AIDS: Immunochemistry, Reductionism and Vaccine Design*, https://doi.org/10.1007/978-3-030-32459-9_21

antigenic repertoire of the organism could be analyzed, demonstrating the enormous potential of RV.

## Genome-Based and Structure-Based Reverse Vaccinology

RV has been very successful for developing new bacterial vaccines (Rappuoli and Bagnoli 2011) but has failed in the case of HIV vaccines. The main reason for this is that HIV particles contain only a small number of proteins useful for vaccination purposes, which require a particular tertiary or quaternary conformation to be effective vaccine immunogens. As a result, the RV strategies used in bacteriology and in virology are completely different. In virology, RV refers to the strategy of trying to generate a vaccine by determining the crystallographic structure of a complex between a viral epitope and a neutralizing monoclonal antibody (nMab). Instead of generating neutralizing Abs by immunization with a viral antigen, investigators use a so-called reverse approach by starting from the known structure of nMabs (Burton 2002; Walker and Burton 2010). The nMab is then used as a template to reconstruct, outside the context of the native viral protein, the epitope recognized by the Ab using reverse engineering. The assumption is made that the reconstructed epitope designed to fit the nMab will have acquired the immunogenic capacity to induce a polyclonal Ab response endowed with the same neutralizing capacity as the nMab. Since the RV strategies used in bacteriology and virology are completely different, it has been suggested that they should be differentiated and called genome-based and structure-based RV, respectively (Van Regenmortel 2011a).

This structure-based RV approach has been used in hundreds of attempts to develop an HIV-1 vaccine using as templates broadly neutralizing (bn)Mabs that recognized the major antigenic sites of HIV-1 (Wyatt et al. 1998; Zwick et al. 2003a; Binley et al. 2008). Additional bnMabs have also been obtained from HIV-1 infected individuals by cloning antigen-specific memory B cells (Scheid et al. 2009) and these Mabs made it possible to dissect numerous epitopes located in the HIV-1 Env protein such as the conserved CD4-binding site, the CD-4 induced antigenic site and the V3 antigenic site, as well as epitopes located in the membrane proximate external region (MPER) of the HIV-1 gp41 protein. Numerous strategies were developed for improving the antigenic reactivity of these epitopes (Pejchal & Wilson 2010; Kong and Sattentau 2012) and although some of the engineered epitopes reacted better with the bnMabs used as templates, none of them were effective immunogens able to induce broadly neutralizing antibodies.

## Structural Vaccinology

The large number of protein antigens that can be studied by genome-based RV makes it possible to select native bacterial proteins that will be effective vaccine immunogens. In some cases, knowledge of the 3D structure of certain bacterial proteins helps to improve the efficacy of bacterial vaccines, as demonstrated with the factor H-binding protein of Men B (Scarselli et al. 2011) and the pili of group B Streptococcus (Pajon et al. 2012). This led to the use of the term "structural vaccinology" in the field of bacterial vaccines (Dormitzer et al. 2012).

In virology, structural vaccinology had been introduced about 10 years earlier (Burton 2002) and was called RV, although structure-based RV is probably a better term for it. Following the dramatic world-wide expansion of the AIDS epidemic, considerable funding became available for studying HIV-1, and large teams of investigators embarked on the search for a much needed HIV-1 vaccine.

In view of the considerable knowledge that was available regarding the structure of immunoglobulins and of HIV-1 (Sundberg and Mariuzza 2002; Wyatt and Sodroski 1998; Zwick 2005; Hoxie 2010; Mascola and Montefiori 2010; Klasse 2012), many scientists believed that structure-based RV was a realistic strategy for developing an HIV-1 vaccine. However, after more than a decade of intensive research funded by billions of dollars, the outcome is very disappointing since little tangible progress has been made in the search for a preventive HIV-1 vaccine (Pejchal and Wilson 2010; Kong and Sattentau 2012; Van Regenmortel 2012a; Burton et al. 2012).

There are many reasons why the successful strategies used for developing viral vaccines in the past failed in the case of HIV-1 (Hilleman 1992, 2000; Kusters and Almond 2010; Virgin and Walker 2010). The natural immune response in HIV-1 infected individuals, for instance, does not clear the infection, which means that a successful vaccine must achieve something that the human immune system (IS) is not capable of doing when it encounters the virus. There are other major impediments namely that (1) HIV-1 integrates into the host genome and conceals the virus from immune recognition; (2) the virus exhibits an enormous antigenic variability and progressively destroys the IS; and (3) in order to acquire sufficient neutralizing capacity, HIV-1 Abs need a much longer affinity maturation process than do Abs elicited by other viruses, with the result that protective Abs usually appear only after about 2 years (Chen et al. 2012). It is not possible during such a long Ab maturation process to use the structure-based RV approach since it is not clear which of the many intermediate Abs in the maturation pathway should be selected as templates for engineering candidate vaccine epitopes. Reverse vaccinologists base the design of vaccine immunogens solely on the observed structural complementarity between one epitope and one bnMab and they disregard the fact that numerous complex biological mechanisms and components of the IS are actually involved in the production of antibodies. Once antigens or individual epitopes are introduced in an IS, they become known as immunogens because they trigger the IS to start synthesizing Abs that may or may not recognize the immunogen and related antigens (Van Regenmortel 2014a). When it was established that HIV-1 epitopes recognized by affinity-matured bnAbs derived from HIV-1-infected individuals did not bind the germline predecessors of these antibodies (Xiao et al. 2009; Dimitrov 2010), it became clear that a lengthy process of antibody affinity maturation would have to be taken into account for obtaining neutralizing anti-HIV-1 antibodies. This meant that the design of HIV vaccines using RV was doomed and a huge, novel research effort was started in an attempt to unravel individual Ab affinity maturation pathways that lead from non-neutralizing to neutralizing Abs (Mascola and Haynes 2013; Klein et al. 2013; Prabakaran et al. 2014). The general aim was to drive human immune responses towards highly mutated anti-HIV bnAbs, using sequential immunizations with various Env immunogens (Doria-Rose and Joyce 2015). In view of the stochastic nature of mutations and the huge number of individual maturation pathways that exist (Dimitrov 2010), it is not at all clear whether this would make it possible to identify a small number of effective immunogens that could be employed for large scale human vaccination campaigns. Since this new vaccination strategy (Doria-Rose and Joyce 2015) has nothing in common with the RV approach that aims to transform HIV-1 epitopes of known structure directly into vaccine immunogens, it will not be further discussed in this review, which only analyzes why the RV approach used for several years failed to deliver an effective HIV-1 vaccine.

## Structure and Dynamics of Epitopes and Paratopes

The B cell epitopes of proteins are the regions that are recognized by the binding sites (i.e., the so-called paratopes) of Ab molecules present either in their free form in serum or as membrane-bound B cell receptors (BCRs). B cell epitopes must be distinguished from T cell epitopes, which are proteolytically cleaved peptides of the antigen that interact with T cell receptors. In the present review, B cell epitopes will simply be referred to as epitopes.

Protein epitopes are usually classified as continuous or discontinuous, depending on whether the amino acids present in the epitope are contiguous in the peptide chain or not. X-ray crystallographic analysis of epitopes bound to paratopes revealed that the vast majority of protein epitopes are discontinuous and consist of residues located on two to five separate chain segments that are brought together on the protein surface by the folding of the peptide chain. This distinction between the two types of epitopes may lead one to believe that amino acid residues are the entities involved in Ab-antigen recognition, although it is at the level of individual atoms that interactions take place (Van Oss 1995).

### *Discontinuous Epitopes*

Figure 1 shows the structure of a discontinuous epitope present in the outer surface protein A (OspA) of the spirochete *Borrelia burdorferi*, the etiological agent of Lyme disease. The structure of this epitope was elucidated by X-ray crystallography of OspA complexed with the Fab fragment of the mouse Mab 184 (Li et al. 1997b). The epitope consists of residues 30, 33–35, 42–46, 52, 67–71, 92–95, and 117–119, which are in contact with residues of the paratope. Atomic groups in the residues that form the seven sections of the epitope are not held together by internal chemical

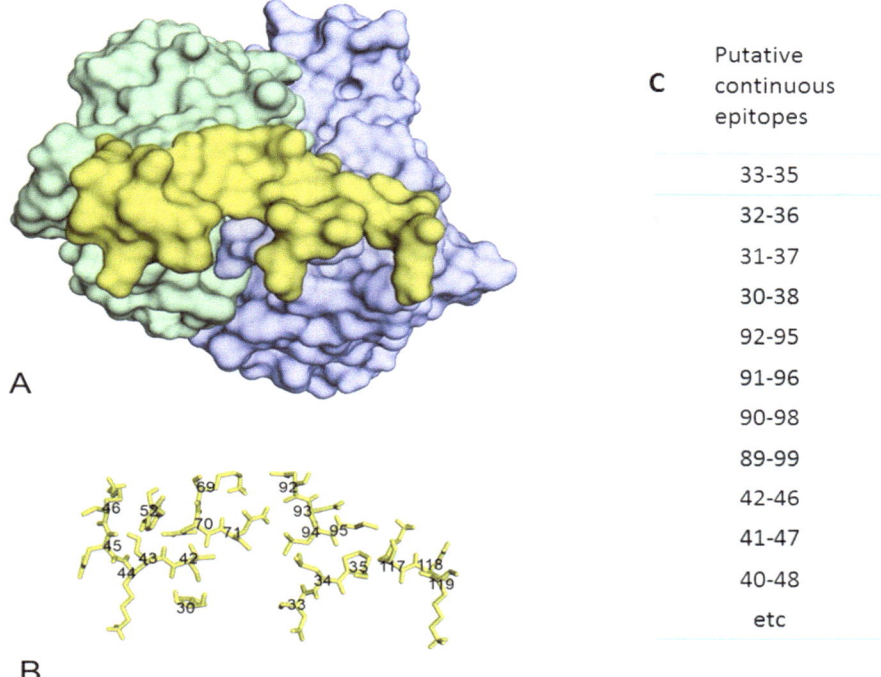

| C | Putative continuous epitopes |
|---|---|
| | 33-35 |
| | 32-36 |
| | 31-37 |
| | 30-38 |
| | 92-95 |
| | 91-96 |
| | 90-98 |
| | 89-99 |
| | 42-46 |
| | 41-47 |
| | 40-48 |
| | etc |

**Fig. 1** Discontinuous epitope of the outer surface protein A of the spirochete *Borrelia burdorferi* elucidated by X-ray crystallography from a complex with Mab 184.1. (**a**) Outline of the epitope in yellow; (**b**) position in space of the residues comprising the epitope. This set of residues cannot be isolated as such from the protein to demonstrate that it possesses binding activity in its own; (**c**) Parts of the discontinuous epitope and other peptide segments of the protein that may be able to bind Mab 184.1, in which case they would be called continuous epitopes (courtesy of Pernille Haste-Andersen, Danish Technical University)

bonds and they are recognized by antibodies only because the entire peptide chain acts as a scaffold. A discontinuous epitope lies in a molecule and acts like a molecule although it is not actually a molecule (Rosen 1991, p. 273). If the tertiary structure is perturbed—for instance, when the protein is denatured—the epitope ceases to exist. Discontinuous epitopes in a native protein are therefore only defined structurally by the set of atoms that make contact with residues of the Ab. A contact is said to occur if the interatomic distance between atoms in the epitope and paratope is 3.5–4.5 Å, depending on the chemical nature of the interaction (Chen et al. 2009).

When the epitope is represented as a disembodied set of residues (Fig. 1b) it is clear that this set of residues cannot be isolated or extracted from the OspA protein as a functional unit possessing binding activity of its own. The functional capacity of a discontinuous epitope of HIV-1 to induce Abs can only be assessed by using as immunogen the native protein in which the epitope is embedded. Since such a procedure always gives rise to a heterogeneous immune response against the many epitopes present in the protein, it is not possible to study the immunogenic activity of a single discontinuous epitope separately from the immunogenicity of the whole protein. 70.

X-ray crystallographic analysis of Ab-antigen complexes provides only static pictures of epitopes and paratopes. A molecular structure is always the result of selective attention to the visual experience of an object at a specific time (Woodger 1967) and thinking in terms of static instead of dynamic structures tends to conceal the role that mutual adaptation in the Ab and antigen plays in facilitating their interaction (Demchenko 2001). Proteins and immunoglobulins are fairly dynamic molecules (Jimenez et al. 2003) and the plasticity of epitopes and paratopes has led to them being described as flexible keys interacting with adjustable locks (Edmundson et al. 1987).

It is well established that the structures visualized in epitope–paratope complexes are often very different from the structures of the binding sites in the free molecules before they were altered by the mutual adaptation and induced fit that occur when the two partners interact (Wilson and Stanfield 1994; Bosshard 2001). The particular epitope structure observed in an epitope–paratope complex therefore does not necessarily correspond to the immunogenic structure recognized by BCRs during the immunization process that gave rise to the Ab; such an epitope identified in a complex may thus not be the best structure that should be mimicked in a vaccine.

Since the entire accessible surface of a protein is a continuum of potential epitopes, the same residues may contribute to several overlapping discontinuous epitopes recognized by different Abs. This is illustrated in Fig. 2 for two Mabs that recognize a very similar epitope in lysozyme by using two completely different paratopes presenting no similarity in their chemical bonding patterns with the epitope. Since the epitope nature of a set of amino acids can only be established when an Ab able to bind to it has been found, an epitope is not an intrinsic structural feature of a protein that could be identified in the absence of a particular interaction with a paratope. Similarly, since a paratope requires the existence of a complementary epitope in order to acquire a recognizable identity, both epitopes and paratopes are relational entities defined by their mutual complementarity (Van Regenmortel 2009b). This means that the two Mabs in Fig. 2 do not recognize the same epitope.

The number of epitopes in a protein can be estimated from the number of different Mabs that can be raised against it. This number was found to be 115 epitopes for insulin (Schroer et al. 1983) and more than a thousand for the BLyS molecule (Edwards et al. 2003). This relational nexus means that analyzing the antigenicity of a protein amounts to analyzing the size of the immunological repertoire of the host immunized with the protein (Van Regenmortel 2009b).

The identical viral subunits that are assembled in virions and in spikes embedded in viral membranes (Liu et al. 2008) give rise to a quaternary protein structure that creates discontinuous epitopes by the juxtaposition of residues from neighboring subunits that Abs recognize as a single epitope. Such epitopes, which have been called neotopes (Van Regenmortel 1966, 1992a), have been shown to be present in the capsids and membrane proteins of many viruses including HIV-1 (Neurath and Rubin 1971; Broder et al. 1994; Gorny et al. 2005). Neotopes can also arise from

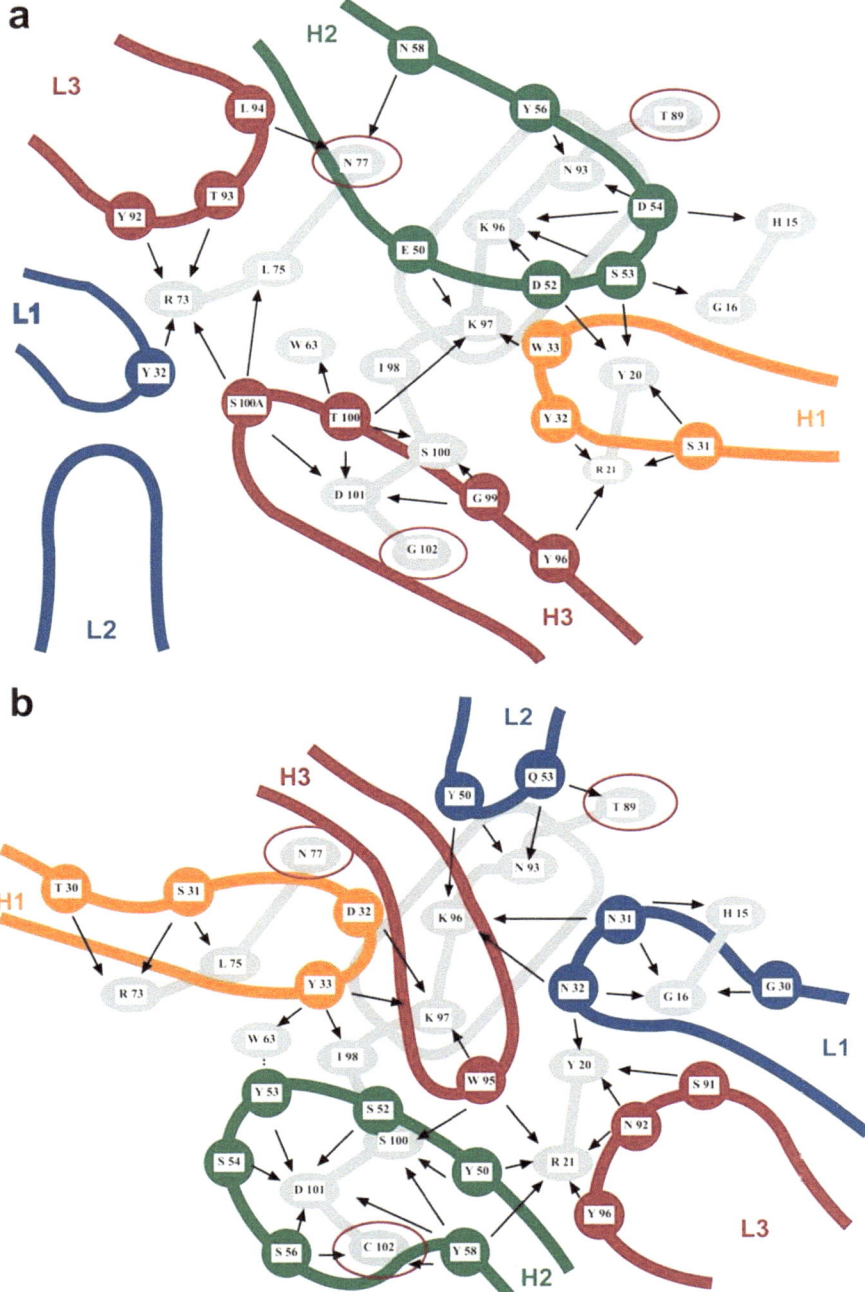

**Fig. 2** The same residues of an antigen can contribute to different overlapping disccntinuous epitopes. Two overlapping discontinuous epitopes of lysozyme are recognized by Mabs F9–13.7 (**a**) and HyHEL10 (**b**) elucidated by X-ray crystallography. Thirteen residues of lysozyme (in gray)

conformational changes induced in the subunits by intersubunit interactions. Since the quaternary structure of viral proteins is sensitive to small changes in the chemical environment (Bothner et al. 1998), neotopes in HIV can be transient epitopes capable of assuming different structures alternating between open and closed conformations (Harris et al. 2011). It is likely that such conformational variability will facilitate induced fit adjustments and BCR recognition although there is also evidence that immunogenicity can be enhanced when epitope flexibility is either increased or decreased (Dey et al. 2009; Moseri et al. 2010; Guenaga et al. 2011).

Since many crystal structures of epitope–paratope complexes have been determined to a resolution of 30 Å or less, the precise nature of the atomic interactions is usually not established. This may partly explain why protein crystallographers for a long time believed that water was actually extruded from epitope–paratope interfaces (Davies et al. 1990). Water of hydration surrounds all protein molecules (Van Oss 1995) but these water molecules are not necessarily expelled from the contact interface because of the imperfect complementarity between epitopes and paratopes. As a result, water molecules may contribute H-bonds to the interaction between the two partners (Goldbaum et al. 1996; Schwartz et al. 1995), although their contribution to the binding energetics is difficult to quantify (Ladbury 1996).

In a seminal study of the complex between lysozyme and the Fab fragment of Mab D 1.3 in its antigen-bound form, compared to its free form, at an improved resolution of 1.8 Å, Roberto Poljak and his collaborators established that 48 water molecules were present at the epitope–paratope interface, which is larger than the number of water molecules associated with the unbound sites (Bhat et al. 1994). Both buried and exposed water molecules establish links between the antigen and Ab and contribute to the chemical complementarity between the two interacting surfaces (Fig. 3), leading to decreased mobility in the Ab structure upon complex formation.

The history of how these results became rapidly accepted is interesting. The journals *Nature* and *Science*, as custodians of scientific orthodoxy, had refused to publish the paper because it contradicted the theory, universally accepted by crystallographers at the time, that water molecules are always extruded from antigen–antibody interfaces, leading to increased entropy. However, for more than a year before the results were finally published, Poljak had presented his data at several international meetings and the quality of his experimental data (Fig. 3) together with his new interpretation of Ab-antigen interactions convinced many crystallographers. Very soon after the 1994 publication (Bhat et al. 1994), such results became

←——————————————————————————————————————————

**Fig. 2** (continued) are recognized by both antibodies, albeit with different bonding patterns. The rounded rectangle in gray represents the lysozyme α-helix. The two sets of CDRs are shown in color and have different orientations on the lysozyme surface. Three residues (N77, T89, and G102 highlighted with red circles) are not shared by the two epitopes. Intermolecular contacts are shown by arrows. Mab HyHEL10 forms a salt bridge between lysozyme K97 and residue D32 of the H1 antibody loop. Mab F9–13.7 forms salt bridges between lysozyme residues K97, K96, and H15 and respectively residues E50, D52, and D54 of the H2 antibody loop (adapted from Lescar et al. 1995)

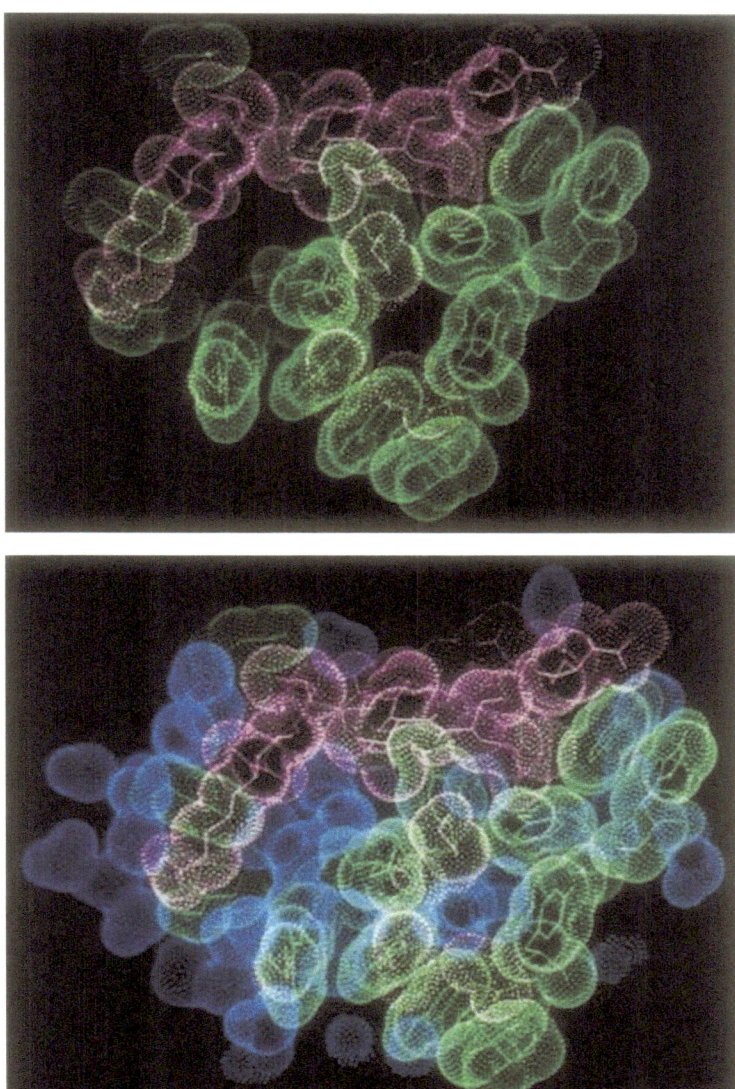

**Fig. 3** 3D structure at 1.8 Å resolution of the anti-lysozyme Fv fragment of Mab D1.3. Upper frame: van der Waals surface of antibody residues (in green) interacting with residues of lysozyme (in pink). Lower frame: the same model showing cavities and channels filled with water molecules (in blue) after complex formation. There are more water molecules at the interface than in the free binding sites and they contribute H-bonds that stabilize the complex. The association is not driven by an hydrophobic entropy effect arising from the extrusion of water from the interface but by a large negative enthalpy arising from H-bonds and van der Waals interactions (Bhat et al. 1994)

acceptable and were published by other groups (Malby et al. 1994; Chacko et al. 1995).

## Continuous Epitopes

Any short linear peptide fragment of a protein able to bind to anti-protein antibodies is called a continuous epitope of that protein. This terminology is unfortunate because it may give the impression that these epitopes are present as such in the native protein although short peptide fragments very rarely retain the conformation present in the corresponding part of the folded protein (Van Regenmortel 2009b). Since short peptides of 3–5 residues corresponding only to portions of a discontinuous epitope are often able to bind weakly to antibodies raised against the discontinuous epitope, such peptides will often be called continuous epitopes (Van Regenmortel 2006). For instance, peptide fragments of OspA corresponding to residues 42–46, 40–48, or 40–50 of the molecule (Fig. 1b) could easily be identified as continuous epitopes because they all contain five contiguous residues of the OspA sequence, which is sufficient to give rise to a weak cross-reaction with the OspA antibody (Van Regenmortel 1992b, pp. 1–27; Roberts et al. 1993). It has been suggested that the majority of continuous epitopes described in the literature may correspond to unfolded regions of denatured proteins and are therefore not genuine epitopes of native proteins (Laver et al. 1990). It can, indeed, never be excluded that the protein antisera used for measuring the antigenic reactivity of peptides contain Abs to denatured molecules that were present in the protein preparation used for eliciting the antisera. In the reciprocal assay when Abs raised against peptides are tested for their ability to react with the cognate protein, it may appear that the Abs also recognize the protein molecules because these have become denatured when they are adsorbed to plastic in a solid phase immunoassay. The initial claim that individual immunizations with 20 residue-long synthetic peptides corresponding to the entire sequence of influenza hemagglutinin were all able to induce antibodies that cross-reacted with the viral protein stimulated considerable interest in the possibility of using synthetic peptides as vaccines (Lerner 1984). However, it was subsequently shown that such misleading results were obtained because the viral antigens used in the solid-phase immunoassay had become denatured by adsorption to plastic (Darst et al. 1988).

More than a thousand linear peptides corresponding to continuous epitopes have been studied over many years as potential synthetic peptide vaccines against numerous pathogens and it is remarkable that by 2006 not a single peptide had passed phased III clinical trials and was marketed as a human vaccine (Hans et al. 2006). It is therefore astonishing that many investigators still try to develop HIV-1 vaccines starting from continuous epitopes, especially because it is well-established that most nAbs against HIV-1 recognize discontinuous epitopes.

In order to ascertain whether antibodies elicited by peptides corresponding to small regions of a protein truly recognize the native form of the protein, it is necessary to test these antibodies in a liquid-phase immunoassay that preserves the native conformation of the protein (Spangler 1991). Thousands of continuous epitopes listed in epitope databases have been identified only on the basis of their ability to cross-react with protein antibodies in a variety of immunoassays and not by showing that each residue in the peptide made contact with residues of a paratope (Greenbaum et al. 2007; Zhang et al. 2008). Since the criterion used for identifying a continuous epitope is only its binding activity, the exact structure of the epitope is never clearly defined (Van Regenmortel 1999a, pp. 1–78). When epitopes are defined structurally by crystallography they comprise about 15 residues whereas if they are defined functionally by binding assays, they appear to be much smaller (Van Regenmortel 1989a). When peptides of decreasing size are tested for binding activity, the smallest peptide that retains a significant level of reactivity is sometimes given the status of epitope although it may be as short as a pentapeptide or a tripeptide (Van Regenmortel 1992b; Trifilieff et al. 1991; Benjamini 1997, pp. 265–310). This agrees with the finding that when small haptens of increasing size are tested for their ability to bind to a range of Mabs, a maximum binding affinity of $10^{10}$ $M^{-1}$ is observed when the hapten molecular weight reaches a value of 450, which corresponds to about four amino acid residues (Chappey et al. 1994; Morel-Montero and Delaage 1994, pp. 357–372). All these finding have led to the conclusion that most studies that analyze the antigenic structure of proteins by binding assays do not focus on antigenicity *per se* but on the phenomenon of cross-reactive antigenicity between proteins and short peptides (Van Regenmortel 1989a).

## Mimotopes

The term mimotope was coined by Mario Geysen et al. (1986) to refer to a peptide that binds a particular Ab but possesses little or no sequence similarity with the protein antigen used to raise the Ab. Mimotopes are usually identified by testing combinatorial peptide libraries and selecting peptides that bind anti-protein Abs. The phenomenon of hydropathic complementarity (see section "Paratopes") explains why peptides with little or no sequence similarity may nevertheless be able to react with the same Ab.

To qualify as a mimotope, the peptide in addition to binding to an anti-protein Ab should also be capable of eliciting Abs that recognize the epitope being mimicked (Van Regenmortel 2009b). The reason why this is a requirement is that a peptide may bind to a paratope in a polyspecific Ab (section "Paratopes") that is not the same as the paratope that initially selected the mimotope from a peptide library.

Many mimotopes of HIV-1 epitopes have been isolated from peptide libraries but none of them induced nAbs when they were used as immunogens (Scala et al. 1999; Zwick et al. 2001a; Dorgham et al. 2005). Peptide analogs that retain the original hydropathic complementarity present in a continuous epitope but no longer possess any sequence similarity with it may still bind to the same Ab, which explains the binding activity of certain mimotopes (see section "Paratopes").

## Paratopes

Paratopes are the binding sites of Abs that recognize epitopes. IgG immunoglobulins possess two identical binding sites formed by six surface loops located at the tip of the two Fab fragments of the IgG (Sundberg 2009, pp. 23–36), and comprising 50–70 hypervariable residues distributed over six complementarity determining regions (CDRs) in the variable domains of the heavy and light immunoglobulin chains. Each of the two Ig binding sites contains several smaller overlapping or non-overlapping subsites of 10–20 residues, called paratopes, which are the regions that bind to different epitopes of an antigen. As a result, an antibody can never be monospecific for a single epitope because the surface of one paratope represents only about 20–35% of the total surface of the entire Ig binding site (Raghunathan et al. 2012). Steric hindrance usually prevents two antigens from binding simultaneously to the same Ig binding cleft, but when two paratope subsites do not overlap, two small antigens or haptens may be able to bind simultaneously to the same antibody molecule (Bhattacharjee and Glaudemans 1978). The overall structure of a paratope built up from residues present in different hypervariable loops is reminiscent of the structure of a discontinuous epitope that always consists of residues located in separate surface regions of a protein molecule. Since small peptides corresponding to a single CDR loop of an antibody are sometimes able on their own to bind the cognate antigen, they can be viewed as continuous paratopes, analogous to continuous epitopes (Laune et al. 1997). The ability of some continuous epitopes to bind continuous paratopes is due to the type of hydropathic complementarity that is usually observed between the sense and antisense peptides encoded by complementary sense and antisense messenger RNAs (Van Regenmortel 2014a; Blalock and Bost 1986; Tropsha et al. 1992; Biro 2007). Hydropathic complementarity arises from an inverted hydropathic pattern in two short peptide sequences and is caused by the attraction that exists between hydrophilic and hydrophobic groups (Van Regenmortel 2014a).

Recent work by the group of Ofran in Israel has considerably improved our understanding of the structural basis of antibody specificity (Sela-Culang et al. 2015; Sela-Culang et al. 2013; Kunik and Ofran 2013). By analyzing 140 antibody–antigen complexes, they demonstrated the contributions of individual Ig binding loops to the overall binding specificity of antibodies and clearly established that there is no noticeable difference between the amino acid residue composition of epitopes and the composition of the entire protein surface (Van Regenmortel 2014a; Kunik and Ofran 2013). This finding explains why innumerable attempts to predict epitopes in proteins using amino acid propensities such as surface accessibility or hydrophilicity were notoriously unsuccessful (Blythe and Flower 2005) because these prediction methods which concentrated on continuous epitopes assumed that certain surface residues possessed a superior intrinsic capacity for binding any potential antibody (Van Regenmortel 1999a, 2014a). Prediction methods for discontinuous epitopes are also very limited because these methods are unable to identify the complete set of residues from distant parts of the protein sequence that needs to be assembled in a precise configuration in order to possess the antigenic and immunogenic activity of the epitope (Van Regenmortel 2014a; Hopp 1993; Van Regenmortel and Pellequer 1994; Ponomarenko and Van Regenmortel 2009). Antibody polyspecifity is the ability of an antibody molecule to bind a large variety of epitopes present in different antigens, and is caused by the presence in each Ab binding site of different paratopes that possess considerable conformational plasticity and flexibility. Several mechanisms of conformational adaptation within the antibody combining site allow a paratope to bind to different epitopes (Mariuzza 2006), which explains why most Abs derived from Ab germline genes are highly polyreactive and are able to react with a variety of self-antigens such as DNA, cytoskeleton proteins, carbohydrates as well as bacterial and viral antigens (Zhou et al. 2007b; Manivel et al. 2002; Mouquet and Nussenzweig 2012; Dimitrov et al. 2013). During somatic hypermutation and the resulting Ab maturation, the flexibility of paratopes tends to decrease as their affinity increases (Jimenez et al. 2003; Thorpe and Brooks 2007). During normal human B cell development, many such Abs are deleted from the Ig repertoire in the bone narrow before they enter the mature B cell compartment (Wardemann et al. 2003) although as many of 20% of the Abs subsequently produced by mature human B cells may still be polyreactive and self-reactive (Notkins 2004). Such findings put into question the validity of the need for absolute self–nonself discrimination as the primary factor that best explains the evolutionary origin of the IS (Sandberg et al. 2000; Leng and Bentwich 2002).

# The So-Called Rational Design of HIV-1 Vaccine Immunogens

The term design refers to the deliberate conceiving and creation of a novel object or process by an intelligent being. It is widely believed today that rational design is the best strategy for developing new drugs and vaccines (Bramwell and Perrie 2005; D'Argenio and Wilson 2010) and that it is applicable to HIV-1 vaccine development (Walker and Burton 2010; Douek et al. 2006; Schief et al. 2009; Sattentau and McMichael 2010; Nabel et al. 2011).

The term "rational" is applied to modern vaccine design in the way the term is used in "rational drug design" i.e., for describing how candidate molecules are designed to fit the 3D structure of a biological target or receptor so that they will bind with high affinity and selectivity to it and inhibit a biological function (Kuntz 1992).

As stated by Amzel in 1998 (Amzel 1998): "One of the goals of research in biotechnology is to transform the process of developing a drug from a trial-and-error empirical operation into a rational structure-based process". This computer-assisted approach based on molecular docking (Gschwend et al. 1996) succeeds because the complementarity between drugs and their target molecules is fairly unique. This explains, for instance, the remarkable success in developing the antiretroviral drugs that inhibit various HIV-1 enzymes (Wlodawer 2002; De Clercq 2009).

In contrast, molecular recognition processes between an antigen and the many Abs that are able to bind to it are much less specific than drug-receptor interactions, mainly because of the conformational flexibility of paratopes, the degeneracy of the immune system and the polyspecificity of antibodies (Jimenez et al. 2003; James et al. 2003). This allows an individual IS to recognize virtually any antigenic motif it may encounter (Sperling et al. 1983; Parnes 2004; Wucherpfennig et al. 2007).

## What Do Vaccinologists Actually Do When They Claim to be Rationally Designing an HIV-1 Vaccine?

Doing something by design is doing it intentionally; believers in "intelligent design" accept, for instance, that an intelligent deity designed all living organisms and that this explains the presence of life on our planet. Scientists today no longer believe that living organisms and their biological functions were designed by the preconceived plan of a designer rather than being shaped by the filter and pressure of Darwinian natural selection (Van Regenmortel 2007). It seems equally inappropriate to use design terminology for describing the activities of vaccinologists when they attempt to develop an HIV-1 vaccine by studying the structure of antibodies able to neutralize the virus. What is of course feasible is to improve the binding complementarity present in one epitope–bnMab pair by either designing an HIV-Env epitope that binds better (Wu et al. 2010) or a paratope that binds better (Diskin et al. 2011). However, in both cases this amounts only to designing improved binders and not immunogens. Although the antigen binding capacity (i.e., the antigenicity) of an Env epitope may sometimes be improved, this is not the case for the ability of that epitope to elicit protective Abs (i.e., its immunogenicity) when it is used for immunization. The claim that what is being rationally designed is actually an HIV-1 vaccine immunogen instead of an antigen (Douek et al. 2006; Nabel al 2011) arises because antigenicity and immunogenicity are confounded and the unwarranted assumption is made that when an epitope reacts strongly with a paratope present in the bnMab, this epitope will also be able to elicit similar neutralizing antibodies in an immunized host. Antigen design is simply masquerading as immunogen design since there is no justification for the belief that antigenic reactivity necessarily entails an immunogenic capacity to produce neutralizing Abs similar to the Mab that was used as template for designing the antigen (Van Regenmortel 2012b, 2015a).

## Rational Vaccine Design Without a Designer Is Only a Metaphor

In order to rationally design an HIV-1 vaccine, a designer would need to know the immune correlates of protection; these, however, are identifiable only after an effective vaccine has been developed. In the case of Darwinian evolution, Ayala (2004) tried to rescue design as an explanation by suggesting that it is possible to have "design without a designer" although this simply encourages the use of metaphors of goal-directed teleology for describing natural biological processes in terms of design and purposes (Van Regenmortel 2007; Ruse 2002).

It has also been claimed that HIV-1 needs to continuously develop new strategies and mechanisms in order to escape IS defences, and that this amounts to a momentous evolutionary battle between virus and host (Burton et al. 2005). However, such metaphoric language implies that the virus is able to develop new goal-directed mechanisms for winning a battle, whereas it is only the occurrence of stochastic mutations in the reverse transcriptase enzyme of the virus that is responsible for the persistent chronic infection (Van Regenmortel 2016a).

Vaccinologists should not claim that they rationally design an HIV-1 vaccine when they are only improving the antigenic binding capacity of viral proteins. The neutralizing efficacy of Abs obtained by vaccination is due to the presence in the host IS of numerous biological and regulatory mechanisms that are unrelated to the actual epitope–paratope recognition process, which is the only parameter that the vaccine designer takes into account and tries to control (Van Regenmortel 2012b; Berzofsky 1985). Since it is not possible by structure-based RV to alter the activity of an IS so that it will produce nAbs, rational vaccine design without a designer, is best relegated to the sphere of poetic, metaphorical language.

## Paradigms in HIV Vaccinology

The term paradigm was introduced by the philosopher and historian of science, Thomas Kuhn, in his 1962 book *The Structure of Scientific Revolutions* (Kuhn 1962), to refer to the conceptual framework within which scientists think and operate at a given time. Paradigms in vaccine research are usually based on recognized past vaccine achievements, since these provide an indication of which experiments might be conducted and what responses could be expected.

Investigators operating within a shared paradigm are often unaware of the many presuppositions, assumptions and conjectures that underlie their research agenda. When they obtain results that differ from what they expected, they will tend to formulate new ad hoc hypotheses to resolve contradictions between their observations and what they had predicted would happen (Bird 2018). However, they rarely conclude that the paradigm has been refuted and must be abandoned. Kuhn argued that when scientists obtain unexpected results, they remain committed to an accepted paradigm because they view it as a coherent conceptual framework that has been successful in the past. Only if a large number of anomalous results keep accumulating over many years may they lose their confidence in a given paradigm and eventually replace it by another one, giving rise to what has been called a paradigm shift. Kuhn (1962) described such a change by saying that a long period of normal science is then followed by a short period of revolutionary science.

When vaccinologists do not obtain the results they expected, they do not abandon a paradigm because they do not consider that the main purpose of their experimental work is to confirm the validity of the presuppositions inherent in the paradigm they adhere to. They will thus not follow the injunction of Karl Popper (1959) that the primary aim of an investigator must be to try to disprove or falsify hypotheses or theories rather than to prove them. It is well known that scientific observations and empirical experimentation are never able to logically prove the correctness of a theory although they can sometimes refute it. When vaccinologists believe that they will achieve a particular research objective by following their chosen paradigm, they will disregard the possibility that some of their assumptions may be incompatible with current immunological theory (see section "An Effective HIV-1 Vaccine Cannot Be Developed by Disregarding Accepted Immunological Theory"). In the field of HIV vaccine research there is evidence that certain paradigms, for instance the assumption that successful results obtained previously with other viruses can be replicated with HIV, have led investigators astray and made them pursue strategies that were unlikely to be successful.

Paradigms come in different forms and some can be applicable to many viruses, whereas others are more restricted in scope. The first paradigm that guided HIV vaccine research in the years 1985–2003 was based on the belief that immunization with HIV-1 Env recombinant proteins would succeed in eliciting neutralizing Abs that would confer protection against infection. When this was found not to be the case, a new paradigm based on cell-mediated immunity was followed for a few years but was also abandoned when vaccine trials showed no efficacy (Buchbinder et al. 2008; Esparza 2013b).

A third paradigm based on the expectation that structure-based RV would be able to identify HIV-1 Env epitopes capable of eliciting broadly neutralizing Ab responses was proposed in 2002 (Burton 2002) and pursued for several years but did not succeed in producing effective vaccine immunogens (Pejchal and Wilson 2010; Van Regenmortel 2012a, b, 2014b; Burton et al. 2012).

The R144 vaccine trial started in Thailand in 2004 combined a poxvirus-vectored gp120 env prime with a gp120 protein boost and followed a mixed paradigm that was widely condemned as ill-conceived (Burton et al. 2004a; Jefferys and Harrington 2004). However, it turned out to be the first human vaccine trial that led to some limited reduction in HIV-1 acquisition (Rerks-Ngarm et al. 2009; Esparza 2015) illustrating the vagaries of paradigm selection.

Another paradigm of a broader nature has become popular in recent years because many investigators believed that we possess too little knowledge of the human IS and of HIV-1 immunopathology to be able to develop an effective HIV vaccine. This led to the view that progress would occur only if large-scale basic research programs in immunology were undertaken (Virgin and Walker 2010; McElrath and Haynes 2010). This paradigm finds its origin in the belief prevalent since the middle of the last century (Bush 1945) that all technological innovations are the result of applied research made possible by previous curiosity-driven basic research that is under-taken without any consideration of potential use. Such a belief is less prevalent today because a tight separation between basic and applied research is considered to be somewhat artificial and it is accepted that research can be driven simultaneously by a quest for fundamental understanding and by considerations of use for solving a practical problem. According to the so-called quadrant model of scientific research (Stokes 1997), a commitment to study various properties of the IS does not exclude a commitment to try to control HIV-1 infection by vaccination (Esparza 2005; Van Regenmortel et al. 2014). In order to understand how a technological innovation such as a new vaccine becomes possible, it is necessary to distinguish between two types of human knowledge (Mokyr 2002a, b):

1. Knowing what is the case, which leads to factual knowledge (also called prop-ositional knowledge) that takes the form of true statements concerning an established scientific fact. New factual knowledge is always *a discovery* that reveals something that existed all along but was unknown to anybody.
2. Knowing how to do something of practical utility that leads to new prescriptive knowledge in the form of an *invention* that makes it possible to do something that was previously not feasible.

Factual knowledge is always derived from basic research whereas prescriptive knowledge, which could take the form of an innovation such as a new vaccine, always requires applied research to demonstrate how nature and natural phenomena can be successfully manipulated and controlled. This is illustrated, for instance, by the fact that factual knowledge concerning the contribution of long CDR-H3 regions of Abs to the exceptional neutralizing ability of certain anti-HIV-1 bnMabs (Yu and Guan 2014; Briney et al. 2012) was of little help for developing a vaccine because there was no prescriptive knowledge available on how such Abs could be elicited by immunization (Van Regenmortel 2015a).

In his book *Representing and Intervening*, Ian Hacking (1983) pointed out that scientists obtain prescriptive knowledge only when they successfully intervene empirically in the system under investigation because this is the only way for them to demonstrate that they are indeed able to control it. The complexity of the IS makes it impossible to fully "understand" it in a reductionist manner since only a few causal links can be revealed between a small number of the numerous individual constit-uents of the system (Van Regenmortel 2004a, b, c; Mazzocchi 2008; Gannon 2007). Biological science does not exhibit the law-like regularities found in the physical sciences and biologists are only able to partly explain phenomena by positing mechanisms and making predictions about their occurrence. However, they do not fully understand the innumerable interconnections that exist between all the parts of a complex system (Van Regenmortel 2012b; Kelley and Scott 2008).

On the other hand, if vaccinologists succeed in intervening in an IS and achieve protective immunity against virus infection, this newly acquired prescriptive knowledge will allow them to make reliable predictions about the likely outcome of a vaccination trial. In other words, the empirical development of an effective vaccine must take place when the mode of action of an ideal HIV-1 vaccine has not yet been elucidated using factual knowledge obtained by previous basic research in immunology. Increasing our factual knowledge of immunological phenomena in no way eliminates the need to develop empirically the prescriptive knowledge required for vaccine development (Van Regenmortel 2012a, 2014b).

## An Effective HIV-1 Vaccine Cannot Be Developed by Disregarding Accepted Immunological Theory

### *Facts, Hypotheses, Laws, Theories, and Empirical Experimentation*

Alan Chalmers, in his popular book translated into 15 languages entitled *What is This Thing Called Science?* (Chalmers 1999) suggested that "science is derived from the facts," the facts being claims about the world that can be directly established by a careful, unprejudiced use of the senses. The term "fact" is somewhat ambiguous because it can refer to a statement that expresses a certain observable state of affairs or to the state of affairs itself. Scientific facts correspond to objective and verifiable observations that are often collected in order to build or reject hypotheses that could explain the occurrence of a certain phenomenon. A hypothesis may be nothing more than guesswork, a hunch or a conjecture, which is usually the case for the assumptions that underlie a paradigm. Scientific experiments are never able to "prove" that a hypothesis is correct although they can sometimes refute a mistaken hypothesis. Experiments may also fail to reject it, which means that the hypothesis could be correct but without any certainty (Popper 1969; Horgan 2006, pp. 32–41). Absolute proof in the mathematical sense or in the logical sense of deduction can never be derived from empirical, experimental science.

It is commonly believed that we derive facts simply by observing the world but this would mean that our visual perceptions require no interpretation and are not influenced by prior-knowledge or by a particular conceptual framework. That this is not the case is shown, for instance, by the belief, prevalent until the seventeenth century, that the earth is stationary. This seemed to be confirmed by the observation that if we jump in the air, the earth does not spin away beneath us. Today, we know that the earth moves, spinning on its axis and orbiting the sun and we accept that the earth moves at more than 100 m per second and that inertia explains why, after jumping, we land where we took off. What has happened is that nowadays we interpret our observation of being apparently stationary in a different way because of a new conceptual framework derived from our improved scientific knowledge. In the same way, we have learned to doubt the validity of certain observations such as the well-known moon illusion that makes the moon appear much smaller in the sky than when it is low on the horizon (Chalmers 1999).

Instead of relying on simple observations for determining what the "facts" are, scientists accept that reliable facts can usually be obtained by performing experiments that are always theory-dependent and require that we make certain inductive inferences. When we extrapolate from a finite number of experimental observations and make a general scientific claim, we accept that this induction procedure only allows us to reach approximate or probable scientific truths but never the type of "proven" truths achievable by logical deduction.

Scientific observations do not only refer to objects, properties and events that we can observe using our senses but also concern the observation of theoretical entities that are less accessible to direct human experience. Certain objects or phenomena may simply be non-observable because they lie outside the range of objects that our senses can directly observe, and therefore require instruments such as microscopes or telescopes. Scientists sometimes also need to observe theoretical entities for instance epitopes or paratopes that can only be defined structurally or functionally on the basis of certain criteria such as a minimum contact distance separating two interacting partners, or the occurrence of a binding reaction between them. In such cases, making a distinction between observable terms and theoretical terms may become difficult and this has led to the claim that all scientific observations may be theory-laden (Klee 1997, pp. 41–61).

Although experimental observations may not on their own justify a belief in the reality of invisible entities such as epitopes, it is our ability to successfully manipulate them empirically that convinces us that they actually exist. As pointed out by Hacking (Hacking 1983), we need to interfere with the material world in order to obtain knowledge about it and our scientific understanding will increase when we are able to intervene in a system and can manipulate it successfully.

A scientific theory refers to a well-substantiated body of explanations, repeatedly tested and confirmed experimentally, which have so far overcome all attempts to falsify them and allow accurate predictions to be made. Theories, therefore, should not be easily dismissed and paradigms are likely to be useful only if they are not contradicted by accepted scientific theories. In everyday conversation, the term theory is often used pejoratively as being a wild guess not to be taken seriously (i.e., this is just a theory), but this is not the case for well-established scientific theories that have become synonymous with well-established scientific facts. For instance, Darwinian evolution is today considered a scientific fact and no longer just a theory, as argued by the adepts of intelligent design.

Scientific explanations usually contain laws in the form of causal explanations that link causes and effects. Causation is a much stronger relation between events than mere temporal succession since the effect is unavoidable and follows the cause necessarily and not only regularly. However, what constitutes the actual link in a causal sequence remains highly mysterious and is an unsolved problem in the philosophy of science. Gravity, for instance, remains mysterious because it does not need any material connection in order to be transmitted.

Causation in the physical and biological sciences possesses very different explanatory power. In physics, only a small number of physical forces exist that cause observable phenomena, and the occurrence of an eclipse, for instance, can be explained using Newton's laws of universal gravitation together with certain initial conditions. These laws on their own suffice to describe the motions of celestial bodies in an interstellar vacuum where there are no external interferences. On Earth, these laws explain physical phenomena only if we add a *ceteris paribus* clause (i.e., other things being equal) to rule out interfering factors such as electrostatic or magnetic forces (Cartwright 1983). The very accurate predictions achievable in astronomy are not possible in the case of complex biological systems that always present an enormous diversity of interactions between their numerous constituents and give rise to network behaviour absent in the isolated components (Alm and Arkin 2003). Complex systems also possess emergent properties that have their own causal powers not found in the individual constituents of the system and these properties cannot be predicted by a reductionist analysis of the individual components of the system (Van Regenmortel 2002a, 2004a, b, c; Kelley and Scott 2008). However, it is nowadays accepted that we need not achieve a full theoretical understanding of a complex biological system in order to be able to make some reliable predictions about how it is likely to behave. This means that we might be able to achieve protective immunity by vaccination without necessarily understanding all the multiple underlying mechanisms at work in a human immune system.

## *Degeneracy of the Immune System and Polyspecificity of Antibodies*

The term degeneracy was introduced in the life sciences to refer to the fact that the genetic code is "degenerate" or "redundant" because there are many more triplet codons than encoded amino acid residues (Jones and Nirenberg 1966). As a result, a huge number of distinct mRNA sequences can be translated to generate the same protein sequence. Degeneracy is a common feature of complex biological networks, where structurally different elements are able to perform the same activity or function. In immunology, degeneracy refers to both the ability of structurally different Abs, BCRs, or T cell receptors (TCRs) to bind to the same ligand and to the ability of a single Ab or T cell receptor to bind many different antigens or peptides (Sperling et al. 1983; Cohn 2005).

Degeneracy is a property of the IS as a whole and not of its individual components (Parnes 2004) whereas the ability of an individual Ab to bind different antigens is described as cross-reactivity, molecular mimicry or promiscuity. This means that there is no unique intrinsic epitope for any Ab molecule but only a diverse group of potential ligands (Wucherpfennig et al. 2007). The degeneracy of the IS makes it possible for a limited number of immune receptors to recognize, albeit often with very low affinity, essentially any molecular motif that an animal's IS is likely to encounter. Specificity has been defined as the exact complementary relationship between an agent and something acted on (Medawar and Medawar 1978) a definition that applies to the stereochemical complementarity between antigen and Ab, enzyme and substrate or receptor and ligand. Specificity in immunology has been difficult to define (Mazumdar 1995; Silverstein 1982) and in the case of Abs, it has been suggested that it may be better to refer to their discriminating capacity rather than to their specificity since it is only the wish of an investigator to distinguish between two antigens that determines whether an Ab is considered to be specific or not (Day 1990, p. 291; Van Regenmortel 1998; Frank 2002, p. 42).

Whereas Ab affinity describes the strength of the binary relationship between one epitope and one paratope, Ab specificity is a ternary relational property that has meaning only with a minimum of three partners, for instance one paratope and two epitopes. There is no necessary link between affinity and specificity because antibodies of low affinity are often able to better discriminate between two antigens than antibodies of high affinity that tend to reveal more cross-reactions (Van Regenmortel 1998). In order to evaluate the efficacy of a vaccine, it may be necessary to measure the affinity of the Abs that it elicits and this can be done, for instance, by calculating Ab avidity indices obtained by disrupting Ab-HIV-1 Env complexes with chaotropes (Klasse 2016). Ab binding to continuous epitopes resists chaotrope-induced disruption more than Ab binding to the more fragile discontinuous epitopes which, however, are the epitopes that induce most bnAbs (Alexander et al. 2015). There is only limited correlation between Ab affinity and chaotrope resistance because of the phenomenon of hysteresis, which occurs because the energy of Ab-antigen dissociation is higher than the energy of association (Van Oss 1992, pp. 99–125).

It is therefore preferable to assess Ab affinity by measuring the kinetics of Ab-HIV-1 Env interactions using surface plasmon resonance biosensors, a technique that also allows the determination of the active concentration of the reactants (Zeder-Lutz et al. 1999) and provides kinetic constants that are better correlated with Ab neutralizing capacity than the avidity indices obtained by chaotrope disruption (Van Cott et al. 1994; Van Regenmortel and Azimzadeh 1994, pp. 805–828).

It may seem counterintuitive that the polyspecificity of Abs is in fact responsible for the greater collective specificity of a polyclonal antiserum induced by a multi-epitopic protein compared to the discriminating capacity of a single Mab (Talmage 1959; Richards et al. 1975). Since all the Abs in a polyclonal antiserum are polyspecific, they will all react with many unrelated epitopes, but as these cross-reactive epitopes are all different for each Ab in the antiserum, the cross-reactive potential present in the serum will be diluted out and will be masked (Fig. 4) Such a dilution effect does not occur in the case of a Mab and its ability to cross-react with other epitopes is therefore more easily observed (Van Regenmortel 2014a). The same phenomenon explains the remarkable specificity of the multi-epitope ligand cartography technique that uses Mabs conjugated to different fluorescent dyes for detecting the topological location of individual proteins in multimolecular assemblies present within cells and tissues (Schubert et al. 2006; Schubert 2014).

In structure-based RV, the Mab used to characterize one HIV-1 epitope by crystallographic analysis of a Mab-Env complex is always polyspecific, which

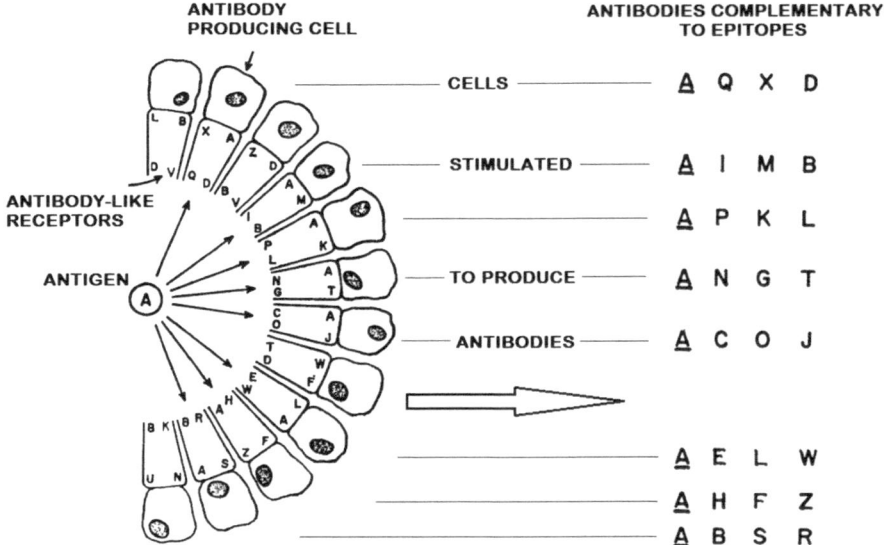

**Fig. 4** Immune serum specificity as a population phenomenon. Individual B cell receptors are shown as having properties similar to immunoglobulin combining regions. For illustrative purposes, these are drawn as being each complementary to four different epitopes or antigens; we suppose that this number is in fact much larger. Stimulation by antigen A causes the cells with A specificity to divide and produce antibodies directed against A. The immune serum produced will therefore react in high titer with antigen A. Each produced immunoglobulin also has other specificities, but because these need not be the same in every molecule, the other specificities, B to Z, will be diluted out and will react only in low titer (Richards et al. 1975)

means that this epitope is only one of the several epitopes that the Mab is able to recognize. When that Mab is present as a BCR in the immunized host, it could also have been selected by other cross-reactive epitopes rather than by the one elucidated by crystallography. This means that this structurally defined epitope does not necessarily correspond to the immunogenic structure believed to be active during the immunization process. Since an Ab is never monospecific for a single epitope, there is no reason to assume that a particular Ab can only be elicited by a single immunogenic, epitope structure. This is one of the reasons why a reductionist structure-based RV approach that concentrates on only one epitope is unlikely to be able to reveal the structure of effective vaccine immunogens capable of inducing polyclonal immune responses (Van Regenmortel 2012b, 2014b).

Since Abs are also heterospecific, i.e., often able to react much better with other antigens than the one that was used in the immunization process used to elicit them, the immunogenicity of an epitope is also not necessarily accompanied by an antigenic reactivity that will enable it to bind to the induced Abs (Van Regenmortel 2014a). Antigenicity and immunogenicity are thus not properties that are always present simultaneously in the same regions of a protein molecule. This underlines the need to distinguish the binding capacity of an antigen or epitope, i.e., its antigenicity, which is a chemical property, from its immunogenicity, which is a biological property determined mainly by mechanisms and factors in the IS that are independent of the chemical basis of epitope–paratope recognition—the only parameter considered by reverse vaccinologists.

## *Structural and Functional Analyses Delineate Different Epitopes*

It has been known for a long time that structural and functional approaches to the study of protein antigenicity lead to different perceptions of the nature of protein epitopes (Van Regenmortel 1989a; Greenspan 1992, pp. 55–79).

Functional studies take the form of Ab-antigen binding measurements that tend not to provide reliable information on which residues in the antigen and Ab molecules make contact with each other since the conformation of the interacting molecules depends considerably on the format and conditions of the binding assay that is used. When the results of liquid-phase and solid-phase immunoassays are compared it may lead to the conclusion that different residues in the two partners participate in the two types of assay (Van Regenmortel and Muller 1999); a similar uncertainly prevails when the antigenic peptides are tested in their free form or conjugated to a carrier (Muller et al. 1986). When longer antigenic peptides are found to bind better in an assay that shorter ones, it is difficult to know if the additional residues present in the longer peptide actually bind to the Ab or simply induce a conformation that is better recognized by the Ab. It is also sometimes observed that when the length of an active peptide is increased, this completely abolishes its ability to bind to an Ab (Muller et al. 1986). It is difficult to interpret in structural terms the results of functional binding assays with both continuous and discontinuous epitopes since such assays are rarely able to clearly establish which amino acids are directly involved in an interaction. These assays therefore are more useful for providing information on the affinity and biological activity of the reactants than on their structure (Van Regenmortel 1995).

It is undeniable that studies of the structure of protein epitopes involving the crystallographic analysis of Ab-antigen complexes provide a clear picture of what epitopes consist of. Initially structural studies were interpreted using a lock-and-key model which assumed that the binding process did not lead to significant conformational changes in the two partners (Mariuzza et al. 1987). This model was later abandoned when structural studies of antigen and Ab molecules in their free and bound state demonstrated that binding usually involves a process of induced mutual complementarity and fit resulting from important side-chain movements and changes in the backbone conformation of the two partners (Wilson and Stanfield 1994; Bosshard 2001; Rini et al. 1992). Since the structures visualized in Ab–antigen complexes usually differ considerably from the structures of the unbound binding sites, it is in fact impossible to know exactly which epitope structure was initially recognized by the Ab molecule when it was in its free form or present as a BCR at the surface of a lymphocyte in the host IS. Although somewhat different criteria can be used for inferring which residues are in contact with each other at an antigen-Ab interface (Chen et al. 2009; Getzoff et al. 1988) crystallographic elucidation of the complexes tends to promote a deceptive perception that epitopes and paratopes are fairly static patches of interacting surface residues. Space-filling models of protein surfaces show many juxtaposed neighboring atoms and this may obscure underlying secondary structure elements and give the impression that two static interacting units are in contact. Such an interpretation of epitopes and paratopes as static interacting surfaces is, however, refuted by the fact that paratopes are constituted of flexible CDR loops that allow them to adapt to a variety of epitopes, which themselves possess considerable flexibility (James et al. 2003; Ramsland et al. 1997; Yin et al. 2003; Khan and Salunke 2012) and are often located in regions of high segmental mobility in the protein sequence (Kong and Sattentau 2012; Berzofsky 1985; Westhof et al. 1984; Tainer et al. 1985; Uversky et al. 2005).

The HIV-1 V3 antigenic site is a particularly good example of the difficulties encountered when a well-delineated structural element of the Env spike is analyzed for its ability to react with neutralizing Abs found in HIV-1 infected individuals or in immunized humans and animals. The semi-conserved and flexible loop V3 plays a major role in the recognition of chemokine co-receptors by the virus and deleting the V3 region is known to completely abrogate virus infectivity (Saunders et al. 2005). The V3 loop has a constant size of about 35 residues and has for many years been considered as the principal neutralizing domain of the virus (Javaherian et al. 1989; Spear et al. 1994). Anti-V3 antibodies are able to block virus infectivity by interfering with co-receptor recognition (Sharon et al. 2003; Gorny et al. 2002) although the value of V3 as a potential vaccine candidate has sometimes been questioned because no anti V3 antibodies appeared to be present in anti-HIV-1 antisera when these were tested with linear V3 peptides (Van Cott et al. 1995a). It was subsequently demonstrated that these antisera did actually contain anti-V3 antibodies that only reacted with constrained cyclized V3 loops possessing the correct conformation not present in linear V3 peptides (Gorny et al. 2002; Zolla-Pazner 2004; Gorny et al. 2006).

The V3 loop is very flexible and can adopt different conformations when it binds to different Mabs (Stanfield et al. 1999), making it difficult to ascertain which conformation is likely to be optimal for a V3 synthetic peptide vaccine. NMR studies of V3 loops bound to Abs specific for different HIV-1 isolates, revealed that the conformation of Ab-bound V3 peptides was dictated by a process of induced fit to each Mab (Rosen et al. 2005) and thus that the observed structure of a bound V3 epitope did not indicate which immunogen conformation was likely to elicit a particular type of neutralizing Ab (Fig. 5). Even cyclic V3 peptides retain significant flexibility and although V3 loops can be chemically constrained in particular conformations (Mor et al. 2009; Mester et al. 2009) there is no evidence whether better vaccine candidates would necessarily be obtained by increasing the rigidity of the V3 loop that is naturally flexible and is therefore more easily recognized by a variety of BCRs (Van Regenmortel 2011b).

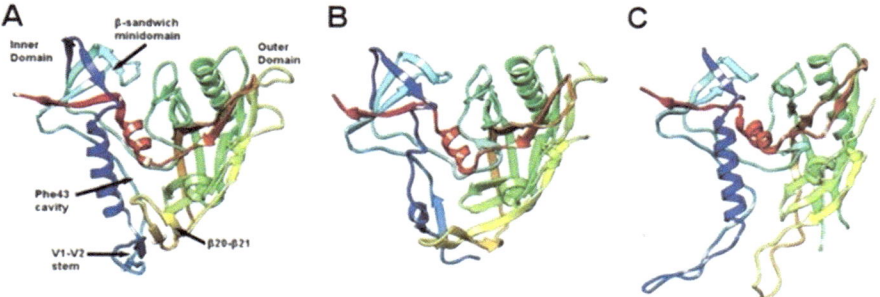

**Fig. 5** Structure of gp120 in different ligation states. (**a–c**) Comparisons of gp120 structures in (**a**) the CD4-bound state; (**b**) the b12 antibody-bound state; and (**c**) the F105 antibody-bound state. Structures are rendered as ribbon diagrams colored spectrally along the sequence from blue at the N-termini to red at the C-termini. The view is into the face occupied by CD4 interactions (Korkut and Hendrickson 2012)

Recent crystallographic analysis of the trimeric pre-fusion HIV-1 Env showed that V3 loop residues are occluded in a pocket, formed by the V1 V2 loop and V1 V2 stem of its own protomer and by the V1 V2 stem of the neighbouring protomer, giving rise to quaternary interprotomer interactions (Pancera et al. 2014). As a result the V3 loop is poorly accessible to anti V3 antibodies (Julien et al. 2013). A recent study (Zolla-Pazner et al. 2016) showed that individual mutations in various regions of the Env-trimer induce an open conformation that exposes the V3 loop to V3 Abs and lead to virus neutralization. It had previously been shown that if V3 Mabs were incubated with the virus for a long time before adding target cells in a neutralization assay, the neutralization capacity of V3 antibodies was considerably enhanced, suggesting that dynamic changes in the V3 loop accessibility were required for optimal neutralization (Upadhyay et al. 2014). There is considerable evidence that the trimeric Env spike possesses considerably more flexibility than previously thought and that this plasticity is also present in the unliganded state of the trimer. Env promotes entry of the virus into target cells by recognizing the cellular CD4 receptor and this leads to conformational rearrangements in Env, which allow binding to the co-receptor binding site and triggers the fusion of viral and cellular membranes and mediates virus entry in the host cell (Klasse 2012; Kwong et al. 2002).

By incorporating fluorophores into unliganded HIV-1 Env trimers on the surface of native virions, it was possible, using fluorescence resonance energy transfer (FRET) methods, to demonstrate that the trimers fluctuated between three distinct conformations (Munro et al. 2014; Munro and Mothes 2015). A low-FRET state reflected the closed ground state conformation of the prefusion Env while two higher-FRET states could be stabilized by soluble CD4 and by the co-receptor-mimicking antibody 17b. Since all three FRET states were also observed in the absence of ligands, the conformations that are stabilized by CD4 and co-receptor could be studied with unliganded HIV-Env, demonstrating an unexpected, considerable intrinsic structural flexibility in the Env protein. These new findings helped to clarify some of the unexplained features of the neutralization behaviour observed with certain nMabs that could be the result of large scale conformational transitions in gp120 (Korkut and Hendrickson 2012; Zolla-Pazner et al. 2016).

Crystal structures of gp120 bound to Fab fragments of Mabs b12 and F105 and to CD4 have demonstrated the extensive rearrangements of gp120 in different ligation states and shown that intermediates generated from antibody-bound states deviate from the initial structure more than those generated from the CD4-bound state, highlighting the crucial role that conformational masking of epitopes plays in allowing the virus to escape the IS (Korkut and Hendrickson 2012). The extreme plasticity of the HIV-1 Env will no doubt continue to pose considerable challenges for developing a vaccine aimed at a virus that exhibits such enormous antigenic variability and structural plasticity.

## The Immunogenicity of a Protein Molecule Is a Complex Biological Property that Cannot Be Predicted from the Molecule's Antigenicity, Which Is Simply Its Ability to Bind Certain Immunoglobulins

An epitope can be defined as any accessible part of a protein that can be recognized by a paratope present in a free Ab molecule or in a BCR at the surface of certain B cells. It must be emphasized that this epitope definition does not include an immunogenic capacity to induce, in different immune systems, the same Ab that the epitope is able to react with; if this were the case, the existence of an epitope in a protein would depend on immunogenetic and immunoregulatory mechanisms of the immunized host. If a molecular biologist engineers a new protein that previously did not exist and therefore was never used to elicit Abs, it would nevertheless possess epitopes able to cross-react with Abs previously induced by other proteins.

Jay Berzofsky in a much cited review published more than 30 years ago (Berzofsky 1985) clearly distinguished between properties intrinsic to protein antigens (such as their chemical and physical structure) and other properties of the host IS that are extrinsic to the antigen molecule but control its immunogenicity, such as the host antibody gene repertoire, the specificity of helper and suppressor T cells, antigen processing, self-tolerance, idiotypic networks, and many additional immunoregulatory mechanisms. Of all the potential epitopes of a protein, only a subset will therefore be immunogenic in any individual host.

The difference between antigens and immunogens is crucial in vaccinology but is often not sufficiently appreciated. Most peptides are immunogenic since they readily elicit Abs that react with the peptide immunogen (Muller 1999a, pp. 113–177). In the context of vaccination, however, what is relevant is so-called cross-reactive immunogenicity which is the ability of epitopes to induce Abs that cross-react with the parent native protein, as well as cross-protective immunogenicity which is the ability of epitopes to induce Abs that neutralize the infectivity of the pathogen harboring the antigen (Van Regenmortel 2006). It is of course the IS that elicits and produces antibodies and not the antigen or immunogen which only plays a triggering role. It is important to appreciate that the chemical environment is not the same when an epitope recognizes either a free Ab molecule or a BCR embedded in a lipid membrane. This means that the antigenic epitope bound to a free Ab may not be identical with the immunogenic epitope that interacts with a BCR (Zwick 2005; Scherer et al. 2010). When an HIV-1 peptide reacts in an immunoassay with an Ab directed to a native protein, the Ab may select one of the many conformations present in the peptide or it may induce a reactive peptide conformation by induced fit. On the other hand, when the same peptide is used as an immunogen, different peptide conformations may be recognized by separate BCRs and there is no reason why the peptide should bind preferentially to those rare receptors that either cross-react with an epitope present in the native protein or that will subsequently lead to the secretion of neutralizing Abs by plasmocytes (Van Regenmortel 2009b, 2015a; Greenspan 2014).

The selection of BCRs during immunization is blind to the presence of neutralizing anti-viral activity in the Abs that will subsequently be secreted since the selective forces operating during immunization depend mainly on reactant concentration and on the probability of lymphocyte stimulation occurring above a certain triggering threshold (Foote and Eisen 1995; Batista and Neuberger 1998).

## Conclusions

Many hypotheses and theories have been proposed to explain the evolutionary origin of the IS. Initially, the IS was considered to be a defense system against pathogens and invaders and subsequently this gave rise to the concept of the Protecton (Cohn and Langman 1990). Later on, it became generally accepted that the main purpose of the IS was to discriminate between self and non-self, although there is no generally accepted definition of immune selfhood (Tauber 1997). In 1974 Niels Jerne proposed his idiotypic network theory (Jerne 1974) according to which the IS is a network of anti-idiotypic Abs that recognize epitopes called idiotopes present on the variable regions of other Abs in the same individual. Such Abs are produced in an IS because the initial Abs1 induced by an external antigen act as functional self-antigens that induce secondary Abs2 directed against idiotopes present in the Abs1. An Ab2 could mimic the epitope recognized by Ab1 and may correspond to an internal image of the external antigen. Subsequently, tertiary Abs3 directed against the Abs2 may appear leading to a cascade of anti-idiotypes which transforms the IS into a self-regulatory network of anti-idiotypes (Greenspan 1992, pp. 55–79). Such a cognitive system acts as a regulatory network of anti-idiotypic Abs, which stabilizes the antigenic integrity of the body and prevents the immune response from getting out of control or producing autoimmune disease (Parnes 2004; Foote and Eisen 1995; Tauber 1997). Once it was recognized that all Abs, BCRs and TCRs are polyspecific and that autoimmunity is a natural, normal immune phenomenon

(Dimitrov et al. 2013; Tauber 1997), other theories favoring immune incorporation instead of interception were developed (Parnes 2004) which gave rise to intense debates between the proponents of various interpretations of immune activity (Cohen et al. 2004).

In the title of the present review, reference was made to the fact that an approach or procedure used to develop a vaccine should not be incompatible with accepted immunological theories. However, none of the theories about the origin and nature of the IS mentioned above are universally accepted and immunologists disagree about which theory will eventually gain general acceptance. On the other hand (see section "Facts, Hypotheses, Laws, Theories, and Empirical Experimentation"), other immunological theories have become widely accepted and since they have overcome all attempts to falsify them, such theories should not be ignored by vaccine developers. For instance, the degeneracy of the IS and the polyspecificity of Abs and BCRs are scientific facts that must accepted by vaccine developers and, since they are not compatible with the presuppositions underlying structure-based RV, it is not astonishing that this approach failed to yield an effective HIV-1 vaccine. The failure of structure-based RV can be explained as follows (Van Regenmortel 2014b, 2015a):

1. The epitope structure observed in a complex when it is bound to a neutralizing, polyspecific Mab is only one of the many epitopes that can be accommodated by that Ab and there is no reason to assume that it must correspond to the immunogenic epitope that elicited the Mab.
2. The structural delineation of HIV-1 epitopes bound to neutralizing Mabs cannot tell us which vaccine immunogens will be able to induce a protective polyclonal immune response. Improving the antigenic binding capacity of an epitope does not constitute vaccine design.
3. Short continuous epitopes of HIV-1 that are mostly only a part of more complex discontinuous epitopes of the virus are not a suitable starting material for developing effective vaccine immunogens.
4. Rational design can be used for optimizing the antigenic binding capacity of one HIV-1 epitope for one nMab or for improving the reactivity of one nMab intended for passive immunotherapy. However, structure-based rational design for improving the binding capacity of one epitope or one paratope does not constitute vaccine design and it cannot elevate vaccine research from an empirical exercise to a scientific discipline. Empirical experimentation aimed at manipulating the immune system successfully is entirely rational and is the approach that was successfully used in the past for developing effective vaccines.
5. Since a complex biological system like the IS possesses many emergent properties that are not present in the individual components of the system, it is not possible to fully understand and control the IS using the reductionist strategy of dissecting and analyzing separately the structure, activity, and interactions of all its numerous constituents.

The RV paradigm has guided a considerable part of HIV-1 vaccine research for about a decade although it completely failed to produce an effective vaccine immunogen. Hundreds of RV research projects aimed at increasing our basic knowledge of HIV-1 immune responses and immunopathology were funded which produced a remarkable harvest of basic science publications that described in detail the structure of many new HIV Env epitopes and their complementary Ab paratopes. This large body of new factual knowledge may one day be useful for passive immunotherapy purposes in spite of the fact that it did not help the development of a vaccine intended for active immunization. This situation once more illustrates the principle that the prescriptive knowledge that vaccinologists require in order to successfully intervene in an immune system and achieve protective immunity is not simply derived from previously acquired factual knowledge obtained by basic research. As explained in section "Paradigms in HIV Vaccinology", a new HIV-1 vaccine is always an invention that necessarily involves an empirical, trial-and-error step that inherently differs from a rational design approach; in the case of HIV vaccines the so-called rational design approach was also ineffectual because it confounded immunogenicity with antigenicity.

Referring to HIV-1 epitopes as vaccine immunogens able to elicit bnAbs probably also added to the confusion since it encouraged investigators not to focus on the intrinsic complex features of the host immune system which hold the key to the development on any vaccine. The past unproductive RV episode that did not help the development of an HIV-1 vaccine could possibly be valuable if it reminds future investigators that they should not disregard established scientific theories in their search for the prescriptive knowledge needed for developing any vaccine.

# Immune Systems Rather than Antigenic Epitopes Elicit and Produce Protective Antibodies Against HIV

## Letter to the Editor

It is common parlance to refer to viral antigens and their epitopes as immunogens capable of producing antibodies (Abs) against the virus that harbours them. Words have an insiduous capacity to fashion our thinking and terms like immunogen and immunogenicity do suggest that epitopes are able to generate immune responses, although they only trigger in the host a series of reactions with B-cell receptors that eventually leads to the immune system (IS) producing a variety of antibodies. Although everyone in the field is well aware that antigens are different from immunogens, it seems that in many cases investigators do not use appropriate experimental tools for studying and controlling the immunogenicity of proteins rather than their antigenicity.

For many years, much of HIV vaccine research concentrated on elucidating the structure of HIV epitopes present on the virus glycoprotein spikes because these epitopes were considered to be potential vaccine immunogens capable of inducing protective Abs against virus infection. The approach known as structure-based reverse vaccinology (SBRV) analysed the structure of complexes between HIV epitopes and neutralizing monoclonal Abs (nMabs) in an attempt to design epitopes by reverse molecular engineering that would elicit nAbs when used as vaccine immunogens (Van Regenmortel 2011a). Investigators using this approach called it vaccine design because they assumed that if an antigenic epitope did bind strongly to a nMab, it would also be able to induce similar neutralizing Abs when used as a vaccine (Van Regenmortel 2012b). They also assumed that when an antigenic epitope binds to a free Ab molecule, the recognition process is exactly the same as

---

Vaccine, 2017, 35, 1985–1986.
Marc H V Van Regenmortel

© Springer Nature Switzerland AG 2019
Marc H V Van Regenmortel, *HIV/AIDS: Immunochemistry, Reductionism and Vaccine Design*, https://doi.org/10.1007/978-3-030-32459-9_22

when the same epitope (which is then an immunogen) binds to the cognate B-cell receptor embedded in a lipid membrane. There is, however, evidence that lipids present in B-cell membranes may contribute to the binding observed when a protein antigen interacts with a B-cell receptor, which means that the same antigen is likely to bind more weakly to the corresponding free Ab molecule. It is also known that when an Ab binds to a free viral peptide, the binding may be weaker than when that Ab binds to the same peptide embedded in a viral membrane, because of additional hydrophobic interactions with membrane lipids (Scherer et al. 2010). An additional problem with the SBRV approach was that it ignored the fact that all Abs are polyspecific or even heterospecific and that the antigenic and immunogenic properties of a protein antigen may be located in different parts of the molecule (Van Regenmortel 2014a). Since they were focusing on only one of the epitopes recognized by a polyspecific nMab, the investigators assumed that this epitope was the immunogen that elicited the nAb and should therefore be able to elicit similar nAbs when used as a vaccine (Van Regenmortel 2014b). Although many structures of nMabs-epitope complexes were elucidated in this manner, the insights derived from these immunochemical studies did not lead to an effective HIV vaccine (Pejchal and Wilson 2010; Van Regenmortel 2016b). It was found, for instance, that strongly neutralizing Abs often possessed very long CDR-H3 but it was not feasible to reverse engineer that property in a vaccine immunogen so that the IS would elicit such nAbs (Yu and Guan 2014).

Many properties of the IS are known to control the types of Abs that are produced, for instance the host Ab gene repertoire, the presence of helper and suppressor T cells, self tolerance and numerous other immunoregulatory mechanisms in the host (Berzofsky 1985). However, vaccine designers using the SBRV approach tended to ignore these factors because they focused their attention on recognition processes between single epitope-paratope pairs which made it impossible for them to figure out how to intervene in the IS so that it would produce neutralizing Abs. Although the multiple shortcomings of SBRV have been repeatedly pointed out (Van Regenmortel 2016b) its proponents have never acknowledged that the paradigm they followed was therefore unlikely to lead to an effective HIV vaccine. When it was discovered after 10 years that HIV Env epitopes recognized by affinity-matured Abs obtained from HIV-infected individuals did not bind the germline predecessors of these Abs (Xiao et al. 2009), it became obvious that potential vaccine immunogens would only be discovered if one took into account the slow but extensive Ab maturation that is required for obtaining nAbs. Finding potential HIV vaccine immunogens by SBRV appeared no longer feasible and a huge research effort was initiated to analyze the innumerable maturation pathways that allow individual immune systems to transform non-neutralizing germline Abs into protective Abs (Mascola and Haynes 2013). Our understanding of antibody maturation processes increased dramatically but the complexity that emerged was daunting because it indicated that several sequential immunizations with particular Env immunogens would be needed to drive the immune responses of genetically heterogeneous human vaccinees towards highly mutated anti-HIV bnAbs (Stamatatos et al. 2017).

Since rational vaccine design based on antigenicity studies was unable to reveal which immunogens may lead to a protective immune response (Van Regenmortel 2016b), it became increasingly evident that only in vivo empirical, experimental immunogenicity studies might be able to provide the required information. Unfortunately, most animal models are poorly predictive of responses in humans and testing a huge variety of HIV vaccine immunogens in humans is not a practical alternative. The most promising approach for evaluating HIV vaccine immunogens developed in recent years is the use of human Ig knockin (KI) mice that have incorporated human Ig variable regions in the corresponding mouse loci (Verkoczy et al. 2017). The remarkable results that have been obtained indicate that KI mice are the best model system for testing HIV vaccine immunogens as well as novel adjuvants and T cell help. The use of KI mice might also be able to conclusively establish whether the existence of self-reactivity in nAbs leads to host tolerance and controls the immune responses induced by particular Env immunogens (Verkoczy et al. 2017).

Immune correlates of protection are often believed to be essential information for guiding HIV vaccine development although they can be inferred only retrospectively after an efficacious vaccine has been obtained empirically (Van Regenmortel 2012b). At the moment, only correlates of HIV-1 decreased transmission risk (i.e. not protection) involving multiple immune responses observed during the RV144 vaccine trial have been investigated (Tomaras and Plotkin 2017) and their usefulness for further improving vaccine regimens will need to be confirmed after a more effective vaccine has been developed. This reminds us that the prescriptive, empirical knowledge required for producing an efficacious HIV vaccine cannot be obtained beforehand by deducing immunogenicity from antigenicity (Van Regenmortel 2016b). The current, huge investment in curiosity-driven basic research aimed at understanding HIV-Ab interactions will no doubt greatly increase our factual knowledge of the enormous complexity of the IS and of HIV. However, it also cannot be excluded, as was often observed in vaccinology in the past, that applied vaccine research may succeed even if we lack a complete understanding of the complex immunological mechanisms that would underlie the mode of action of an ideal HIV vaccine.

# Development of a Preventive HIV Vaccine Requires Solving Inverse Problems Which Is Unattainable by Rational Vaccine Design

*Science is the acceptance of what works and the rejection of
what does not. That needs more courage than we might think.*
Jacob Bronowski

*The average scientist unequipped with the powerful lenses of
conceptual insight, is a nearsighted creature, who cheerfully
attacks each difficulty in the hope that it will be the last.*
Gilbert Lewis

## Introduction

Most scientists have heard of the scientific method but usually have rather vague
ideas about what it might be. In his popular book, translated into 15 languages,
entitled *What is this Thing Called Science?* Chalmers (1999) knew better than to
formulate a philosophically articulated and rigorous account of a universal scientific
method that would fully explain the success of science. It has also been argued
convincingly that a universal scientific method applicable to all scientific disciplines is actually a myth (Bauer 1992) and there is, indeed, no scientific recipe book
that scientists could use to make either discoveries (i.e., revealing something
that existed all along but was unknown to anybody) or inventions (i.e., achieving
something new that was not previously feasible, for instance, successfully vaccinating people against HIV infection). Furthermore, it is no longer considered
feasible to simply derive a desirable invention such as an effective HIV vaccine
from prior discoveries obtained by curiosity-driven basic research in immunology
(Stokes 1997; Van Regenmortel 2016b).

Frontiers in Immunology, 2018, volume 8, 2009
Marc H V Van Regenmortel

© Springer Nature Switzerland AG 2019
Marc H V Van Regenmortel, *HIV/AIDS: Immunochemistry, Reductionism and
Vaccine Design*, https://doi.org/10.1007/978-3-030-32459-9_23

The scientific method is often misrepresented as an invariable sequence of steps going from observations to hypothesis formulation, theory building and experimental verification rather than as a highly variable, complex and creative process (Gauch 2012). Attempts by philosophers of science to describe the methods used by scientists for obtaining new, reliable scientific knowledge often bear little resemblance to the way scientists behave in the laboratory. In their book entitled *Theories of Scientific Method* (Nola and Sankey 2007, p. 2) the authors quoted the physicist Richard Feynman who said: "Philosophy of science is about as useful to scientists as ornithology is to birds," as well as the philosopher of science Imre Lakatos who quipped: "Most scientists tend to understand little more about science than fish about hydrodynamics."

A major difficulty for explaining the success of the scientific method is due to the fact that scientific experiments in biology, for instance, produce results that regularly require continual self-correction because the data that are obtained are always approximations consisting of mixtures of "real" meaningful signals (relevant to what is being investigated) together with spurious "noise" due to random errors (Tsimring 2014). This prevents absolutely certain and reliable knowledge to be obtained and explains why the scientific method has frequently failed to solve problems in the past. A satisfactory purely logical description of the scientific method should therefore be able to explain how the same method can either succeed or fail, a feat that has not yet been possible (Chalmers 1999; Bauer 1992, p. 51).

The present review will discuss a number of invalid assumptions that have been detrimental to the search for a preventive HIV vaccine. Although several scientific journals are entirely devoted to the different approaches that can be used to solve inverse problems (*Journal of Inverse and Ill-posed Problems, Inverse Problems in Science and Engineering, Inverse Problems and Imaging*) many vaccinologists are unaware that the problems they are trying to solve are mostly inverse problems which consist in imagining what could bring about a desired outcome, for instance immunological protection against virus infection (Van Regenmortel 2012b). Investigators may not clearly differentiate between antigens and immunogens and often think of viral antigens as immunogens, implying that viral antigens and epitopes are able to trigger and generate immune responses although it is obvious that it is the immune system (IS) that produces antibodies (Van Regenmortel 2017).

This review will describe the reductionist mindset (Van Regenmortel 2004b; Mazzocchi 2008), which has dominated molecular biology for half a century as well as a number of research paradigms that have been counterproductive for developing an HIV vaccine (Esparza 2013a, 2015; Van Regenmortel 2015b). It appears that the widespread expectation that the rational design of an HIV vaccine is likely to be more successful than tentative, small scale trial-and-error vaccine experimentation may have made it more difficult for vaccinologists to solve the inverse problems they are faced with.

# What are Inverse Problems?

Solving an inverse problem consists in first proposing a theoretical model for explaining, for instance, the absence of deleterious HIV infection in elite controllers, and subsequently demonstrating that by adjusting experimentally certain parameters in human ISs which the model predicts are crucial, it is indeed possible to obtain the desired outcome. Solving inverse problems therefore consists in identifying what are the causes that produced an observed effect, which requires reasoning by induction from a particular instance to a tentative generalization. Solving a direct problem, on the other hand, involves reasoning by deduction from a reliable premise or from an accepted physicochemical law to a particular instance, a procedure that in principle leads to the correct solution (Kell and Oliver 2004).

Many scientific problems in biology are direct, forward or downstream problems that can be solved by experimentally determining what are the effects that follow certain causes. Usually this involves elucidating particular causal mechanisms that are responsible for the occurrence of certain biological phenomena. A specific mechanism consists in interactions between the numerous parts of a complex biological system that are linked by direct causal relations and make it possible to systematically predict a result by providing an explanation for the occurrence of a given phenomenon (Bechtel and Abrahamsen 2005). Solving a direct problem, therefore, always starts with known causes and makes the investigator observe, analyze, or calculate the results (Allen 2001).

In physics and astronomy only a small number of physical forces exist that cause observable phenomena. For instance, the laws of universal gravitation together with certain initial conditions suffice to predict the occurrence of an eclipse because there are no external interferences in an interstellar vacuum.

The situation is completely different in the biological sciences because there are no universal laws in biology (Dupré 1993). No single event in biology can be said to be *the* cause of another event because it was necessary and sufficient on its own for the effect to occur (Van Regenmortel 2007, p. 969). Since any observed effect in complex biological systems always results from a network of causal interactions and internal regulations, an analysis in terms of a single causal factor instead of a multitude of contributory causes is never satisfactory (Berger 1998).

Solving an inverse or upstream problem in biology requires inferring from a set of observations what are the multiple causes that produced certain effects. Medical diagnosis, for instance, is a typical inverse problem which consists in guessing the causes of a disease from some of its symptoms (Anger and Moritz 2003). An inverse problem thus starts with a result and requires that the investigator must try to imagine what could be the causes that produced it. This means that it is always necessary to first conceptualize a theoretical model that would account for what has been observed and subsequently to demonstrate experimentally that what the model predicts actually occurs. This is based on less reliable inductive reasoning which starts with the observation of particular instances in order to derive a general principle or model that unavoidably will only have a certain probability of being correct (Allen 2001). Contrary to the solution of a direct problem which is easily obtained by deductive reasoning from correct premises or from known mechanisms or laws, solving an inverse problem is usually extremely difficult and may often be impossible.

In view of the complexity of the IS, a multiplicity of unknown causes may be responsible for what is observed when a patient has been infected or when a vaccine has been administered, and the inverse problem of developing an HIV vaccine could therefore have numerous solutions or possibly none at all. HIV vaccinologists, for instance, need first to conceptualize and articulate what are the multiple causes that sometimes allow the IS in a small number of individuals to elicit immune responses that protect against HIV infection. Subsequently, they need to make these rare causal chains occur regularly in large populations of genetically and physiologically heterogeneous human vaccinees. If the desired result is not achieved when the theoretical model is implemented, the inverse problem of developing a vaccine has not been solved!

Unfortunately, vaccinologists cannot follow the procedure followed by X-ray crystallographers when they derive the structure of a protein from the set of concentric rings or parallel bands observed in the X-ray diffraction pattern of the protein crystal. This procedure involves calculating by Fourier analysis the diffraction patterns produced by irradiating conceptual crystals with an imaginary beam of X-rays and choosing which of these theoretical diffraction patterns best matches the pattern observed with the protein of interest (Bunge 2017, p. 98). Such a procedure transforms the insoluble inverse problem posed by the uninterpretable diffraction pattern of the protein crystal into a series of hypothetical direct problems that are easily solved. Unfortunately, such an approach is not applicable for solving inverse problems in vaccinology.

## Can Systems Biology Help Solve Inverse Problems in Vaccinology?

Living organisms and biological systems are always organized into successive levels of increasing complexity from genes to RNAs, proteins, subcellular organelles, cells, tissues, and organs. There is no privileged level of causal determination at any of these levels since both bottom-up and top-down causal determination occur, in addition to horizontal causation operating within the boundaries of each of the different levels (Noble 2012). There is, however, no single causal pathway that links gene sequence to protein sequence and conformation or to protein binding and function and which further extends to the upper levels of cells, tissues, organs and organisms (Van Regenmortel 2012b). At each successive level of biological organization, innumerable interactions between genetic, epigenetic, biochemical, physiological and environmental factors occur and it is necessary to first study each level separately as an independent system (Mahner 2017) before the different levels can all be integrated to give rise to the notion of systems biology. Whereas top-down systems biology analyzes how higher levels of organization determine events at lower levels, bottom-up systems biology analyzes the functional properties, that emerge from a subsystem that has been characterized to a high level of mechanistic detail, in an attempt to predict its behavior and the occurrence of certain biological phenomena (Bruggeman and Westerhoff 2006). Unfortunately, few kinetic parameters are currently measurable in vitro which severely limits the usefulness of the models that can be proposed at the different levels. Systems biology is therefore rarely capable of developing theoretical models that encompass several different levels and could form the basis for investigating experimentally whether integrated immunological reactions associated with vaccine activity occur as predicted in models.

In an often-quoted review, Brenner (2010) claimed that systems biologists are unable to solve inverse problems of physiology by deriving how a complex system works by simply observing its behavior, because a hypothetical model can never be more than one tentative, imperfect representation of reality. The human IS consists of an extremely intricate network of specialized cells, tissues and organs which is probably the most complex biological system that exists, with the exception of the human brain. In recent years, systems vaccinology which arose from systems biology, has employed high-throughput microarray technologies such as genomics, transcriptomics, proteomics, and metabolomics for describing the innumerable interactions that occur between all the parts of an IS in an effort to present a more holistic view of the entire system than is feasible when the constituents of each level are studied separately (Nakaya and Pulendran 2012). This makes it possible, for instance, to identify gene signatures that may be able to predict the immunogenicity of certain vaccines, as well as the changes that occur in important cell subsets isolated from human blood, for instance CD4+ T cells, CD8+ T cells, and regulatory T cells. However, identifying signatures that predict vaccine efficacy does not necessarily provide a mechanistic insight for explaining how a particular vaccine stimulates protective immunity (Pulendran 2014).

Transcriptional profiling of human immune responses to vaccination using peripheral blood mononuclear cells obtained from whole blood of vaccinees has demonstrated that immunization leads to up-or downregulation of many heterogeneous gene clusters, such as genes associated with innate immunity, dendritic cell activation, production of different cytokines, T cell receptor signaling, antibody titers, cell cycle progression, inflammatory responses, protein folding, mitochondrial dysfunction, etc. (Wang et al. 2012). Although some of these gene signatures are considered to be predictive of specific adaptive immune responses (Pulendran et al. 2010), the considerable measurement noise that is observed in the assays makes it difficult to distinguish correlations from causal relationships that may be linked to protection. Since a large number of up or downregulated gene products act in combination to generate a biological function such as protection, it is extremely difficult to elaborate an integrated theoretical model that includes all the relevant causal factors and which could be tested in an attempt to solve a particular inverse problem. Furthermore, in non-linear dynamic systems, a complex network of internal regulations together with negative feedback and feed-forward control is always present, making it difficult to identify which causal factors are responsible for a biological effect (Van Regenmortel 2012b; Noble 2012).

In his comprehensive overview of the promises of systems vaccinology for identifying effective vaccines, especially when correlates of protection are unknown, Pulendran (2014) pointed out that several subsystems present in a completely integrated human IS are still insufficiently understood to provide the necessary knowledge required for developing a protective HIV vaccine. The inverse problem of solving each subsystem must first be achieved by developing appropriate theoretical models for each of them and showing experimentally that these models are correct. This is particularly difficult in the case of HIV because genetic, immunological and gut microbiome characteristic of the host must be considered together with the enormous antigenic variability of the virus. Since suitable theoretical models are not yet available for many of the IS subsystems, the problem of solving the more complex inverse problem arising in an integrated IS that incorporates all these subsystems is even more intractable. Such an holistic inverse problem can be called an ill-posed inverse problem which is defined as a problem for which insufficient information is available to make it possible to derive a plausible explanatory model that could be validated experimentally (Engl et al. 2009). Our ignorance of all the different parameters that control the IS subsystems does not allow us to elaborate a testable theoretical model that incorporates all the individual constituents and all the bottom-up and top-down causal links in each subsystem. It seems that several inverse problems will first have to be solved at the level of individual components of the IS before it would become possible to test the validity of an integrated theoretical model of immune protection against HIV infection.

This situation is somewhat reminiscent of the conundrum we face when trying to explain human consciousness (Mahner 2017). Consciousness is an emergent, systemic property and activity of the brain that cannot be attributed to a specific region of the brain but results when innumerable neuronal systems undergo many different processes. When we loose consciousness, following the administration of an anesthetic, we are unable to localize this phenomenon to a specific region of the brain, in the way sensory perceptions of vision or hearing originate in particular brain regions (Tonini and Koch 2008). Consciousness is thus an holistic and emergent brain property that results from the integration of a large number of individual brain processes. Similarly, the IS achieves protection against infection by an integrated combination of all its subsystems that cannot be explained by the activity of a single individual subsystem.

## Reductionist Thinking and Invalid Paradigms Have Hampered Attempts to Develop an HIV-1 Vaccine

Methodological reductionism has guided the development of molecular biology for half a century and was widely accepted because it was able to describe biological systems in terms of its physicochemical properties, giving rise to the belief that biology could be reduced to chemistry and physics (Van Regenmortel 2004b; Mazzocchi 2008, 2012).

Such a belief was consistent with the assertion of Crick (1966): "The ultimate aim of the modern movement in biology is to explain all biology in terms of physics and chemistry." When the double-helical structure of DNA was discovered in 1953, some commentators maintained that this discovery solved the mystery of life. Today, we recognize that the DNA structure only allowed us to understand in considerable detail, the molecular mechanisms involved in gene replication and expression and that it did not give us any insight on how genes actually lead to phenotypes.

For many years, biologists have had misgivings about the validity of reductionist explanations in biology (Bock and Goode 1998; Van Regenmortel and Hull 2002) because they were aware that biological systems possess so-called emergent properties that are not present in their constituent parts taken in isolation. Since dissecting the IS into its numerous components inevitably severs the connections that link the various parts together in a functionally integrated manner, essential features that regulate the IS are destroyed and it is no longer feasible to account for the behavior of the system as a whole (Van Regenmortel 2002a).

The advent of Mabs revolutionized immunochemistry because it allowed immunologists to investigate the chemical and functional properties of polyclonal antibody responses by analyzing single epitope–paratope interactions. The use of Mabs, however, also promoted a reductionist approach because vaccinologists tended to focus on individual epitopes as elicitors of nAbs instead of studying larger antigenic surfaces that harbor a continuum of overlapping epitopes. They also neglected the fact that most protective immune responses are polyclonal and often involve synergistic effects arising from the collective neutralizing activities of Abs directed to different epitopes. Relying on a single epitope, identified by reverse vaccinology using one Mab for inducing a protective vaccine response is known to be less effective than using cocktails of several Env epitopes (Klein et al. 2012).

Reductionist thinking also made many vaccinologists accept that if an epitope of HIV Env binds to a broadly neutralizing (bn) Mab, this epitope should also be able to induce similar bnAbs in an immunized human host. The *chemical* nature of antigenicity was thereby confounded with the *biological* nature of immunogenicity and this led investigators to claim that they were designing a vaccine immunogen capable of generating protective antibodies whereas they were actually only improving the binding reactivity, i.e., the antigenicity, of a single HIV epitope (Van Regenmortel 2016b). Another reductionist paradigm assumed that antibody protection against HIV infection could be analyzed at the level of individual epitope–paratope interactions although protection against infection is only meaningful in the biological context of particular cells and tissues and at the level of individual organisms. Additional methods for detecting emergent immunological patterns that are not detected by classical reductionist approaches have been described recently (Rivas et al. 2017).

The concept of scientific paradigm was introduced by Kuhn (1962) in his influential book *The Structure of Scientific Revolutions*. He argued that scientific investigators are always guided by assumptions and theoretical presuppositions, which he called paradigms, that determine the type of research and the experimental approaches they will use when trying to solve a particular problem. Since prevailing paradigms are usually not explicitly stated, investigators are often not aware of the tacit assumptions underlying their choice of paradigm and when they obtain results that are not consistent with these assumptions, they may fail to realize that their paradigm has been refuted and should be abandoned. This could then lead them to pursue unfruitful lines of investigation that may impede scientific progress. There is, indeed, evidence that in the field of HIV vaccine research several invalid paradigms have had a detrimental effect (Van Regenmortel 2015b; Van Regenmortel et al. 2014). Paradigms are sometimes followed because they appear to have been useful for achieving earlier vaccine successes and they may then continue to be used even if they are incompatible with more recent immunological knowledge (Van Regenmortel 2016b; Esparza 2015). Other invalid paradigms used in HIV vaccinology have been discussed previously (Van Regenmortel 2009a, 2014b, 2016b; Esparza 2013b).

## What Is Structure-Based Reverse Vaccinology (SBRV)?

Two types of reverse vaccinology should be distinguished. Genome-based reverse vaccinology, introduced by Rino Rappuoli (2001) predicts potential vaccine immunogens using bioinformatics analyses of entire bacterial genomes and has allowed the development of several effective bacterial vaccines (Rappuoli and Bagnoli 2011). A completely different approach was proposed by Burton (2002) for developing viral vaccines and it has been suggested that this should be called SBRV to clearly differentiate it from genome-based reverse vaccinology (Van Regenmortel 2011a). Such a nomenclature is more explicit than calling the two procedures "reverse vaccinology 1" and "reverse vaccinology 2," respectively (Burton 2017).

Structure-based reverse vaccinology attempts to generate a vaccine by first determining the X-ray crystallographic structure of a complex between, for instance, an HIV Env and a bnMab, and then using the 3D structure of the Mab as a template to reconstruct the epitope by reverse engineering outside the context of the Env protein in which it was originally embedded. This approach was called vaccine design because it was assumed that the reconstructed epitope designed to fit the Mab might be able to acquire the immunogenic capacity to induce a polyclonal Ab response endowed with the same neutralizing capacity as the bnMab. However, what was being designed was an antigen able to bind the Mab and not a viral immunogen able to induce a protective immune response (Van Regenmortel 2012a, 2014b).

The SBRV approach only focuses on the recognition process between a single epitope–paratope pair and it does not take into account that Abs are produced by the IS and not by the viral immunogen and that T cell help always play a crucial role.

Properties of the IS such as the Ab gene repertoire, self-tolerance, the presence of appropriate helper and suppressor T cells, the production of cytokines, and various immunoregulatory mechanisms determine whether protective Abs will actually be elicited. Since SBRV investigators made no attempt to control these IS parameters, it is not astonishing that they did not succeed in developing an effective HIV vaccine (Van Regenmortel 2012a, 2016b, 2017; Pejchal and Wilson 2010; Ho et al. 2005; Burton 2010).

Another reason for the lack of success of SBRV is that many investigators did not fully appreciate the extent of degeneracy of the IS. Degeneracy refers to the ability of many different Abs, B cell and T cell receptors to bind to the same ligand as well as to the ability of a single Ab or T cell receptor to bind many different antigens or peptides. This means that there is never a unique intrinsic epitope for any Ab molecule but always a diverse group of potential ligands (Wucherpfennig et al. 2007; Eisen and Chakraborty 2010). The reason for this is that every combining site of an Ig, consisting of 50–70 residues distributed over 6 hypervariable loops, always harbors numerous, overlapping smaller paratope subsites of 10–20 residues that possess considerable conformational plasticity and flexibility, allowing them to bind a variety of epitopes (Mariuzza 2006; Sundberg 2009). Since each paratope represents only about 20–35% of the entire Ig combining site, the one epitope structure identified in an SBRV experiment does not necessarily correspond to the immunogenic structure that elicited the Mab used as template. When that Mab was present as a B cell receptor in the original immunized host, it could have been selected by other cross-reactive epitopes than by the one identified by X-ray crystallography in SBRV. Since an Ab is never monospecific for a single epitope, there is actually no reason to assume that the particular Mab used as template could only have been elicited by a single immunogenic structure, a fact that clearly invalidates the SBRV approach (Van Regenmortel 2014b).

Another somewhat unexpected consequence of the polyspecificity of Abs is that the collective specificity of a polyclonal antiserum induced by a multiepitopic protein is always greater than the specificity of a single Mab raised against that protein. Since each Ab in a polyclonal antiserum is able to cross-react with a variety of different epitopes, these various cross-reactivities present in the antiserum are diluted out and will usually be masked. In contrast, there will be no such dilution effect in the case of a Mab and its ability to cross-react will therefore be observed more easily (Van Regenmortel 2014a).

Although the crystallographic structure of many anti-HIV Env bnMabs has been elucidated in SBRV studies, it is evident that this knowledge has not made it possible to manipulate the IS so that it will elicit protective Abs. For instance, it has been known for many years that strongly neutralizing anti-HIV-1 Mabs often possess a very long heavy chain complementarity-determining region 3 (HCDR3) of 20–34 residues, compared to the average length of 16 residues of HCDR3 in human B cells (Yu and Guan 2014). These long HCDR3 loops allow the Abs to penetrate the glycan shields of Env trimers and to reach highly conserved hydrophobic residues in the gp41 region of Env (Kwong and Mascola 2012). It has also been established that long HCDR3s are not predominantly generated by somatic hypermutation-related insertions (Briney et al. 2012) but usually arise in humans during VDJ recombination and VH replacement events (Huang et al. 2016). As a result it is not possible to simply rely on stochastic Ab affinity maturation events for increasing the number of immune cells in a human IS that will contain long HCDR3s. One can only regret than humans ISs do not possess the large numbers of Abs with long CDR3s that are present in the IS of cows and allow these animals to produce large quantities of neutralizing anti-HIV bnAbs (Sok et al. 2017).

Most of the Env surface is covered with glycans, but when selective glycans are removed in the vicinity of the CD4 binding site, the underlying peptidic Env region becomes accessible to nAbs which can lead to an 1000-fold increase in neutralization efficacy (Zhou et al. 2017). Certain bnAbs of the VCR38 Ab lineage (Bhiman and Moore 2017) that have HCRD3 loops of a normal length are able to bind neotopes at the V1/V2 apex of Env trimers. It has been known since the 1960s that the presence of identical subunits in viral capsids and spikes always gives rise to a quaternary structure and to new types of epitopes known as neotopes absent in the monomeric subunits (Van Regenmortel 1966; Neurath and Rubin 1971). Neotopes arise from the juxtaposition of residues from neighboring subunits or from conformational changes induced by intersubunit bonds. It is unfortunate that such neotopes capable of inducing neutralizing Abs were detected at the V1 V2 apex and other regions of Env only 30 years later. If neutralizing antineotope antibodies that do not bind to Env protomers had been detected earlier, it might have dissuaded investigators to try to use Env monomers as potential vaccine immunogens (Van Regenmortel 1992a).

Epitope structures visualized by X-ray crystallography in Ab-antigen complexes usually provide little information on which exact structures of the immunogens are recognized by B cell receptors (BCRs) at the surface of lymphocytes. The exceptional plasticity, flexibility, and metastability of the HIV Env as well as the extensive Env conformational changes that occur during the transition from the prefusion to the postfusion state make it extremely difficult to know what structures are recognized by BCRs at different stages of the immunization process (Ward and Wilson 2017). Crystallographic analysis of the trimeric prefusion Env showed that V3 loop residues are not accessible to anti V3 Abs because they are occluded in a pocket formed by the V1 V2 region that gives rise to neotopes. When Abs bind to such a neotope, a conformational change occurs which exposes the V3 loop to V3 Abs (Zolla-Pazner et al. 2016). Extensive conformational transitions are induced by the binding of Env Abs and these present considerable challenges for HIV vaccine development (Van Regenmortel 2016b; Korkut and Hendrickson 2012).

After 15 years of unsuccessful attempts to develop an HIV vaccine by SBRV, this approach has lost its popularity with most HIV vaccinologists who no longer attempt to transform HIV epitopes of known structure directly into effective vaccine immunogens. However, the lure of a rational design approach for developing an HIV vaccine still exists as shown by the recent claim that "reverse vaccinology 2.0" (i.e., SBRV) shows great promise as a powerful vaccine design strategy (Burton 2017).

# The Bottleneck of Stochastic Ab Affinity Maturation

Ab affinity maturation refers to a process whereby BCRs and Abs develop an increased affinity for binding to a particular antigen present in the light zone of germinal centers in secondary lymphoid tissue (Suan et al. 2017). When the local concentration of this initial viral antigen decreases over time, only BCRs that bind to it with the highest affinity may be triggered because the mass action law will allow such BCRs to recognize a partner present at a very low concentration. This process differs from what is observed when new bnAbs appear in the chronic phase of HIV infection, usually after a couple of years, when the virus has been integrated in the host genome. At that stage, intrahost evolutionary pressures lead to the appearance of many mutated viruses that are able to trigger new Abs recognizing novel viral antigens, a process that differs from what occurs when truly affinity matured Abs appear that react better with one particular viral antigen. The presence of increasing numbers of antigenic variants during chronic infection may induce the appearance of novel bnAbs that are not necessarily "affinity" matured since they recognize antigens that were not previously available. The capacity to neutralize the viral immunogen only becomes effective when Abs are subsequently produced by plasma cells and the selecting factor during BCR recognition may be the appearance at low concentrations of novel antigenic variants (Batista and Neuberger 1998). This could explain why chronically HIV-1-infected humans from whom bnAbs are isolated do not benefit from the presence of such Abs for controlling virus replication (Bunnik et al. 2008; Euler et al. 2010).

When it was demonstrated that HIV-1 epitopes recognized by affinity-matured bnMabs derived from infected individuals did not bind germline predecessors of these Abs (Xiao et al. 2009; Dimitrov 2010) it became evident that a lengthy process of Ab affinity maturation would be needed to obtain anti-HIV-1 Abs by vaccination. This meant that it would not be possible to directly engineer HIV epitopes by SBRV using germline-encoded Abs that had not undergone affinity maturation. The SBRV approach that could have worked with other viruses was doomed in the case of HIV because of the Ab affinity maturation bottleneck and of the enormous antigenic variability of the virus. This led many laboratories to initiate a major new research effort aimed at elucidating a huge number of possible stochastic maturation pathways that may lead from germline Abs to mature BCRs and to Abs endowed with a high mutation frequency (Chen et al. 2012; Mascola and Haynes 2013; Klein et al. 2013; Prabakaran et al. 2014). It is generally accepted today that a series of sequential immunizations with HIV epitopes that recognize increasingly mutated Abs will be necessary for obtaining a protective immune response in a genetically heterogeneous human population (Doria-Rose and Joyce 2015; Greenspan 2014), although the logistics of such an approach appear daunting. Such a vaccination strategy departs from the SBRV approach since it requires the unraveling of particular stochastic maturation pathways that need to be transferred to large numbers of genetically heterogeneous vaccinees. Numerous studies have attempted to identify which structural elements of Env recognized by different bnMabs should be present in the sequential immunogens needed for designing an HIV-1 vaccine (Andrabi et al. 2015; Garces et al. 2015; Sok et al. 2016) but it has not yet been possible to demonstrate empirically which structural features in immunogens are needed for eliciting a protective immune response.

Protein antigens and immunoglobulins are fairly flexible and dynamic molecules (Jimenez et al. 2003) and the plasticity of epitopes and paratopes has been compared with flexible keys interacting with adjustable locks (Edmundson et al. 1987). The X-ray crystallographic analysis of many Ab-antigen complexes has demonstrated that the binding process leads to important side-chain movements and changes in the backbone conformation of the two partners (Wilson and Stanfield 1994). The mutual adaptation and induced fit (Bosshard 2001) that occur when the two partners interact often completely modifies the binding site structure originally present in the two free partners, with the result that the epitope structure observed in the complex may show little resemblance with the immunogenic structure that was recognized by the BCR during the immunization process. The recognition process between the epitope and a free Ab molecule may differ considerably from what occurs when this epitope binds to the cognate BCR embedded in a lipid membrane because hydrophobic interactions between membrane lipids and aromatic residues often lead to stronger binding (Scherer et al. 2010). Many HIV Env epitopes are also chemically heterogeneous and may bind to Abs by involving specific peptidic interactions as well as numerous glycans or even lipids (Van Regenmortel 2016b; Sok et al. 2014).

## Can Rational HIV Vaccine Design Overcome the Limits of Bounded Rationality?

The term design refers to the intentional, deliberate conceiving of a novel object or process by an intelligent mind. The designer's task is to pose and solve an inverse problem by imagining a theoretical model or process that would make it possible to obtain the desired outcome. It used to be believed that living organisms were designed by the preconceived plan of an intelligent deity although it is generally accepted today that organisms are shaped by the filter of Darwinian selection. There is, indeed, little anatomical evidence for the intentional design excellence of a mythical designer and, instead, plenty of evidence that utilitarian tinkering processes took place during evolution rather than design (Olshansky et al. 2003; Laubichler 2007). Design terminology also seems inappropriate for describing the activities of vaccine developers since investigators do not know how the IS succeeds in eliciting protective Abs and they need first to imagine a plausible theoretical model and subsequently to test its validity by demonstrating empirically that the predicted outcome actually does occur. Such a procedure hardly qualifies as intentional design and does not really justify the common belief that structure-based design is preferable to the empirical testing of HIV vaccine candidates. It may even seem odd that many authors still maintain that rational design is the best strategy for developing an

HIV vaccine (Douek et al. 2006; Schief et al. 2009; Walker and Burton 2010; D'Argenio and Wilson 2010; Nabel et al. 2011; Karlsson-Hedestam et al. 2008) although they rarely made it clear what they mean by the term "rational." Rational vaccine design is derived from the concept of "rational drug design" which uses the 3D structure of a biological target for designing molecules that will selectively bind to it and inhibit its biological activity. Such a computer-assisted approach based on molecular docking (Gschwend et al. 1996) is also feasible for improving the binding complementarity in an epitope–paratope pair by either designing an HIV Env epitope that binds better (Wu et al. 2010) or a paratope that binds better (Diskin et al. 2011). In both cases, however, this amounts only to designing improved binders and not immunogens. As explained in the section "*What is structure-based reverse vaccinology*," improving the binding capacity or antigenicity of an HIV Env epitope does not constitute rational vaccine design since an epitope that strongly binds a bnMab will not necessarily elicit Abs possessing the same neutralizing capacity as the bnMab. Abs are also often heterospecific, i.e., able to react better with other antigens than the one used in the immunization process that elicited the Ab; this demonstrates that antigenic and immunogenic properties are not necessarily located in the same regions of a protein molecule (Van Regenmortel 2014a). In rational vaccine design, the term "rational" actually does not have the same meaning as in rational drug design that is based on molecular docking and it simply refers to the fact that the investigator is focusing on parts of the system for which molecular information is available.

The economist and Nobel laureate Herbert Simon introduced the concept of "bounded rationality" (Simon 1996) to describe the intrinsic limitations of human cognition that are due to the many unavoidable constraints that always limit our ability to reach rational decisions. Such limitations are due, for instance, to insufficient and inaccurate information, a limited capacity to investigate any real-world complex system in all of its complexity and the limited time and resources that are available to humans. Since only a small number of possibilities can ever be investigated, humans cannot achieve the complete analysis and understanding of highly complex systems that they would require in order to make entirely rational decisions based on a complete knowledge of all relevant parameters. Instead of guaranteeing that correct solutions to a problem can be reached, bounded rationality forces humans to make decisions by following tentative heuristic procedures that cannot provide definitive solutions to inverse problems (Wimsatt 2007; Gannon 2007). The reason for this is that all inductive and discovery procedures in science are heuristic principles that fail as algorithms in part because they do not represent logically valid argument forms (Wimsatt 2007). Multiple solutions to the same problem may also arise from the degeneracy of structure-function relationships present in biological systems which results from nearly equal, energetically available configurations. It is also not clear whether intrinsically stochastic elements can be incorporated into holistic paradigms in order to provide an understanding of the emergent properties characteristic of complex biological systems (Conti et al. 2007).

When all the relevant parameters of a complex system are actually not all known, which is the case, for instance, with the world economic system, totally rational choices and decisions are not feasible which explains why it was not possible to predict an event such as the world financial crisis of 2007–2008. Similarly, long term weather prediction also remains impossible as illustrated by the metaphoric "butterfly effect" which explains that small changes in initial conditions in a complex non-linear system lead to unpredictable consequences at a later stage. It is also impossible for vaccinologists to know in perfect detail the physiological and genetic backgrounds of all the ISs of human vaccines nor importantly their prior immunological exposure. The limitations imposed by bounded rationality does not allow them to predict the outcome of various immunological interventions, and this makes the strategy of rationally designing an HIV vaccine seem totally unrealistic. Only empirical evidence obtained by successfully manipulating and controlling an IS in order to achieve a certain degree of immune protection may allow investigators to reach plausible, but nevertheless tentative conclusions, on how best to proceed for attempting to develop an effective HIV vaccine. Since the aim of vaccinologists is to control immunological phenomena, they always need to interfere in the IS in order to gain knowledge about how to control it (Hacking 1983; Leonelli 2009). Increasing our knowledge of basic immunology and of HIV-1 antigenic structure is unlikely on its own, to allow us to rationally design an effective HIV-1 vaccine (Van Regenmortel 2016b), although it may be useful for elaborating better theoretical models when trying to solve the numerous inverse problems that have until now impeded the development of an HIV vaccine.

## The Use of Transmitted/Founder Virus as HIV Vaccine Immunogen

The preceding sections of this review may seem to some to be excessively negative and devoid of constructive suggestions about alternative approaches that could be investigated for developing an HIV vaccine. In this last section, it will be argued that most HIV vaccinologists have so far mainly studied the rather ineffective immune responses that occur after several years of chronic HIV infection. Since only a few studies have been devoted to the immune responses that occur during the initial acute phase of HIV infection, there clearly is a need to further investigate these responses that may be highly relevant for developing a protective vaccine.

Despite the high diversity of HIV populations found in infected individuals, it is remarkable that about 80% of heterosexually transmitted HIV-1 infections are caused by a single, so-called transmitted/founder (T/F) virus which is not the predominant variant in the donor. This phenomenon is known as the transmission bottleneck of HIV-1 (Kariuki et al. 2017). Compared to viruses present in chronic HIV infections, T/F viruses have about twice more Env spikes per virion, have shorter variable loops and fewer N-linked glycosylation sites, and are more efficiently captured by dendritic cells and transmitted to CD4+ T cells (Keele et al. 2008; Parrish et al. 2013). T/F viruses also utilize the chemokine receptor 5 as coreceptor and bind to an integrin receptor that transfers the virus from the genital tract to the gut-associated lymphoid tissue (GALT). Following infection by a single T/F virus, an acute phase of HIV infection occurs in the following month(s), which is characterized by the presence of infected CD4+ T cells in lymph nodes and intense viremia. Subsequently, the HIV proviral DNA is integrated in the host genome, seroconversion occurs as infected cells move from the GALT to the blood and chronic HIV-1 infection sets in, leading eventually to the progressive destruction of the host IS (Rios et al. 2016). In the chronic phase, Ab-based selection and virus escape occur continuously although this does not allow infected individuals to control virus replication (Euler et al. 2010). The bnAbs that appear during chronic infection 2–4 years after the initial infection are the ones that have been studied most extensively in SBRV experiments, although the initial immune response against T/F viruses, which also generates some neutralizing Abs, may be more relevant when trying to developing a preventive HIV-1 vaccine (McMichael et al. 2010). It is somewhat paradoxical that most crystallographic studies of bnMabs were done with human anti HIV antibodies that are present after several years of chronic infection since these Abs are usually unable to control HIV infection in the infected individuals from whom they had been obtained.

Recently, the antigenicity and immunogenicity in guinea pigs of recombinant gp140 Env glycoproteins from T/F viruses and from chronic viruses have been compared in order to estimate their relative efficacy as potential vaccine immunogens (Liao et al. 2013a, b). For this study, Env glycoproteins were selected that reacted with the largest number of available bnMabs, although a pronounced cross-reactive antigenicity may not necessarily entail a superior immunogenic efficacy against T/F viruses (Van Regenmortel 2014a). The T/F Envs were found to induce Abs with the greatest neutralization breadth which suggests that T/F viruses may be suitable vaccine immunogens. The T/F Envs also induced Abs against the V3 loop.

In recent years, several vaccinologists have suggested that chemically inactivated HIV virions should be reconsidered as potential vaccine immunogens (Rios et al. 2016; Sheppard 2005; Sheppard and Dorman 2015). Inactivated HIV vaccines had already been investigated more than 20 years earlier but these attempts failed because harsh and inappropriate inactivation methods were used. Recently, new and more reliable inactivation methods have been developed that no longer suffer from the defects of earlier methods (Van Regenmortel 2011b). A particular attractive method consists in irreversibly destroying the infectivity of HIV by targeting the viral reverse transcriptase (RT) using UV irradiation of a photo-labeled RT inhibitor, a procedure that fully preserves the viral antigenic structure (Rios et al. 2015, 2016). Such a targeted inactivation of RT could be used for inactivating various T/F viruses and could lead to the development of a whole-virus polyvalent vaccine that would possess double the amount of viral Env spikes per virion. Such a vaccine may possibly trigger an immune response in the acute infection phase and since the RT is inactivated, the vaccine may be able to prevent HIV integration and subsequent chronic infection (Rios et al. 2016). It is the neutralization of T/F viruses that are present initially and easily spread from cell to cell that may be the most relevant for preventing rapid viral dissemination. Improving the immunogenic properties of T/F viruses may thus be more important for developing a preventive HIV-1 vaccine than concentrating on the immunogens that give rise to bnAbs during the chronic phase of viral infection (Ensoli et al. 2016a, b).

## Conclusion

It is often claimed that the IS is unable to control HIV infection because the virus continuously evolves new strategies and mechanisms in order to evade IS defenses. Such metaphoric language seems to attribute to HIV a goal directed capacity to defeat the host IS, although it is only the highly error-prone activity of the viral RT that is responsible for the enormous structural plasticity and antigenic variability of the Env glycoprotein, allowing the virus to evade immune control (Zolla-Pazner et al. 2016; Van Regenmortel 2016a). Since nAbs appear only very slowly during chronic infection, the Ab response lags behind the rapidly diversifying virus and the IS is unable to control HIV infection. Numerous additional, unsolved challenges have further contributed to our inability to control HIV-1 infection by vaccination (Greenspan 2014). It also seems unrealistic to expect that passive immunization could provide the enormous quantities of recombinant Abs that would be needed to combat a world-wide pandemic. In spite of an immense, high caliber research effort backed by very considerable funding over several decades, it seems increasingly likely that the multiplicity of unsolved inverse problems that defy a rational vaccine design approach may prevent us from developing an effective HIV-1 vaccine. Although the time-honored vaccine approach of using chemically inactivated virions should urgently be tested with T/F viruses, we might have to accept that the HIV pandemic will not be brought under control by what would have been the ideal solution: an effective, preventive HIV-1 vaccine (Corey and Gray 2017).

# Viral Species, Viral Genomes and HIV Vaccine Design: Is the Rational Design of Biological Complexity a Utopia?

Marc H V Van Regenmortel

## Virus Species and Classification

Classifying objects is a human prerogative based on the capacity of the mind to conceptualize and recognize the presence of similar properties in individual ob ects. Properties and classes are related abstract entities, and when a property is ascribed to an object, the object becomes a member of a particular class defned by that property. The property is distinct from the object that possesses it, in the sense that a red object is said to have the property of redness. Redness is not a thing but is a concept. The classifcation of viruses consists of demarcating taxonomic classes such as species or families and allocating individual viruses to these classes in order to be able to group similar viruses together. A virus species is thus an abstract concept fabricated by the mind and not a real object located in space and time. The term "species" has two different meanings. It can refer to the innumerable species taxa created by virologists that have viruses as their members, or it can refer to the lowest species category in the hierarchical virus classifcation, below the categories "genus" and "family". The category "virus species" is thus the class of all the virus species taxa that virologists have demarcated. The ongoing debate about the correct defnition of virus species concerns the defnition of the species category (the class of all species taxa) and not the defnition of an individual species using the properties of the viruses that are its members. For this reason, the original 1991 ICTV defnition of the species category did not make it any easier for virologists to decide which viruses should be allocated to a particular species taxon. Virologists readily accept that the categories of genus,

Archives of Virology, 2018, volume 63, 2047–2054
Marc H V Van Regenmortel

Marc H V Van Regenmortel
School of Biotechnology, University of Strasbourg, Illkirch, France

© Springer Nature Switzerland AG 2019
Marc H V Van Regenmortel, *HIV/AIDS: Immunochemistry, Reductionism and Vaccine Design*, https://doi.org/10.1007/978-3-030-32459-9_24

family or order in virus classifcation correspond to conceptual constructions of the mind. They are fully aware, for instance, that a virus family cannot be sequenced, purifed by centrifugation or visualized by electron microscopy, since it is an imaginary creation of the mind and not a physical entity. It is therefore surprising that many virologists think of a virus species as a kind of virus that is physically real and able to infect a particular host rather than as a member of a conceptual class. In the case of the HIV species, only two highly variable viruses, HIV-1 and HIV-2 have been recognized. It should be obvious that it is not possible to develop a vaccine against the HIV species. The Entrez forms utilized to enter viral sequences in the GenBank database used to refer to them as being sequences of new virus species instead of new virus isolates. This confusion between a species as a concrete kind of object and as an abstract category in a classifcation is actually prevalent in the whole of biology, which in part may be due to the fact that the vast majority of organisms in microbiology, botany and zoology do not have common names that difer from the Latin name of the species to which the organism belongs. One commonly reads that somebody has been infected by Escherichia coli as if a conceptual species class would be able to infect a physical organism. Virologists are the only biologists who could easily avoid this type of logical confusion, since every infecting virus has a common name, different from the species name to which the virus belongs, which could be used to refer to the infectious agent. It is therefore strange that this logical confusion between object and concept has been encouraged by a virologist (Simmonds 2018) when he advocated the use of genetic metrics for allegedly defining viral species "logically" on the basis of specific genome-based inclusion and exclusion criteria. He argued that virologists should abandon their unique linguistic capacity to use diferent names for distinguishing between an abstract virus species and a concrete viral object on the grounds that "it would eliminate a whole range of problems and confusion with a terminological usage in virology that simply does not exist in other areas of biological classification". Simmonds (2018) even recommended that the ICTV should reconsider the need for making a rigid distinction between a conceptual species and a population of real viral objects because this would bring virology more in line with the use of the term species in the rest of biology, which accepts that a person can be infected by Escherichia coli! Such a recommendation probably arose because of the unfortunate rule that the ICTV had adopted in 1998 to use the common English name of a virus (written in Roman) also for designating the species to which the virus belongs, but writing the species name in italics as is done with all other viral taxa. Authors now had the choice to refer to either the virus "measles virus" or to the species "Measles virus", but they often got it wrong because the diference between a virus and a species was often not clear to them. Having to use the same word, of course, did not help clear thinking, and the ICTV has been widely criticized for introducing this rule, especially because the ICTV in 1998 had rejected the alternative, more cogent proposal of using nonLatinized binomial names (NLBNs) as species names (Van Regenmortel 2001a, b). Such names had been widely used since the 1970s in plant virology papers and books as well as in early ICTV reports but were less popular with animal virologists. Italicized species NLBNs are formed by replacing the terminal word "virus" occurring in all common English virus names with the genus name to which the virus belongs, which also ends in virus. Measles virus, for instance, then becomes a member of the species *Measles morbillivirus*. Such a system is widely applicable to all viruses and does not require the creation of new species names, since it combines known virus names with known genus names. It would also avoid the creation, advocated by some virologists, of thousands of new unpronounceable Latin species names that are unlikely to be popular with most virologists who had never accepted that they should follow the rules of the Biological Nomenclature Code regarding the use of Latin and of italics in taxon names.

Since binomial names in biology have always been associated with taxonomic entities, the species name *Measles morbillivirus* is unlikely to be confused with the virus name "measles virus", and including the genus name in the species name is a useful reminder of some of the characteristics of the viruses that are members of the species. NLBNs could have been immediately introduced in 1997 for more than 90% of the species that were recognized at the time, and in recent years, many such species names have been introduced in numerous virus families (Amarasingham et al. 2015).

## Are Virus Species Polythetic Classes or Monophyletic Groups?

The ICTV in 1991 endorsed the following defnition of virus species: "*A virus species is a polythetic class of viruses that constitute a replicating lineage and occupy a particular ecological niche*".

The notion of polythetic class, commonly used in taxonomy, has been extensively defined numerous times (Van Regenmortel 2011a, b). However, this did not prevent a few virologists (Gibbs and Gibbs 2006) from objecting to virus species being called polythetic classes because they claimed that the term "polythetic" was obscure and that the term "class" should only be used to denote a category in an hierarchical classifcation, i.e., the one situated above the category "order" and below the category "phylum". A few virologists even claimed that the terms "polythetic", "replicating lineage" and "ecological niche" used in the 1991 species defnition were meaningless (King 2012).

While monothetic classes are defned by one or a few properties that are both necessary and sufficient for membership in the class, polythetic classes are defned by a variable combination of properties, none of which is a defining property necessarily present in every member of the class. This means that 1) each member of a polythetic class shares a certain number of properties, 2) each property is present in a large but unspecifed number of members, and 3) no property is necessarily present in all the members of the class and absent in the members of other classes. It must be emphasized that a concept such as a species class cannot have physical or material properties, since only its members do. A species thus cannot be described and can only be defned by listing the intrinsic properties of its members, such as chemical composition and relational properties that arise by virtue of a relation between a virus and other objects such as a host or a vector. Since many biological properties used for defining virus species, such as their natural host range, symptomatology, pathogenicity or mode of transmission, can be altered by a few mutations, these species-defning properties can vary considerably in diferent members of the same species, which is one reason why it is useful to defne species polythetically. The continuous nature of biological variation often leads to the absence of clear-cut discontinuities between closely related species, and these may then be viewed as fuzzy classes with blurred boundaries. In formal bivalent logic, vagueness in the descriptors of reality is prohibited, and logical dichotomies are mandatory; for instance, a glass is said to be either full or not full. Such neat distinctions, however, are rarely feasible when describing biological objects that possess inherent variability. Whether the symptoms induced by a virus infection are mild or severe is inherently not quantifable and, likewise, describing the degree of genome sequence similarity between two viruses is going to depend on which genome characteristics are arbitrarily considered to be most signifcant. It is thus unavoidable that the classes that will be conceptualized on this basis will be fuzzy and devoid of precise boundaries. Membership in a biological species class is therefore more a matter of stipulation than of logical necessity, and polythetic classes correspond to fuzzy sets with hazy boundaries. However, this would not justify abandoning the species concept, since the continuous nature of electromagnetic radiation, for instance, does not prevent us from recognizing different colours in a rainbow. The question whether virus species exist as real biological entities independently of any human conceptualization is still the object of intense debate. However, if one accepts that real, phenomenal and conceptual modes of existence are actually possible (Bunge 2016), it is possible to say that virus species at least have a conceptual existence in human minds. Classes, like all concepts, can only be defned, while objects can only be described. Virus species are sometimes viewed as natural kinds, similar to the chemical elements that exist in nature independently of their discovery and naming by humans. In a chemical context, a chemical species always consists of identical molecules. It is therefore unfortunate that virologists borrowed this chemical species category and used it to describe clones of RNA viruses as being imperfect quasispecies, since obviously no virus species can have members that are all chemically identical. Taxa are defned intensionally by listing the qualifications for membership in the class. The so-called intension of a class is its meaning, which, however, does not give it any reality outside the realm of conceptual constructions. The so-called extension of the virus species class is the set of all the real viruses that are the members of the species, and they are the concrete referents of the class. Since the intension of a class determines its extension, the extension of a species class can only be determined by taxonomists after they have previously demarcated such a species taxon (Mahner and Bunge 1997). Before this is done, it is not possible to ascertain if a sufficient number of the species-defning properties are present in an individual virus to make it a member of the species. The proposal that a monothetic species class can be established by relying on a single defning property such as a particular nucleotide motif found in viral genomes overlooks the fact that it would be necessary to know beforehand that this motif is present in all the members of the species and absent in other species. The extension of the species class would thus need to precede its intension, which is not possible (Van Regenmortel 2011a, b).

When Gibbs et al. (2004) showed that the RNA genome sequences of members of diferent species in the *Tobamovirus* genus possessed unique nucleotide combination motifs (NC-motifs) of up to 47 nucleotides in the viral polymerase gene, they suggested that such NC-motifs could be used as diagnostic markers for identifying members of different species in that genus. However, the authors had confused a diagnostic marker used for recognizing members of an existing species with a species defning property that taxonomists could use for demarcating a new species. Gibbs and Gibbs (2006) nevertheless made the proposal that a virus species could be defned monothetically by the presence in all its members of a common NC-motif. This led to the claim that a single characteristic feature in a viral genome was sufficient to establish a new species and to the conclusion that species were not polythetic classes defned by a combination of biological and phenotypic properties. In 2013, the ICTV then replaced the defnition of a virus species as a polythetic class by the following new defnition: "A species is a monophyletic group of viruses whose properties can be distinguished from those of other species by multiple criteria" (Adams et al. 2013). However, since every species and genus or even a family may be considered to be monophyletic groups, this new definition is both a definition of a taxon and of a species. Monophyly is also not an adequate criterion for species demarcation because of the common occurrence of recombination and reassortment among parts of many viral genomes, which produces chimeric viruses with poly-phyletic genomes. This makes it impossible to represent such multidimensional phylogenies in a monophyletic scheme (Ball 2005).

The new ICTV species defnition allowed virologists to establish virus species as monothetic groups of concrete viral objects that share only one common character-istic such as a nucleotide motif or a certain percentage of genome similarity. The classical view that properties, classes and virus species are related conceptual entities was thus abandoned, and virus species were given the status of physical objects. Viruses became linked to species taxa by part-whole relations instead of by the membership relation that is the foundation of all hierarchical biological classifcations and makes lower taxa also to be "included" in the higher taxa above it (Buck and Hull 1966).

Class membership is the relation that makes it possible to establish a link between the two logical categories of abstract classes and of their concrete members that are located in space and time. This membership relation differs from the part-whole relationship that exists between two concrete objects, one being a part of the other in the way a limb is part of a body. A viral object obviously cannot be part of a conceptual species construct, nor can a thought or a concept be part of a material object. Since a lower taxon such as a species is "included" in the higher genus taxon immediately above it, the properties required for classifying a virus as a member of the species Measles virus include, besides others, all the properties required for classifying it as a member of the genus *Morbillivirus*. Class inclusion in the classifcation hierarchy obviates the need to repeat the properties used for defining the higher taxon as properties of the lower taxon, although the genus properties that are still present in members of the lower species taxon are not species-defning properties that could allow species to be differentiated. Since lower taxa always have fewer members than higher taxa, they require more properties to meet the qualifcations for membership. This illustrates the logical principle that reducing the number of required qualifcations increases membership, whereas increasing the number of required qualifcations decreases membership (Buck and Hull 1966). It is amusing to note that this principle totally invalidates the currently popular claim of many viral taxonomists that a virus species can be defined by a single property and therefore cannot be a polythetic class. It also has been argued that if a species corresponds to a replicating lineage, it cannot be a polythetic class because all its members must have inherited one or more properties from a common ancestor, which must therefore make the class a monothetic one [20]. The notion that viruses can form a polythetic class was also rejected by Simmonds (2018) on the basis that "two members of a polythetic class may conceivably share no properties at all that differentiate them from other species". However, as discussed above, a polythetic class is defined by a variable distribution of properties in the members of the class together with the absence of a single common species-defining property in all its members. This combination of properties in a polythetic class does not itself constitute a single common property shared by all the members of the class, since it is actually a characteristic of the class and not a property of its members. Virus classes only admit viruses as members but cannot admit themselves as members. The adjective "long" denotes the class of long things, but since it is not a long adjective, it is a non-selfdenoting property of the class. If one fails to appreciate that the non-self-membership condition does not determine the class, one lands with logical para-doxes like the well-known Russell paradox (Baldwin and Lessmann 1998) or with the paradox that a polythetic class is also a monothetic one because its members share a particular distribution of properties. The argument that polythetic species may contain members with no common species-defining properties (Gibbs and Gibbs 2006) is based on the misconception that a part of an object is a property instead of being only a smaller part of the object, and it contradicts the fact that members of a polythetic class always share a combination of properties. A glaring example of the pernicious consequence of allowing virus species to be established only on the basis of a single percentage of genome divergence in Pairwise Sequence Comparisons (PASC) are the 307 species of begomoviruses that were demarcated on the basis of a 90% identity percentage in their viral DNA-A genomes. This produced 42 different species of *Tomato leaf curl virus* whose members had very similar biological properties and were diferentiated only by including in the species name the geographical location of the first isolation of the virus. These 42 so-called species could all have been considered to be strains of one virus if a lower cut of percentage in the PASC peaks had been chosen for separating viruses from strains (Van Regenmortel 2018c).

## A Philosophical Interlude: Bionominalism Regards Species as Concrete Individuals Rather than Conceptual Classes

The school of thought known as bionominalism, which has many adepts, considers that since species change during evolution, giving rise to new species, they must be evolving historical entities with a temporal dimension rather than immutable and timeless classes. This changed the ontological status of species that no longer were considered to have viral objects as their members, since viruses were now viewed as being part of a material species. This new ontology of "species as individuals" (SAI) has been analyzed in considerable detail by Mahner and Bunge (1997, pp. 232–270) and readers interested in this philosophical debate are advised to consult the three entries on Ontology, Properties and Reifcation in Wikipedia. Whereas ontology deals with the nature of being and of what exists, reification is the fallacy of treating an abstraction as if it were a concrete, physical entity. The SAI thesis holds that species are lineages of ancestral-descendant populations with a spatiotemporal location and that all taxa are so-called historical entities forming cohesive wholes. However, viewing historical entities as concrete individuals is an instance of reification. The relational concepts of ancestry, progeny and lineage are not real objects that are linked by causal relations, since the ancestry and progeny of a population cannot act upon each other unless they exist at the same time. Conceptual classes also cannot descend from each other, since only concrete viruses can do so. Species have been recognized as fundamental units of biological classifcation and taxonomy long before the theory of evolution had been formulated. When species are viewed as both conceptual constructs and as units of classifcation, they cannot be said to evolve, because only units of evolution evolve in the form of lineages.

Since a classifcation is only a conceptual construct, taxa can only be interpreted to be real individuals when concepts are conflated with their referents. The major shortcomings of bionominalism is that it does not distinguish a thing from it conceptual representation and does not accept that it is possible to conceive of conceptual classes and taxa that exist on Earth for only limited amounts of time, instead of being eternal and timeless (Ruse 1998). Mahner and Bunge (1997) conclude that the mistaken ontology that underlies bionominalism is responsible for its inability to provide an adequate philosophical framework for any hierarchical biological classification.

# Is it Possible to Incorporate Viral Metagenomic Data in the Current ICTV System of Virus Classification Established Using the Chemical and Biological Properties of Viruses?

The current ICTV system of virus classifcation comprising 9 orders, 131 families, 46 subfamilies, 803 genera and 4853 species was initially developed because virologists needed a practical scheme for recognizing and naming the viruses they were working with. The properties that were most relevant for classifcation were therefore the pathogenic and biological properties of viruses and the structural, biochemical and antigenic properties of virus particles that are useful for diagnosing or controlling viral infections. Viruses were allocated to orders, families and sub-families on the basis of a small number of stable properties such as the type and strandedness of the viral nucleic acid, the presence of a reverse transcription process and of diferent replication strategies, the virion morphology and presence of an envelope. These properties were sufficient for defining whether a virus was a member of any of these three categories. Membership in the lower categories of genera and species required a larger number of defning biological properties, for instance host range, vectors, pathogenicity and epidemiology, which frequently difered among the individual members of genera and species. The resulting classifcation is still very useful for laboratory virologists who investigate the biology, pathogenicity, immune responses and epidemiology of the viruses and the diseases they cause.

With the development of more efficient and rapid methods of sequencing genomes, it became feasible to study in considerable detail the genetic diversity of viruses. Highthroughput sequencing then made it possible to study virus evolution and to reconstruct comprehensive virus phylogenies with the aim of having taxonomy reflect phylogeny. The evolutionary history of viruses on our planet became a topic of considerable scientifc interest, in spite of the fact that viruses do not have a single evolutionary origin that links them to the history of the Universal Tree of Life. It must also be pointed out that virus evolution is not necessarily the major scientific concern of laboratory virologists who study present-day virus diseases, because they tend to be more concerned with the phenotypic and biological properties of viruses than with evolutionary considerations pertaining to the entire virosphere. It is sometimes claimed that the development of the Global Virome Project (Daszak et al. 2016) and our increased understanding of virus evolution is of practical importance because it will allow us to predict the future emergence of virus diseases in humans, but this seems unlikely because the time-scale of evolutionary events encompasses millions of years (Geoghenan and Holmes 2017).

When viral species are represented as only physical nucleotide sequences present in metagenomic datasets, the absence of any viral phenotypic traits makes it problematic to incorporate them in the current ICTV classifcation established using phenotypic properties. Since it has been estimated that more than 99% of the eukaryotic virosphere remains undiscovered, it would be necessary to devise a hybrid classifcation consisting of the current 4853 ICTV species together with huge numbers of viral nucleotide sequences. Since the viral genome sequence is part of the chemical composition of a virion, a classification relying only on genome sequences, nucleotide motifs and inferred genes could still be viewed as a phenotypic classification based on sequences instead of on other viral phenotypic properties, although this would produce a classifcation of sequences and of viral genomes and not a classifcation of viruses (Van Regenmortel 2016a, b). Although it is believed by many virologists that viral genomes identifed from metagenomic datasets should become part of a greatly extended future ICTV classifcation (Simmonds et al. 2017), it is currently not clear how this could be achieved in practice if no phenotypic properties can be associated with the nucleotide sequences assumed to "represent" real viruses. Although it is also widely believed that millions of bacterial species are still undiscovered, it is possible that these bacterial genomes in metagenomic data sets refect the accumulation of countless adopted neutral mutations with no concomitant phenotypic diversification. This has led to the suggestion that what is termed species in this context might be more accurately described as bacterial DNA sequences that are not yet in a database (Finlay and Esteban 2009).

## It Is Rarely Possible to Infer the Phenotypic Properties of a Virus from its Genome Sequence

It is often claimed that the properties of a virus are mostly encoded by its genome, although there are many reasons why the phenotypic and biological properties of a virus cannot be deduced or predicted from its genome. Alternative splicing and discarding of introns always produce unpredictable RNA transcripts, and the resulting proteins interact with various host and vector gene products through mechanisms that have not been elucidated. Furthermore, no causal links have been established between particular nucleotide sequences and most of the complex phenotypic traits of the members of virus species.

It is sometimes possible to identify a viral genome as belonging to a member of an established virus family because certain stable structural features of virions are correlated with particular nucleotide sequences. In the case of species, however, their members always possess a combination of variable phenotypic properties that cannot be correlated with particular nucleotide regions because 1) epigenetic factors still play an important role, 2) it is impossible to infer the hosts or vectors of a virus from the viral genome sequence, and 3) the full pangenome of a virus (i.e., the entire set of genes found in all the members of a species) is not necessarily shared by all the members of a given species (Brito et al. 2015). It seems, therefore, unrealistic to expect that virus species and genus taxa can be established in the future only by analyzing viral nucleotide sequences obtained from metagenomic datasets (Van Regenmortel 2016a, b).

# Determinism, Contingency and Complexity

Although determinism assumes that every phenomenon in nature is entirely determined by pre-existing causes, in biological systems no single event can ever be said to be the necessary and sufficient cause for a particular biological efect to occur. Numerous contributory causes are always present, and one causal explanation is inadequate when one factor is singled out and given undue explanatory weight on its own. The notion of contingency is crucial to appreciate that determinism is entirely compatible with unpredictability, because innumerable individual deterministic processes will follow separate trajectories that very rarely impede on each other, leading therefore to unpredictable results.

Biological systems always consist of hierarchical compartments of increasing complexity, which all communicate with each other via translocating molecules. This organization leads to upward as well as downward causation, for instance when the constituents at the lower levels are controlled by new emerging properties that arise at the higher levels (Van Regenmortel 2012b).

One of the aims of systems biology is to understand how emergent biological phenomena arise from the multiple interactions that occur in complex biological systems such as the immune system. At each successive level of biological organization from genes to proteins and to cellular tissues, innumerable interactions occur between genetic, epigenetic, biochemical and environmental factors, and these must first be analyzed separately before they can be integrated into systems biology or systems vaccinology (Pulendran 2014, Van Regenmortel 2018a). Our understanding of the entire integrated system, however, remains severely limited because of our inability to measure accurately the on- and of-rates of the individual interactions (Zeder-Lutz et al. 1999).

Most biological interactions are reversible, and when increasing amounts of an antigen are added to antibodies immobilized on a chip, a dynamic plateau is eventually reached because the number of antigen molecules that bind to the chip is equal to the number of molecules that dissociate. Whereas the kinetic on-rate of a reaction is mainly controlled by the diffusion constant of proteins, which varies very little, it is a slower off-rate that usually brings about an increased biological effect such as enhanced infectivity neutralization by certain antibodies. Since the kinetic and equilibrium constants of a reaction depend on the concentration of biologically active molecules, the active concentration must be determined, for instance, with surface plasmon resonance biosensors (Richalet-Sécordel et al. 1997) and not by spectrophotometry, which measures both correctly and incorrectly folded molecules. The active concentration is often much lower than the nominal chemical concentration and may vary with the ligand used in the biosensor assay because the degree of incorrect folding may not be the same in different regions of a protein (Azimzadeh and Van Regenmortel 1991). It is our inability to measure accurately the active concentrations of the numerous partners that interact at the diferent levels of a complex biological system that prevents us from analyzing the kinetics of the individual interactions as well as of the fully integrated system.

# HIV Vaccine Design

The term "design" refers to the intentional, deliberate conceiving of a novel object or process by an intelligent mind. The designer's task is to pose and solve an inverse problem by imagining a theoretical model that would make it possible to obtain the desired outcome. Rational drug design, which uses the 3D structure of a biological target for designing molecules that will selectively bind to it and inhibit its biological activity, is a computer-assisted approach that has been very successful in pharmacology. This has led many vaccinologists to believe that rational vaccine design should also be used for developing effective vaccines. Such claims, however, are mostly based on a misunderstanding regarding the capacity of rational design to solve the inverse problems encountered by vaccinologists (Van Regenmortel 2018a). A large number of unsuccessful attempts to rationally design HIV vaccines have been made that consisted of improving the capacity of viral epitopes present at the surface of the HIV Env protein to bind to various neutralizing monoclonal antibodies derived from patients during the chronic phase of HIV infection (Van Regenmortel 2018a). This sometimes led to an improved chemical antigenicity (i.e., binding capacity) of the HIV epitopes, but such results did not constitute rational vaccine design, because the ability of the epitopes to induce neutralizing antibodies was not enhanced (Van Regenmortel 2014a, b). In this approach, called structure-based reverse vaccinology, the investigators always concentrated on certain structural features of the epitopes that were thought to enhance epitope-paratope binding. What was needed, however, was to improve the biological capacity of the immune system to induce neutralizing antibodies, and this would have required acting on the intrinsic properties of various components of the immune system (Ig genes, cytokines, T regulatory cells, etc.) and not simply on the degree of fit between epitope and paratope (Van Regenmortel 2017). Attempts to unravel the pathways of antibody maturation during an immunization process that lead to highly potent anti-HIV antibodies could also identify antigens rather than immunogens. Recent advances in the elucidation of such maturation pathways that lead to the identification of germline-targeting HIV antigens have shown that these trajectories are often extremely complex (Umotoy et al. 2019). Such maturation pathways will therefore be difficult to re-enact for the rational design of an HIV vaccine since such antigens may turn out to be actually ineffective immunogens for eliciting protective antibodies.

The concept of bounded rationality developed by the economist Herbert Simon (Simon 1996) has been used to explain the intrinsic limitations of human cognition that arise because of our insufficient information and available time for achieving a complete understanding of complex systems. In order to make truly rational decisions, we would need to know all the relevant parameters that infuence the behaviour of a complex system, a knowledge that simply is not available. According to Herbert Simon, our bounded rationality would make us unable to predict a world financial crisis (as we failed to do in 2008). In spite of our enormous modern computational power, our bounded rationality together with chaos theory also does not allow us to make long term weather predictions. Physicists and chemists used to believe that the universe was ruled by mathematical laws that would make it possible to predict the future behavior of a system if one had an intimate knowledge of all its initial conditions. That assumption became untenable in the last century with the development of chaos theory, which demonstrated that extremely small diferences in the initial conditions of a dynamic complex system have a huge impact on the subsequent state of the system (Du Sautoy 2016). Since making very precise quantitative measurements of the innumerable initial conditions of a complex biological system is even more unattainable than with physical systems, it should be evident that making accurate predictions about any future state of a dynamic biological system is completely intractable (Green 2015).

Since the problems that HIV vaccinologists have to solve are not direct problems but are mainly inverse problems (Van Regenmortel 2018a), i.e., determining what are the past causes that would lead to a currently observed (or anticipated) phenomenon (for instance the absence of deleterious HIV infection in human elite controllers), it seems that the rational design of an HIV vaccine will forever remain a utopia. In biology, trying to predict the past by solving inverse problems (i.e., the many causes that gave rise to a desired effect) is actually more diffcult than predicting the future, since only direct problems can be investigated experimentally by determining what effects follow from certain causes.

**Compliance with Ethical Standards**

**Conflict of Interest**   The author declares that he has no conflict of interest.

# Conclusion: The Rational Design of a Preventive HIV Vaccine is a Utopia

There are many reasons why it has not been possible to rationally design a preventive HIV vaccine. Since HIV-infected individuals usually do not mount the type of protective immune response that occurs with many other virus infections, past successful vaccines that mimicked such responses have offered little guidance in the case of HIV/AIDS. HIV infection progressively destroys the host immune system and the virus possesses an extremely high variability linked to its error-prone reverse transcriptase enzyme which incorporates at least one mutation in every transcribed DNA copy. Since a billion HIV particles may be produced in an infected patient every day, the probability of mutations arising that are functional for the virus is extremely high and will lead to escape mutants capable of evading the immune responses of the host (Nowak and McMichael 1995).

As discussed in the chapters "Basic Research in HIV Vaccinology Is Hampered by Reductionist Thinking" and "Paradigm Changes Are Required in HIV Vaccine Research", the reductionist mindset as well as other research paradigms followed by HIV vaccinologists made them pursue strategies that were not well adapted to the unique biological and immunological features of the virus. Many studies were devoted to analyzing the structure of the Mabs that appear after several years of chronic HIV infection although these antibodies are not effective in controlling the infection in the patients from whom the antibodies had been obtained. Hundreds of nMabs bound to various epitopes of HIV Env have been studied by X-ray crystallography in an attempt to reconstruct the epitope by reverse molecular engineering, in the hope that the epitope designed to fit a neutralizing Mab would have acquired the immunogenic capacity to elicit a protective immune response. This approach failed to produce an HIV vaccine because the investigators focused only on optimizing the complementarity between single epitope-paratope pairs and did not investigate which intrinsic properties of the immune system control the formation of neutralizing antibodies (Van Regenmortel 2017).

Most protective immune responses against viruses are polyclonal, and in the case of HIV Env, the major antigenic sites always consist of a continuum of overlapping epitopes, each of which is recognized by different antibodies that often possess

© Springer Nature Switzerland AG 2019

Marc H V Van Regenmortel, *HIV/AIDS: Immunochemistry, Reductionism and Vaccine Design*, https://doi.org/10.1007/978-3-030-32459-9

synergistic neutralizing properties. Since most HIV immunogenicity studies use individual Mabs as models, very little information is obtained regarding the neutralization potential that **could** result if several unrelated epitopes within the same engineered HIV antigenic site were used simultaneously for immunization and gave rise to a synergistic, polyclonal neutralization effect. Another difficulty is that the enormous dynamic plasticity of the HIV Env trimer makes the dissection of its numerous transient epitopes during the immunization process extremely difficult (Zolla-Pazner et al. 2016).

The polyspecific functionalities of proteins and their considerable plasticity and flexibility, that are particularly evident in the multiple HIV Env conformations (Korkut and Hendrickson 2012), have changed our perception of proteins as being much more dynamic and adaptable than previously thought.

In a series of insightful reviews, Kurakin (2007, 2009, 2010) argued that the dynamic and adaptive plasticity of protein molecules does challenge the conventional, mechanistic assumptions embodied in the design charts of cellular circuitry, and favours instead a re-interpretation of biological molecular recognition and cellular organization within an expanded framework of scale-invariant self-organization. A protein in solution always consists of a population of conformers continuously interconverting on multiple timescales and capable of accommodating various ligands in different environmental contexts (Ma et al. 2002) and a protein binding site is no longer viewed as a static entity that can exist or be defined in the absence of a binding partner.

Large numbers of proteomics investigations have revealed that many proteins are frequently found in organelles and parts of cells where they are not expected to occur, indicating that protein localization is more random than commonly believed. Although these findings may seem odd, they are in agreement with the accepted view that the intracellular localization and roles of proteins are specified by biological evolution according to a predefined pattern of molecular recognition (Kurakin 2007). Although such apparent mislocalizations of individual protein molecules may not correspond to our expectations, they can be explained if one accepts that individual proteins tend to have multiple functionalities that arise from the presence in individual proteins of several, unrelalted binding sites in individual proteins. These multifunctionalities allow proteins to perform several, unrelated functions in diverse cellular locations, reflecting their adaptive plasticity and continuously variable physicochemical organization. These promiscuous interactions demonstrate that proteins are inherently much more dynamic and flexible than would appear for instance from the static crystallographic representations of protein-ligand complexes which do not reveal that these complexes result from processes of mutual adaptation between sterically dynamic entities. Numerous so-called moonlighting proteins have been discovered that are natively unfolded and acquire a particular structure only after interacting with a number of ligands, providing another striking example of protein plasticity (Tompa et al. 2005; Dunker et al. 2008). This has led to the suggestion that the prevailing paradigm: *structure determines function* should be replaced by *binding determines function* (Van Regenmortel 2002b).

Protein-protein interaction networks comprise large numbers of poorly connected nodes and a smaller number of highly connected nodes known as hubs. The promiscuity of hub proteins allows them to interact with multiple protein partners which is due to the presence of numerous binding sites and disordered regions in the hub proteins. This polyreactivity arises because a hub protein does not consist of a single amino acid sequence but of a large number of related gene products that arise from epigenetic modifications, alternative splicing and the discarding of introns during gene transcription. The hub does contain a large number of unpredictable RNA transcripts and protein isoforms with numerous conformations that display a variety of binding sites. This allows the hubs to interact with many different protein partners (Tsai et al. 2009).

Although it is commonly accepted that living cells and organisms are open, nonequilibrium systems characterized by a continuous flow of energy and matter, many biologists nevertheless assume that molecular and cellular biology can be interpreted within the conceptual framework of classical mechanics and equilibrium thermodynamics.

This contradiction has been highlighed by Kurakin (2009, 2010) who pointed out that biological macrostructures can also emerge in far-from-equilibrium systems because of configurational and flow dynamics. In this case, the order and macroorganization present within the cell are not predetermined by a pre-existing design but are driven by the same physical principles that determine self-organization in open, inorganic nonequilibrium thermodynamic systems. At each level of biological organization, the observed structure and dynamics are shaped by the filter of evolutionary selection and it is no longer relevant to introduce the preconceived plan of a factitious designer. In human bodies, for instance, there is little anatomical evidence for the intentional design excellence of a mythical designer (Olshansky et al. 2003) but plenty of evolutionary evidence for the occurrence of utilitarian tinkering processes (Laubichler 2007). Kurakin (2009) concluded that the critical parameters that define the state of a living system are the flow rates of its contituents rather than their concentrations as is usually assumed in equilibrium thermodynamics. Macromolecular complexes then become steady-state organizations that respond to the needs of cells, in a self-organized and self-regulated manner that is independent of a pre-existing design.

Doing something by design is doing it intentionally and involves the deliberate conceiving of a novel object or process by an intelligent mind. Rational drug design, for instance, utilizes the 3D structure of a biological target for designing molecules that bind to it selectively and inhibit its biological activity. In the narrow context of a single epitope-paratope pair, it is justified to speak of rational design when one tries to improve the ability of the epitope or the paratope to bind to its partner. However, improving the binding complementarity of an HIV Env epitope for a particular nMab does not amount to rationally designing an HIV immunogen better capable of generating protective antibodies.

The economist and Nobel laureate, Herbert Simon, introduced the concept of "bounded rationality" to describe the intrinsic limitations of human cognition for achieving a complete analysis and understanding of any complex system (Simon 1996). These limitations always exist because our information is always incomplete or inaccurate, our ability to investigate countless numbers of interacting components in a complex system is limited and we lack the time and the resources to reach a truly rational decision based on a complete knowledge of the investigated system and all its interactions with the environment (Gigerenzer and Selten 2002).

After the human brain, the human immune system is the most complex system in the human body. In order to make fully rational decisions about how to control the immune system, we would need to know all the relevant parameters that influence the innumerable interactions between the many compartments of the immune system which should also include the on-and-off kinetic constants of all the interactions which is not a realistic requirement (Van Regenmortel 2018b). The limitations of our bounded rationality explain why we have been unable to develop a comprehensive theoretical model of the entire human immune system that would have allowed us to conceive plausible testable hypotheses for trying to solve the multiple inverse problems that HIV vaccinologists are faced with (Van Regenmortel 2018b).

In their research, scientists are mostly confronted to direct problems that start with certain causes and analyze experimentally the mechanisms that produce a wanted result. In vaccinology, scientists are often confronted with inverse problems that require determining the past causes of a presently observed or wanted beneficial outcome (for instance the absence of deleterious HIV infection in human elite controllers). However, predicting something that happened in the past is considerably more difficult than predicting and analyzing what is going to happen in the future, which is what scientists actually do when they try to solve direct problems by investigating experimentally what are the effects that follow certain causes.

The complexity of the immune system and our inability to solve inverse problems are actually responsible for our inability to rationally design a preventive HIV vaccine.

Although the hypothetical capacity of rational design to solve problems is sometimes attributed to intelligent minds or metaphorically to the creative filter of natural selection and evolution, the available scientific evidence indicates that relying on design for solving inverse problems encountered in complex biological systems is a utopia. However, this should not prevent scientists from pursuing the empirical explorations that have allowed them to develop effective vaccines so may times in the past. It is to be welcomed that some of the novel research strategies pursued in HIV vaccinology today no longer claim to follow rational design principles but rely on time-honoured empiricism (Andrieu and Lu 2018; Rios 2018).

**Acknowledgments** Several chapters of this book were written when the author was on three occasions a resident fellow at the Stellenbosch Institute for Advanced Studies (STIAS) which is affiliated to an international consortium of Institutes for Advanced Studies in the US, Europe and Israel. I am grateful to STIAS for having provided an ideal intellectual environment for stimulating interactions with other STIAS fellows and scientific colleagues. I am grateful to Florence Diemer (CNRS, Strasbourg, France) for highly competent secretarial help over many years as well as to Emily Filmer for her help with compiling the extensive bibliography of this eBook.

# References

AcAleer WJ, Bynak EB, Maigeter RZ, Wampler DE, Miller WJ, Hilleman MR. Human hepatitis B vaccine from recombinant yeast. Nature. 1984;307:178–80. https://doi.org/10.1038/307178a0.

Acharya R, Fry E, Stuart D, Fox G, Rowlands D, Brown F. The three-dimensional structure of foot-and-mouth disease virus at 2.9 A resolution. Nature. 1989;337:709–16.

Achinstein P. The nature of explanation. Oxford University Press: New York; 1983.

Achour A, Biquard JM, Krsmanovic V, M'bika JP, Ficheux D, Sikorska M, Cozzone AJ. Induction of human immunodeficiency virus (HIV-1) envelope specific cell-mediated immunity by a non-homologous synthetic peptide. PLoS One. 2007;2(11):e1214.

Ackerman M, Alter G. Mapping the journey to an HIV vaccine. N Engl J Med. 2013;369:389–91.

Adams MJ, Lefkowitz EJ, King AM, Carstens EB. Recently agreed changes to the international code of virus classification and nomenclature. Arch Virol. 2013;158:2633–9.

Aderem A, Smith KD. A systems approach to dissecting immunity and inflammation. Semin Immunol. 2004;16:55–67.

Adu-Bobie J, Arico B, Giuliani MM, Seruto D. The first vaccine obtained through reverse vaccinology: the serogroup B meningococcus vaccine. In: Rappuoli R, Bagnoli F, editors. Vaccine design: innovative approaches and novel strategies. Norfolk: Caister Academic Press; 2011. p. 225–41.

Agarwal A, Hioe CE, Swetnam J, Zolla-Pazner S, Cardozo T. Quantitative assessment of masking of neutralization epitopes in HIV-1. Vaccine. 2011;29:6736–41.

Ahlers JD, Dunlop N, Pendleton CD, Newman M, Nara PL, Berzofsky JA. Candidate HIV type 1 multideterminant cluster peptide-P18MN vaccine constructs elicit type 1 helper T cell, cytotoxic T cells, and neutralizing antibody, all using the same adjuvant immunization. AIDS Res Hum Retrovir. 1996;12:259–72.

Al Moudallal Z, Briand JP, Van Regenmortel MHV. Monoclonal antibodies as probes of the antigenic structure of tobacco mosaic virus. EMBO J. 1982;1:1005–10.

Alam SM, McAdams M, Boren D, Rak M, Scearce RM, Gao F, et al. The role of antibody polyspecificity and lipid reactivity in binding of broadly neutralizing anti-HIV-1 envelope human monoclonal antibodies 2F5 and 4E10 to glycoprotein 41 membrane proximal envelope epitopes. J Immunol. 2007;178:4424–35.

Alam SM, Morelli M, Dennison SM, Liao HX, Zhang R, Xia SM, Rits-Volloch S, Sun L, Harrison SC, Haynes BF, Chen B. Role of HIV membrane in neutralization by two broadly neutralizing antibodies. Proc Natl Acad Sci U S A. 2009;106:20234–9.

Alberts B, Kirschner MW, Tilghman S, Varmus H. Rescuing US biomedical research from its systemic flaws. Proc Natl Acad Sci U S A. 2014;111:5773–7. https://doi.org/10.1073/pnas.1404402111.

Alexander H, Alexander S, Getzoff ED, Tainer JA, Geysen HM, Lerner RA. Altering the antigenicity of proteins. Proc Natl Acad Sci U S A. 1992;89:3352–6.

Alexander MR, Ringe R, Sanders RW, Voss JE, Moore JP, Klasse PJ. What do chaotrope-based avidity assays for antibodies to HIV-1 envelope glycoproteins measure? J Virol. 2015;89:5981–95.

Allen JF. Bioinformatics and discovery: induction beckons again. BioEssays. 2001;23:104–7.

Allen C, Bekoff M. Biological function, adaptation, and natural design. Philos Sci. 1995;62:609–22.

Alm E, Arkin AP. Biological networks. Curr Opin Struct Biol. 2003;13:193–202.

Almagro JC. Identification of differences in the specificity determining residues of antibodies that recognize antigens of different size: implications for the rational design of antibody repertoires. J Mol Recognit. 2004;17:132–43.

Altschuh D, Braun W, Kallen J, Mikol V, Spitzfaden C, Thierry JC, Vix O, Walkinshaw MD, Wuthrich K. Conformational polymorphism of cyclosporin A. Structure. 1994;2:963–72.

Amarasingham A, Geman S, Harrison MT. Ambiguity and nonidentifiability in the statistical analysis of neural codes. Proc Natl Acad Sci U S A. 2015;112:6455–60. https://doi.org/10.1073/pnas.1506400112.

Amela I, Cedano J, Querol E. Pathogen proteins eliciting antibodies do not share epitopes with host proteins: a bioinformatics approach. PLoS One. 2007;2:e512.

Amzel LM. Structure-based drug design. Curr Opin Biotechnol. 1998;9:366–9.

Andersson K, Areskoug D, Hardenborg E. Exploring buffer space for molecular interactions. J Mol Recognit. 1999;12:310–5.

Andersson K, Choulier L, Hamalainen MD, Van Regenmortel MHV, Altschuh D, Malmqvist M. Predicting the kinetics of peptide-antibody interactions using a multivariate experimental design of sequence and chemical space. J Mol Recognit. 2001;14:62–71.

Andrabi R, Voss JE, Liang CH, Briney B, McCoy LE, Wu CY, et al. Identification of common features in prototype broadly neutralizing antibodies to HIV envelope V2 apex to facilitate vaccine design. Immunity. 2015;43:959–73. https://doi.org/10.1016/j.immuni.2015.10.014.

Andreopoulos A, Tsotsos JK. 50 Years of object recognition: directions forward. Comput Vis Image Underst. 2013;117:827–91. https://doi.org/10.1016/j.cviu.2013.04.005.

Andrieu J-M, Lu W. A 30-year journey of trial and error towards a tolerogenic AIDS vaccine. Arch Virol. 2018;163:2025–31.

Andrieu JM, Chen S, Lai C, Guo W, Lu W. Mucosal SIV vaccines comprising inactivated virus particles and bacterial adjuvants induce CD8+ T-regulatory cells that suppress SIV positive CD4+ cell activation and prevent SIV infection in the macaque model. Front Immunol. 2014;5:297. https://doi.org/10.3389/fimmu.2014.00297.

Andris JS, Johnson S, Zolla-Pazner S, Capra JD. Molecular characterization of five human anti-human immunodeficiency virus type1 antibody heavy chains reveals extensive somatic mutation typical of an antigen-driven immune response. Proc Natl Acad Sci U S A. 1991;88:7783–7.

Anger G, Moritz H. Inverse problems and uncertainties in science and medicine. Proc Leibnitz Soc. 2003;61:171–212.

Anuforom O, Wallace GR, Piddock LV. The immune response and anti-bacterial therapy. Med Microbiol Immunol. 2015;204:151–9. https://doi.org/10.1007/s00430-014-0355-0.

Appel JR, Pinilla C, Niman H, Houghten RA. Elucidation of discontinuous linear determinants in peptides. J Immunol. 1990;144:976–83.

Arnaout R, Lee W, Cahill P, Honan T, Sparrow T, Weiand M, Nusbaum C, Rajewsky K, Koralov SB. High-resolution description of antibody heavy-chain repertoires in humans. PLoS One. 2011;6:e22365. https://doi.org/10.1371/journal.pone.0022365.

Arneson R. Egalitarianism. The Stanford Encyclopedia of Philosophy; 2013.

Arnold GF, Velasco PK, Holmes AK, Wrin T, Geisler SC, Phung P, Tian Y, Resnick DA, Ma X, Mariano TM, Petropoulos CJ, Taylor JW, Katinger H, Arnold E. Broad neutralization of human immunodeficiency virus type 1 (HIV-1) elicited from human rhinoviruses that display the HIV-1 gp41 ELDKWA epitope. J Virol. 2009;83:5087–50100.

Arnon R. Synthetic vaccines. Boca Raton, FL: CRC Press; 1987.

Arnon R, Ben Yedidia T. Old and new vaccine approaches. Int Immunopharmacol. 2003;3:1195–204.

Arrhenius S. Immunochemistry: the applications of the principles of physical chemistry to the study of the biological antibodies. New York: Macmillan; 1907.

Arthur LO, Bess JW Jr, Chertova EN, Rossio JL, Esser MT, Benveniste RE, Henderson LE, et al. Chemical inactivation of retroviral infectivity by targeting nucleocapsid protein zinc fingers: a candidate SIV vaccine. AIDS Res Hum Retrovir. 1998;14(Suppl 3):S311–9.

Atassi MZ, Smith JA. A proposal for the nomenclature of antigenic sites in peptides and proteins. Immunochemistry. 1978;15:609–10.

Atassi MZ, Van Oss CJ, Absolom DL. Molecular immunology. New-York: Dekker; 1984.

Auffray C, Imbeaud S, Roux-Rouquié M, Hood L. From functional genomics to systems biology: concepts and practices. CR Biol. 2003;326:879–92.

Auffrey C, Nottale L. Scale relativity theory and integrative systems biology 1. Founding principles and scale laws. Prog Biophys Mol Biol. 2008;97:79–114. https://doi.org/10.1016/j.pbiomolbio.2007.09.002.

Ayala FJ. In: Dembski WA, Ruse M, editors. Debating design: from Darwin to DNA. Cambridge: Cambridge University Press; 2004. p. 55–80.

Azimzadeh A, Van Regenmortel MHV. Measurement of affinity of viral monoclonal antibodies by ELISA titration of free antibody in equilibrium mixtures. J Immunol Methods. 1991;141:199–208.

Azoitei ML, Correia BE, Ban YE, Carrico C, Kalyuzhniy O, Chen L, Schroeter A, Huang PS, McLellan JS, Kwong PD, Baker D, Strong RK, Schief WR. Computation-guided backbone grafting of a discontinuous motif onto a protein scaffold. Science. 2011;334:373–6.

Bacarese-Hamilton T, Mezzasoma L, Ardizzoni A, et al. Serodiagnosis of infectious diseases with antigen microarrays. J Appl Microbiol. 2004;96:10–7.

Bachler BC, Humbert M, Palikuqi B, Siddappa NB, Lakhashe SK, et al. Novel biopanning strategy to identify epitopes associated with vaccine protection. J Virol. 2013;87:4403–16.

Baden LR, Walsh SR, Seaman MS, Tucker RP, Krause KH, Patel A, et al. First-in-human evaluation of the safety and immunogenicity of are combinant adenovirus serotype 26 HIV-1 Env vaccine (IPCAVD001). J Infect Dis. 2013;207:240–7. https://doi.org/10.1093/infcis/jis670.

Bahk YY, Kim SA, Kim JS, et al. Antigens secreted from Mycobacterium tuberculosis: identification by proteomics approach and test for diagnostic marker. Proteomics. 2004;4:3299–307.

Baldwin JT, Lessmann O. What is Russell's paradox? Scientific American; 1998. http://www.scientificamerican.com/article/whatis-russells-paradox/. Accessed 23 July 2018.

Ball LA. The universal taxonomy of viruses in theory and practice. In: Fauquet CM, et al., editors. Eighth ICTV report. Amsterdam: Elsevier; 2005. p. 11–6.

Bamford DH, Gilbert RJC, Grimes JM, Stuart DI. Macromolecular assemblies: greater than their parts. Curr Opin Struct Biol. 2001;11:107–13.

Barabasi AL, Oltvai ZN. Network biology: understanding the cell's functional organization. Nat Rev Genet. 2004;5:101–13.

Barlow DJ, Edwards MS, Thornton JM. Continuous and discontinuous protein antigenic determinants. Nature. 1986;322:747–8.

Batista FD, Neuberger MS. Affinity dependence of the B cell response to antigen: a threshold, a ceiling, and the importance of off-rate. Immunity. 1998;8:751–9.

Bauer HH. Scientific literacy and the myth of the scientific method. Urbana, IL: University of Illinois Press; 1992. p. 1–180.

Bechtel W, Abrahamsen A. Explanation: a mechanist alternative. Stud Hist Phil Biol Biomed Sci. 2005;36:421–41. https://doi.org/10.1016/j.shpsc.2005.03.010.

Bęczkowski PM, Harris M, Techakriengkrai N, Beatty JA, Willett BJ, Hosie MJ. Neutralising antibody response in domestic cats immunised with a commercial feline immunodeficiency virus (FIV) vaccine. Vaccine. 2015;33:977–84.

Behe MJ. Darwin's black box: the biochemical challenge to evolution. New York: Free Press; 1996.

Bellino S, Francavilla V, Longo O, Tripiciano A, Paniccia G, et al. Parallel conduction of the phase I preventive and therapeutic trials based on the Tat vaccine candidate. Rev Recent Clin Trials. 2009;4:195–204.

Bellino S, Tripiciano A, Picconi O, Francavilla V, Longo O, et al. The presence of anti-Tat antibodies in HIV-infected individuals is associated with containment of CD4+ T-cell decay and viral load and with delay of disease progression: Results of a 3 year cohort study. Retrovirology. 2014;11:49.

Belshe R, Franchini G, Girard MP, et al. Support for the RV144 HIV vaccine trial. Science. 2004;305:177–80.

Ben Khalifa M, Weidenhaupt M, Choulier L, Chatellier J, Rauffer-Bruyère N, Altschuh D, Vernet T. Effects of interaction kinetics of mutations at the VH-VL interface of Fabs depend on the structural context. J Mol Recognit. 2000;13:127–39.

Benito A, Van Regenmortel MHV. Biosensor characterization of antigenic site A of foot-and-mouth disease virus presented in different vector systems. FEMS Immunol Med Microbiol. 1998;21:101–15.

Benjamin DC, Berzofsky JA, East IJ, Gurd FRN, Hannum C, Leach SJ, Margoliash E, Michael JG, Miller A, Prager EM, Reichlin M, Sercaz EE, Smith-Gill SJ, Todd PE, Wilson AC. The antigenic structure of proteins: a reappraisal. Annu Rev Immunol. 1984;2:67–101.

Benjamini E. Immunochemistry of the tobacco mosaic virus. In: Atassi MZ, editor. Immunochemistry of proteins, vol. 2. New York: Plenum Press; 1997. p. 265–310.

Berek C. Maturation of the immune response. In: Van Regenmortel MHV, editor. Structure of antigens, vol. 1. Boca Raton: CRC Press; 1992. p. 149–57.

Berek C, Milstein C. Mutation drift and repertoire shift in the maturation of the immune response. Immunol Rev. 1987;96:23–41. https://doi.org/10.1111/j.1600-065X.1987.tb00507.x.

Berger R. Understanding science: why causes are not enough. Philos Sci. 1998;65:306–32.

Berger C, Weber-Bornhauser S, Eggenberger J, Hanes J, Plückthun A, Bosshard HR. Antigen recognition by conformational selection. FEBS Lett. 1999;450:149–53.

Bernardes AT, Zorzenon Dos S. Immune network at the edge of chaos. J Theor Biol. 1997;186:173–87.

Bertuglia CS, Vaio F. Nonlinearity, chaos and complexity. Oxford: Oxford University Press; 2005.

Berzofsky JA. Intrinsic and extrinsic factors in protein antigenic structure. Science. 1985;229:932–40.

Berzofsky JA, Berkover IJ. Novel approaches to peptide and engineered protein vaccines for HIV using defined epitopes: advances in 1994-1995. AIDS. 1995;9:S143–57.

Berzofsky JA, Schechter AN. The concepts of crossreactivity and specificity in immunology. Mol Immunol. 1981;18:751–63.

Berzofsky JA, Pendleton CD, Clerici M, Ahlers J, Lucey DR, Putney SD, Shearer GM. Construction of peptides encompassing multideterminant clusters of HIV envelope to induce in vitro T-cell responses in mice and humans of multiple MHC types. J Clin Invest. 1991;88:876.

Bhalla US, Iyengar R. Emergent properties of networks of biological signaling pathways. Science. 1999;283:381–7.

Bhat TN, Bentley GA, Boulot G, Greene MI, Tello D, Dall'Acqua W, Souchon H, Schwarz FP, Mariuzza RA, Poljak RJ. Bound water molecules and conformational stabilization help mediate an antigen-antibody association. Proc Natl Acad Sci U S A. 1994;91:1089–93.

Bhattacharjee AK, Glaudemans CP. Dual binding specificities in MOPC 384 and 870 murine myeloma immunoglobulins. J Immunol. 1978;120:411–3.

Bhiman JN, Moore PL. Size doesn't matter: shorter antibody loops can infiltrate HIV's env apex defenses. Immunity. 2017;46:762–4. https://doi.org/10.1016/j.immuni.2017.04.012.

Bickle J. Philosophy and neuroscience: a ruthlessly reductive account. Dordrecht: Kluwer Academic; 2003.

Binder H, Blettner M. Big data in medical science – a biostatistical view. Dtsch Arztebl Int. 2015;112:137–42. https://doi.org/10.3238/arztebl.2015.0137.

Binley J. Specificities of broadly neutralizing anti-HIV-1 sera. Curr Opin HIV AIDS. 2009;4:364–72.

Binley JM, Lybarger EA, Crooks ET, Seaman MS, Gray E, Davis KL, Decker JM, Wycuff D, Harris L, Hawkins N, et al. Profiling the specificity of neutralizing antibodies in a large panel of plasmas from patients chronically infected with human immunodeficiency virus type 1 subtypes B and C. J Virol. 2008;82:11651–68.

Binley JM, Ban YE, Crooks ET, Eggink D, Osawa K, Schief WR, et al. Role of complex carbohydrates in human immunodeficiency virus type 1 infection and resistance to antibody neutralization. J Virol. 2010;84:5637–55.

Bird A. Thomas Kuhn. In: Zalta EN, editor. The Stanford Encyclopedia of Philosophy (Winter 2018 Edition); 2018. https://plato.stanford.edu/archives/win2018/entries/thomas-kuhn/

Biro JC. Seven fundamental, unsolved questions in molecular biology. Cooperative storage and bi-directional transfer of biological information by nucleic acids and proteins: an alternative to "central dogma". Med Hypotheses. 2005;63:951–62.

Biro JC. The proteomic code: a molecular recognition code for proteins. Theor Biol Med Model. 2007;4:45.

Bittle JL, Hougten RA, Alexander H, et al. Protection against foot-and-mouth disease by immunization with a chemically synthesized peptide predicted from the viral nucleotide sequence. Nature. 1982;298:30–3.

Blake J. Bio-ontologies: fast and furious. Nat Biotechnol. 2004;22:773–4.

Blalock JE. Complementarity of peptides specified by "sense" and "antisense" strands of DNA. Trends Biotechnol. 1990;8:140–4.

Blalock JE, Bost KL. Binding of peptides that are specified by complementary RNAs. Biochem J. 1986;234:679–89.

Blalok J. Complementarity of peptides specified by "sense" and "antisense" strands of DNA. Trends Biotechnol. 1990;8:140–4.

Blythe MJ, Flower DR. Benchmarking B cell epitope prediction: underperformance of existing methods. Protein Sci. 2005;14:246–8.

Bock G, Goode J. The limits of reductionism in biology. Novartis Foundation Symposium no. 213. Chichester: Wiley; 1998.

Bock WJ, Von Wahlert G. Adaptation and the form-function complex. Evolution. 1965;19:269–99.

Boeyé A, Rombaut B. The proteins of poliovirus. Prog Med Virol. 1992;62:139–66.

Bomsel M, Tudor D, Drillet AS, Alfsen A, Ganor Y, et al. Immunization with HIV-1 gp41 subunit virosomes induces mucosal antibodies protecting nonhuman primates against vaginal SHIV challenges. Immunity. 2011;34:269–80.

Boots LJ, McKenna PM, Arnold BA, et al. Anti-human immunodeficiency virus type 1 human monoclonal antibodies that bind discontinuous epitopes in the viral glycoproteins can identify mimotopes from recombinant phage peptide display libraries. AIDS Res Hum Retrovir. 1997;13:1549–59.

Boquet D, Déry O, Forbert Y, Grassi J, Couraud JY. Is hydropathic complementarity involved in antigen-antibody binding? Mol Immunol. 1995;32:303–8.

Bork P, Serrano L. Towards cellular systems in 4D. Cell. 2005;121:507–9.

Bork P, Dandekar T, Diaz-Lazcoz Y, Eisenhaber F, Huynen M, Yuan Y. Predicting function: from genes to genomes and back. J Mol Biol. 1998;283:707–25.

Borsetti A, Baroncelli S, Maggiorella MT, Moretti S, Fanales-Belasio E, et al. Containment of infection in tat vaccinated monkeys after rechallenge with a higher dose of SHIV89.6P(cy243). Viral Immunol. 2009;22:117–24.

Bosshard HR. Molecular recognition by induced fit: how fit is the concept? News Physiol Sci. 2001;16:171–3.

Bothner B, Dong XF, Bibbs L, Johnson JE, Siuzdak G. Evidence of viral capsid dynamics using limited proteolysis and mass spectrometry. J Biol Chem. 1998;9:673–6.

Bottaro A, Inlay MA, Matzke NJ. Immunology in the spotlight at the Dover 'Intelligent Design' trial. Nat Immun. 2006;7:433–5.

Bou-Habib DC, Roderiquez G, Oravecz T, et al. Cryptic nature of envelope V3 region epitopes protects primary monocytotropic human immunodeficiency virus type 1 from antibody neutralization. J Virol. 1994;68:6006–13.

Bouvet JP, Stahl D, Rose S, Quan CP, Kazatchkine MD, Kaveri SV. Induction of natural autoantibody activity following treatment of human immunoglobulin with dissociating agents. J Autoimmun. 2001;16:163–72.

Boyd WC, Bernard H. Quantitative changes in antibodies and globulin fractions in sera of rabbits injected with several antigens. J Immunol. 1937;33:111–22.

Braden BC, Poljak RJ. Structural features of the reactions between antibodies and protein antigens. FASEB J. 1995;9:9–16.

Bradie M. Review of what genes can't do by Lenny Moss. A Bradford book. The MIT Press, Cambridge, MA. Hum Nat Rev. 2003;3:317–20.

Bramwell VW, Perrie Y. The rational design of vaccines. Drug Discov Today. 2005;10:1527–34.

Brenner S. Sequences and consequences. Philos Trans R Soc Lond B Biol Sci. 2010;365:207–12.

Brentani RR. Biological implications of complementary hydropathy of amino acids. J Theor Biol. 1988;135:495–9.

Briand JP, Benkirane N, Guichard G, Newman JFE, Van Regenmortel MHV, Brown F, Muller S. A retro-inverso peptide corresponding to the GH loop of foot-and-mouth disease virus elicits high levels of long lasting protective neutralizing antibodies. Proc Natl Acad Sci U S A. 1997;94:12545–50.

Briney BS, Willis JR, Crowe JE. Human peripheral blood antibodies with long HCDR3s are established primarily at original recombination using a limited subset of germline genes. PLoS One. 2012;7:e36750.

Brito AF, Braconi CT, Weidmann M, Dichler M, Alves JM, Gruber A, Zanotto PM. The Pangenome of the Anticarsia gemmatalis multiple nucleopolyhedrovirus (AgMNPV). Genome Biol Evol. 2015;8:94.

Broder CC, Earl PL, Long D, Abedon ST, Moss B, Doms RW. Antigenic implications of human immunodeficiency virus type 1 envelope quaternary structure: oligomer-specific and -sensitive monoclonal antibodies. Proc Natl Acad Sci U S A. 1994;91:11699–703.

Brown F. Foot-and-mouth disease. In: Nicholson BH, editor. Synthetic vaccines. Oxford: Blackwell Scientific Publication; 1994. p. 416–32.

Brown F. Inactivation of viruses by aziridines. Vaccine. 2002;20:322–7.

Bruggeman FJ, Westerhoff HV. The nature of systems biology. Trends Microbiol. 2006;15:45–50. https://doi.org/10.1016/j.tim.2006.11.003.

Brunel FM, Zwick MB, Cardoso RMF, Nelson JD, Wilson IA, Burton DR, Dawson PE. Structure-function analysis of the epitope for 4E10, a humanimmunodeficiency virus type 1 (HIV-1) broadly neutralizing antibody. J Virol. 2006;80:1680–7.

Bublil EM, Freund NT, Mayrose I, Penn O, Roitburd-Berman A, Rubinstein ND, Pupko T, Gershoni JM. Stepwise prediction of conformational discontinuous B-cell epitopes using the Mapitope algorithm. Proteins. 2007;68:294–304.

Buchak L. Why high-risk, non-expected-utility-maximising gambles can be rational and beneficial: the case of HIV cure studies. J Med Ethics. 2016;2015:103118.

Buchbinder SP, Mehrotra DV, Duerr A, Fitzgerald DW, Mogg R, Li D, Gilbert PB, Lama JR, Marmor M, Del Rio C, et al. Efficacy assessment of a cell-mediated immunity HIV-1 vaccine (the Step Study): a double-blind, randomised, placebo-controlled, test-of-concept trial. Lancet. 2008;372:1881–93.

Buck RC, Hull DL. The logical structure of the Linnaean hierarchy. Syst Zool. 1966;15:97–111.

Bunge M. Philosophical dictionary. Amherst, MA: Prometheus Books; 2003.

Bunge M. Modes of existence. Rev Metaphys. 2016;70:225.

Bunge M. Science in the light of philosophy. Singapore: World Scientific Publishers; 2017. p. 1–225.

Bunnik EM, Pisas L, van Nuenen AC, Schuitemaker H. Autologous neutralizing humoral immunity and evolution of the viral envelope in the course of subtype B human immunodeficiency virus type 1 infection. J Virol. 2008;82:7932–41. https://doi.org/10.1128/JV1.00757-08.

Bunnik EM, van Gils MJ, Lobbrecht MS, et al. Changing sensitivity to broadly neutralizing antibodies b12, 2G12, 2F5, and 4E10 of primary subtype B human immunodeficiency virus type 1 variants in the natural course of infection. Virology. 2009;390:348–55.

Burggren W, Monticino MG. Assessing physiological complexity. J Exp Biol. 2005;208:3221–32. https://doi.org/10.1242/jeb.01762.

Burke V, Williams C, Sukumaran M, Kim SS, Li H, Wang XH, Gorny MK, Zolla-Pazner S, Kong XP. Structural basis of the cross-reactivity of genetically related human anti-HIV-1 monoclonal antibodies: implications for design of V3-based immunogens. Structure. 2009;17:1538–46.

Burton DR. Antibodies, viruses and vaccines. Nat Rev Immunol. 2002;2:706–13.

Burton DR. Scaffolding to build a rational vaccine design strategy. Proc Natl Acad Sci U S A. 2010;107:17859–60.

Burton DR. What are the most powerful immunogen design strategies? Reverse vaccinology 2.0 shows great promise. Cold Spring Harb Perspect Biol. 2017;9:a030262. https://doi.org/10.1101/cshperspect.a030262.

Burton DR, Moore JP. Why do we not have an HIV vaccine and how can we make one? Nat Med. 1998;4:495–8.

Burton DR, Weiss RA. AIDS/HIV. A boost for HIV vaccine design. Science. 2010;329:770–3. https://doi.org/10.1126/science.1194693.

Burton DR, Pyati J, Koduri R, Sharp SJ, Thornton GB, Parren PW, et al. Efficient isolates of HIV-1 by a recombinant human monoclonal neutralization of primary antibody. Science. 1994;266:1024–7. https://doi.org/10.1126/science.7973652.

Burton DR, Desrosiers RC, Doms RW, Feinberg MB, Gallo RC, Hahn B, Hoxie JA, Hunter E, Korber B, Landay A, et al. Public health. A sound rationale needed for phase III HIV-1 vaccine trials. Science. 2004a;303:316.

Burton DR, Desrosiers RC, Doms RW, Koff WC, Kwong PD, Moore JP, Nabel GJ, Sodroski J, Wilson IA, Wyatt RT. HIV vaccine design and the neutralizing antibody problem. Nat Immunol. 2004b;5:233–6.

Burton DR, Stanfield RL, Wilson IA. Antibody vs HIV in a clash of evolutionary titans. Proc Natl Acad Sci U S A. 2005;102:14943–8.

Burton DR, Ahmed R, Barouch DH, Butera ST, Crotty S, Godzik A, Kaufmann DE, McElrath MJ, Nussenzweig MC, Pulendran B, et al. A blueprint for HIV vaccine discovery. Cell Host Microbe. 2012;12:396–407.

Bush V. Science, the endless frontier: a report to the president on a program for hostwar scientific research. Washington, DC: National Science Foundation; 1945.

Byerly H. Reductionism: analysis and synthesis in biological explanations. Q Rev Biol. 2003;78:336–42.

Cafaro A, Caputo A, Fracasso C, Maggiorella MT, Goletti D, et al. Control of SHIV-89.6P-infection of cynomolgus monkeys by HIV-1 Tat protein vaccine. Nat Med. 1999;5:643–50.

Cafaro A, Bellino S, Titti F, Maggiorella MT, Sernicola L, et al. Impact of viral dose and major histocompatibility complex class IB haplotype on viral outcome in Mauritian Cynomolgus monkeys vaccinated with Tat upon challenge with simian/human immunodeficiency virus SHIV89.6P. J Virol. 2010;84:8953–8.

Cafaro A, Tripiciano A, Sgadari C, Bellino S, Picconi O, et al. Development of a novel AIDS vaccine: the HIV-1 transactivator of transcription protein vaccine. Expert Opin Biol Ther. 2015;15(Suppl 1):S13–29.

Caioli SE. Benchmarking B-cell epitope prediction for the design of peptide-based vaccines: problems and prospects. J Biomed Biotechnol. 2010;2010:910524. https://doi.org/10.1155/2010/910524.

Canady MA, Larson SB, Day J, MacPherson A. Crystal structure of turnip yellow mosaic virus Nat Struct Biol. 1996;3:771–80.

Caputo A, Gavioli R, Bellino S, et al. HIV-1 Tat-based vaccines: an overview and perspectives in the field of HIV/AIDS vaccine development. Int Rev Immunol. 2009;28:285–334.

Cardoso RM, Zwick MB, Stanfield RL, Kunert R, Binley JM, Katinger H, et al. Broadly neutralizing anti-HIV antibody 4E10 recognizes a helical conformation of a highly conserved fusion-associated motif in gp41. Immunity. 2005;22:163–73.

Carnathan DG, Mackel JJ, Sweat SL, Enemuo CA, Gebru EH, Dhadvai P, Gangadhara S, Hicks S, Vanderford TH, Amara RR, Esparza J, Lu W, Andrieu JM, Silvestri G. Intragastric administration of Lactobacillus plantarum and AT-2-inactivated SIV does not protect Indian rhesus macaques from intra-rectal SIV challenge nor reduce virus replication after transmission. J Virol. 2018;92:e02030-17.

Carneiro J, Rethinking SJ. 'Shape space': evidence from simulated docking suggests that steric shape complementarity is not limiting for antibody-antigen recognition and idiotypic interactions. J Theor Biol. 1994;169:391–402.

Cartwright N. How the laws of physics lie. New York, NY: Oxford Uduniversity Press; 1983. p. 1–221.

Carver JA, Esposito G, Viglino P, et al. Structural comparison between retro-inverso and parent peptides: molecular basis for the biological activity of a retro-inverso analogue of the immunodominant fragment of VP1 coat protein from foot-and-mouth disease virus. Biopolymers. 1997;41:569–89.

Casadevall A, Pirofski L. Host-pathogen interactions: redefining the basic concepts of virulence and pathogenicity. Infect Immun. 1999;67:3703–13.

Casadevall A, Pirofski L. Host-pathogen interactions: the basic concepts of microbial commensalism, colonization, infection, and disease. Infect Immun. 2000;68:6511–8. https://doi.org/10.1128/IAI.68.12.6511-6518.2000.

Casadevall A, Pirofski LA. What is a host? Incorporating the microbiota into the damage-response framework. Infect Immun. 2015;83:2–7. https://doi.org/10.1128/IAI.02627-14.

Casadevall A, Fang FC, Pirofski LA. Microbial virulence as an emergent property: consequences and opportunities. PLoS Pathog. 2011;7:e1002136. https://doi.org/10.1371/journal.ppat.1002136.

Casti JL. Complexiication. New York: Harper Collins; 1994.

Cebere I, Dorrell L, McShane H, Simmons A, McCormack S, et al. Phase I clinical trial safety of DNA- and modified virus Ankara-vectored human immunodeficiency virus type 1 (HIV-1) vaccines administered alone and in a prime-boost regime to healthy HIV-1-uninfected volunteers. Vaccine. 2006;24:417–25.

Cedersund G, Roll J. Systems biology: model based evaluation and comparison of potential explanations for given biological data. FEBS J. 2009;276:903–22. https://doi.org/10.1111/j.1742-4658.2008.06845.x.

Chacko S, Silverton E, Kam-Morgan L, Smith-Gill S, Cohen G, Davis D. Structure of an antibody–lysozyme complex unexpected effect of a conservative mutation. J Mol Biol. 1995;245:261–74.

Chakraborty AK, Dustin ML, Shaw AS. In silico models for cellular and molecular immunology: successes, promises and challenges. Nat Immunol. 2003;4:933–6.

Chakraborty K, Durani V, Miranda ER, Citron M, Liang X, Schleif W, Joyce JG, Varadarajan R. Design of immunogens that present the crown of the HIV-IV3 loop in a conformation competent to generate 447-52 D-like antibodies. Biochem J. 2006;399:483–91.

Chalmers AF. What is this thing called science. 3rd ed. Maidenhead: Open University Press; 1999. p. 1–288.

Changela A, Wu X, Yang Y, Zhang B, Zhu J, Nardone GA, O'Dell S, Pancera M, Gorny MK, Phogat S, Robinson JE, Stamatatos L, Zolla-Pazner S, Mascola JR, Kwong PD. Crystal structure of human antibody 2909 reveals conserved features of quaternary structure-specific antibodies that potently neutralize HIV-1. J Virol. 2011;85:2524–35.

Chappey O, Debray M, Niel E, Schermann JM. Association constants of monoclonal antibodies for hapten: heterogeneity of frequency distribution and possible relationship with hapten molecular weight. J Immunol Methods. 1994;172:219–25.

Chatellier J, Van Regenmortel MHV, Vernet T, Altschuh D. Functional mapping of conserved residues located at the VL and VH domain interface of a Fab. J Mol Biol. 1996;264:1–6. https://doi.org/10.1006/jmbi.1996.0618.

Chatzipanagiotou S, Ioannidis A, Trikka-Graphakos E, Charalampaki N, Sereti C, Piccinini E, et al. Detecting the hidden properties of immunological data and predicting the mortality risks of infectious syndromes. Front Immunol. 2016;7:217. https://doi.org/10.3389/fimmu.2016.00217.

Chen ZJ, Wheeler CJ, Shi W, Wu AJ, Yarboro CH, Gallagher M, et al. Polyreactive antigen-binding B cells are the predominant cell type in the newborn B cell repertoire. Eur J Immunol. 1998;28:989–94. https://doi.org/10.1002/(SICI)1521-4141(199803)28:03<989::AID-IMMU989>3.0.CO;2-1.

Chen J, Liu H, Yang J, Chou KC. Prediction of linear B cell epitope using amino acid pair antigenicity scale. Amino Acids. 2007;33:423–8.

Chen SW, Van Regenmortel MHV, Pellequer JL. Structure-activity relationships in peptide-antibody complexes: implications for epitope prediction and development of synthetic peptide vaccines. Curr Med Chem. 2009;16:953–64.

Chen W, Streaker ED, Russ DE, Feng Y, Prabakaran P, Dimitrov D. Characterization of germline antibody libraries from human umbilical cord blood and selection of monoclonal antibodies to viral envelope glycoproteins: implications for mechanisms of immune evasion and design of vaccine immunogens. Biochem Biophys Res Commun. 2012;417:1164–9.

Cho MW, Lee MK, Chen CH, Matthews T, Martin MA. Identification of gp120 regions targeted by a highly potent neutralizing antiserum elicited in a chimpanzee inoculated with a primary human immunodeficiency virus type 1 isolate. J Virol. 2000;74:9749–54.

Chothia C, Lesk AM. Canonical structures for the hypervariable regions of immunoglobulins. J Mol Biol. 1987;196:901–17.

Choulier L, Andersson K, Hamalainen MD, Van Regenmortel MHV, Malmqvist M, Altschuh D. QSAR studies applied to the prediction of antigen-antibody interaction kinetics as measured by Biacore. Prot Eng. 2002;15:101–10.

Churchill MEA, Stura EA, Pinilla C, Appel JR, Houghten RA, Kono DH, Balderas RS, Fieser GG, Schulze-Gahmen U, Wilson IA. Crystal structure of a peptide complex of antiinfluenza peptide antibody Fab 26/9. Comparison of two different antibodies bound to the same peptide antigen. J Mol Biol. 1994;241:534–56.

Coëffier E, Clément JM, Cussac V, Khodaei-Boorane N, Jehanno M, Rojas M, et al. Antigenicity and immunogenicity of the HIV-1 gp41 epitope ELDKWA inserted into permissive sites of the MalE protein. Vaccine. 2001;19:684–93.

Cohen J. The cost of dichotomization. Appl Psychol Meas. 1983;7:249–53. https://doi.org/10.1177/014662168300700301.

Cohen J. Infectious disease. Obstacles loom along path to the end of AIDS. Science. 2016;353:432–3.

Cohen J, Stewart L. The collapse of chaos. New York: Penguin Books; 1994. p. 495.

Cohen GH, Sheriff S, Davies DR. Refined structure of the monoclonal antibody HyHEL-5 with its antigen hen egg-white lysozyme. Acta Crystallogr D Biol Crystallogr. 1996;52:315–26.

Cohen IR, Hershberg U, Solomon S. Antigen-receptor degeneracy and immunological paradigms. Mol Immunol. 2004;40:993–6.

Cohen J, Vincent JL, Adhikari NKJ, Machado FR, Angus DC, Calandra T, et al. Sepsis: a roadmap for future research. Lancet Infect Dis. 2015;15:581–614. https://doi.org/10.1016/S1473-3099(15)70112-X.

Cohn M. A new concept of immune specificity emerges from a consideration of the self-nonself discrimination. Cell Immunol. 1997;181:103–8.

Cohn M. Degeneracy, mimicry and cross-reactivity in immune recognition. Mol Immunol. 2005;42:651–5.

Cohn M, Langman RE. The protecton: the evolutionary selected unit of humoral immunity. Immunol Rev. 1990;115:1–131.

Collis AV, Brouwer AP, Martin AC. Analysis of the antigen combining site: correlations between length and sequence composition of the hypervariable loops and the nature of the antigen. J Mol Biol. 2003;325:337–54.

Conti F, Valerio MC, Zbilut JP, Giuliani A. Will systems biology offer new holistic paradigms to life sciences? Syst Synth Biol. 2007;1:161–5. https://doi.org/10.1007/s11693-008-9016-1.

Corey L, Gray GE. Preventing acquisition of HIV is the only path to an AIDS free generation. Proc Natl Acad Sci U S A. 2017;114:3798–800. https://doi.org/10.1073/pnas.1703236114.

Corey L, Nabel GJ, Dieffenbach C, Gilbert P, Haynes BF, et al. HIV-1 vaccines and adaptive trial designs. Sci Transl Med. 2011;3:79ps13.

Corey L, Gilbert PB, Tomaras GD, Haynes BF, Pantaleo G, et al. Immune correlates of vaccine protection against HIV-1 acquisition. Sci Transl Med. 2015;7:310rv7.

Cornish-Bowden A. Putting the systems back into systems biology. Perspect Biol Med. 2006;49:475–89.

Cornish-Bowden A. Systems biology. How far has it come? Biochemist. 2011;33:16–8.

Correia BE, Ban YE, Holmes MA, Xu H, Ellingson K, Kraft Z, Carrico C, Boni E, Sather DN, Zenobia C, Burke KY, Bradley-Hewitt T, Bruhn-Johannsen JF, Kalyuzhniy O, Baker D, Strong RK, Stamatatos L, Schief WR. Computational design of epitope-scaffolds allows induction of antibodies specific for a poorly immunogenic HIV vaccine epitope. Structure. 2010;18:1116–26.

Correia B, Bates JT, Loomis RJ, et al. Proof of principle for epitope-focused vaccine design. Nature. 2014;507:201–6.

Corti D, Lanzavecchia A. Broadly neutralizing antiviral antibodies. Annu Rev Immunol. 2013;31:705–42. https://doi.org/10.1146/annurev-immunol-032712-095916.

Corti D, Langedijk JP, Hinz A, Seaman MS, Vanzetta F, Fernandez-Rodriguez BM, Silacci C, Pinna D, Jarrossay D, Balla-Jhagjhoorsingh S, Willems B, Zekveld MJ, Dreja H, O'Sullivan E, Pade C, Orkin C, Jeffs SA, Montefiori DC, Davis D, Weissenhorn W, McKnight A, Heeney JL, Sallusto F, Sattentau QJ, Weiss RA, Lanzavecchia A. Analysis of memory B cell responses and isolation of novel monoclonal antibodies with neutralizing breadth from HIV-1-infected individuals. PLoS One. 2010;5:e8805. https://doi.org/10.1371/journal.pone.0008805.

Coutinho A, Kazatchkine MD, Avrameas S. Natural autoantibodies. Curr Opin Immunol. 1995;7:812–8.

Crick FHC. Of molecules and men. Seattle, WA: University of Washington Press; 1966.

Crooks ET, Moore PL, Franti M, Cayanan CS, Zhu P, Jiang P, de Vries RP, Wiley C, Zharkikh I, Schülke N, Roux KH, Montefiori DC, Burton DR, Binley JM. A comparative immunogenicity study of HIV-1 virus-like particles bearing various forms of envelope proteins, particles bearing no envelope and soluble monomeric gp120. Virology. 2007;366:245–62.

Crotty S, Ahmed R. Immunological memory in humans. Semin Immunol. 2004;16:197–203.

Csete ME, Doyle JC. Reserve engineering of biological complexity. Science. 2002;295:1664–9.

Cunningham BC, Wells JA. Comparison of a structural and a functional epitope. J Mol Biol. 1993;234:554–63. https://doi.org/10.1006/jmbi.1993.1611.

D'Argenio DA, Wilson CB. A decade of vaccines: integrating immunology and vaccinology for rational vaccine design. Immunity. 2010;33:437–40.

Dakappagari NK, Lute KD, Rawale S, Steele JT, Allen SD, Phillips G, Reilly RT, Kaumaya PT. Conformational HER-2/neu B-cell epitope peptide vaccine designed to incorporate two native disulfide bonds enhances tumor cell binding and antitumor activities. J Biol Chem. 2005;280:54–63.

Daniels DA, Lane DP. Phage peptide libraries. Methods. 1996;9:494–507.

Darst SA, Robertson CR, Berzofsky JA. Adsorption of the protein antigen myoglobin affects the binding of conformation-specific monoclonal antibodies. Biophys J. 1988;53:533–9.

Daszak, et al. The global virome project. Intern J Infect Dis. 18.001. 2016; https://doi.org/10.1016/ijid.2016.11.096.

Davies DR, Cohen GH. Interactions of protein antigens with antibodies. Proc Natl Acad Sci U S A. 1996;93:7–12.

Davies DR, Padlan EA, Sheriff S. Antibody-antigen complexes. Annu Rev Biochem. 1990;59:439–73.

Dawkins R. River out of eden. London: Phoenix Orion Books; 1995.

Day ED. Advanced immunochemistry. 2nd ed. New York, NY: Wiley; 1990. p. 1–291.

De Clercq E. The history of antiretrovirals: key discoveries over the past 25 years. Rev Med Virol. 2009;19:287–99.

De Vos-Cloetens C, Minsart-Baleriaux V, Urbain-Vansanten G. Possible relationships between antibodies and non-specific immunoglobulins simultaneously induced after antigenic stimulation. Immunology. 1971;20:955–8.

Decker JM, Bibollet-Ruche F, Wei X, Wang S, Levy DN, Wang W, et al. Antigenic conservation and immunogenicity of the HIV coreceptor binding site. J Exp Med. 2005;201:1407–19.

DeLano WL. Unraveling hot spots in binding interfaces: progress and challenges. Curr Opin Struct Biol. 2002;12:14–20.

Delmastro P, Meola A, Monaci P, Cortese R, Galfre G. Immunogenicity of filamentous phage displaying peptide mimotopes after oral administration. Vaccine. 1997;15:1276–85.

Dembski WA, Ruse M, editors. Debating design: from Darwin to DNA. Cambridge, UK: Cambridge University Press; 2004.

Demchenko AP. Recognition between flexible protein molecules: induced and assisted folding. J Mol Recognit. 2001;14:42–61.

Denisova GF, Denisov DA, Bramson JL. Applying bioinformatics for antibody epitope prediction using affinity-selected mimotopes – relevance for vaccine design. Immunome Res. 2010;6 (Suppl. 2):S6.

Denton G, Hudecz F, Kajtar J, Murray A, Tendler SJB, Price MR. Sequential order of T and B cell epitopes affects immunogenicity but not antibody recognition of the B cell. Pept Res. 1994;7:258–64.

Deutschman CS, Tracey KJ. Sepsis: current dogma and new perspectives. Immunity. 2014;40:463–75. https://doi.org/10.1016/j.immuni.2014.04.001.

Dey B, Pancera M, Svehla K, Shu Y, Xiang SH, Vainshtein J, Li Y, Sodroski J, Kwong PD, Mascola JR, Wyatt R. Characterization of human immunodeficiency virus type1 monomeric and trimeric gp120 glycoproteins stabilized in the CD4-bound state: antigenicity, biophysics, and immunogenicity. J Virol. 2007;81:5579–93.

Dey B, Svehla K, Xu L, Wycuff D, Zhou T, Voss G, Phogat A, Chakrabarti BK, Li Y, Shaw G, et al. Structure-based stabilization of HIV-1 gp120 enhances humoral immune responses to the induced co-receptor binding site. PLoS Pathog. 2009;5:e1000445. https://doi.org/10.1371/journal.ppat.1000445.

Dhillon AK, Donners H, Pantophlet R, Johnson WE, Decker JM, Shaw GM, et al. Dissecting the neutralizing antibody specificities of broadly neutralizing sera from human immunodeficiency virus type 1-infected donors. J Virol. 2007;81:6548–62.

DiMarchi R, Brooke G, Gale C, Cracknell V, Doel T, Mowat N. Protection of cattle against foot-and-mouth disease by a synthetic peptide. Science. 1986;232:639–41.

Dimitrov DS. Therapeutic antibodies, vaccines and antibodyomes. MAbs. 2010;2:347–56.

Dimitrov JD, Planchais C, Kang J, Pashov A, Vassilev TL, Kaveri SV, Lacroix-Desmazes S. Heterogeneous antigen recognition behavior of induced polyspecific antibodies. Biochem Biophys Res Commun. 2010;398:266–71.

Dimitrov JD, Planchais C, Roumenina LT, Vassilev TL, Kaveri SV, Lacroix-Desmazes S. Antibody polyreactivity in health and disease: Statuvariabilis. J Immunol. 2013;191:993–9.

Dimmock NJ. Neutralization of animal viruses. Berlin: Springer Verlag; 1993. p. 149.

Diskin R, Scheid JF, Marcovecchio PM, West AP Jr, Klein F, Gao H, Gnanapragasam PN, Abadir A, Seaman MS, Nussenzweig MC, et al. Increasing the potency and breadth of an HIV antibody by using structure-based rational design. Science. 2011;334:1289–93.

Ditzel HJ, Itoh K, Burton DR. Determinants of polyreactivity in a large panel of recombinant human antibodies from HIV-1 infection. J Immunol. 1996;157:739–49.

Doan LX, Li M, Chen C, Yao Q. Virus-like particles as HIV-1 vaccines. Rev Med Virol. 2005;15:75–88.

Doniach S. Antibody-antigen binding: structure-function relationships viewed at atomic scale resolution. In: Herzenberg LA, Weir DM, Hertzenberg LA, Blackwell C, editors. Weir's handbook of experimental immunology. 5th ed. Cambridge: Blackwell Science; 1997. p. 12.1–12.10.

Dorgham K, Dogan I, Bitton N, Parizot C, Cardona V, Debré P, Hartley O, Gorochov G. Immunogenicity of HIV type 1 gp120 CD4 binding site phage mimotopes. AIDS Res Hum Retrovir. 2005;21:82–92.

Doria-Rose NA. HIV neutralizing antibodies: clinical correlates and implications for vaccines. J Infect Dis. 2010;201:981–3. https://doi.org/10.1086/651143.

Doria-Rose NA, Joyce MG. Strategies to guide the antibody maturation process. Curr Opin Virol. 2015;11:137–47.

Doria-Rose NA, Klein RM, Manion MM, O'Dell S, Phogat A, Chakrabarti B, et al. Frequency and phenotype of human immunodeficiency virus envelope-specific B cells from patients with broadly cross-neutralizing antibodies. J Virol. 2009;83:188–99.

Doria-Rose NA, Klein RM, Daniels MG, O'Dell S, Nason M, Lapedes A, et al. Breadth of human immunodeficiency virus-specific neutralizing activity in sera: clustering analysis and association with clinical variables. J Virol. 2010;84:1631–6. https://doi.org/10.1128/JVI.01482-09.

Doria-Rose NA, Louder MK, Yang Z, O'Dell S, Nason M, Schmidt SD, McKee K, Seaman MS, Bailer RT, Mascola JR. HIV-1 neutralization coverage is improved by combining monoclonal antibodies that target in dependent epitopes. J Virol. 2012;86:3393–7.

Doria-Rose NA, Schramm CA, Gorman J, et al. Developmental pathway for potent V1V2-directed HIV-neutralizing antibodies. Nature. 2014;509:55–62.

Dormitzer PR, Grandi G, Rappuoli R. Structural vaccinology starts to deliver. Nat Rev Microbiol. 2012;10:807–12.

Douek DC, Kwong PD, Nabel GJ. The rational design of an AIDS vaccine. Cell. 2006;124:677–81.

Dougherty W, Willis L, Johnston RE. Topographic analysis of tobacco etch virus capsid protein epitopes. Virology. 1985;144:66–72.

Drews J. Strategic trends in the drug industry. Drug Discov Today. 2003;8:411–20.

Du Sautoy M. What we cannot know. New York: Harper Collins Publishers; 2016. p. 36–73.

Du SX, Idiart RJ, Mariano EB, Chen H, Jiang P, Xu L, Ostrow KM, Wrin T, Phung P, Binley JM, Petropoulos CJ, Ballantyne JA, Whalen RG. Effect of trimerization motifs on quaternary structure, antigenicity, and immunogenicity of an oncleavable HIV-1 gp140 envelope glycoprotein. Virology. 2009;395:33–44.

Dubé K, Henderson GE, Margolis DM. Framing expectations in early HIV cure research. Trends Microbiol. 2014;22:547–9.

Dubé K, Ramirez C, Handibode J, Taylor J, Skinner A, et al. Participation in HIV cure-related research: a scoping review of the proxy literature and implications for future research. J Virus Erad. 2015;1:250–6.

Duenas-Decamp MJ, Peters P, Burton D, Clapham PR. Natural resistance of human immunodeficiency virus type 1 to the CD4bs antibody b12 conferred by a glycan and an arginine residue close to the CD4 binding loop. J Virol. 2008;82:5807–14.

Duerr A, Huang Y, Buchbinder S, Coombs RW, Sanchez J, del Rio C, et al. Extended follow-up confirms early vaccine-enhanced risk of HIV acquisition and demonstrates waning effect over time among participants in a randomized trial of recombinant adenovirus HIV vaccine (stepstudy). J Infect Dis. 2012;206:258–66. https://doi.org/10.1093/infdis/jis342.

Dunker AK, Silman I, Uversky VN, Sussman JL. Function and structure of inherently disordered proteins. Curr Opin Struct Biol. 2008;18:756–64.

Dupré J. The disorder of things. Metaphysical foundations of the disunity of science. Cambridge, MA: Harvard University Press; 1993.

Ebensen T, Guzman CA. Immune modulators with defined molecular targets: cornerstone to optimize rational vaccine design. Hum Vaccin. 2008;4:13–22.

Edelman GM, Gally JA. Degeneracy and complexity in biological systems. PNAS. 2001;98:13763–8.

Edmundson AB, Ely KR, Herron JN, Cheson BD. The binding of opioid peptides to the MCG light chain dimer: flexible keys and adjustable locks. Mol Immunol. 1987;24:915–35.

Edwards BM, Barash SC, Main SH, Choi GH, Minter R, Ullrich S, Williams E, DuFou L, Wilton J, Albert VR, et al. The remarkable flexibility of the human antibody repertoire; isolation of over one thousand different antibodies to a single protein, BLyS. J Mol Biol. 2003;334:103–18.

Efroni S, Cohen IR. Simplicity belies a complex system: a response to the minimal model of immunity of Langman and Cohn. Cell Immunol. 2002;216:23–30.

Eisen HN. Specificity and degeneracy in antigen recognition: Yin and Yang in the immune system. Annu Rev Immunol. 2001;19:1–21.

Eisen HN, Chakraborty AK. Evolving concepts of specificity in immune reactions. Proc Natl Acad Sci U S A. 2010;107:22373–80. https://doi.org/10.1073/pnas.1012051108.

El-Manzalawy Y, Honavar V. Recent advances in B cell epitope prediction methods. Immunome Res. 2010;6(Suppl 2):1–9. https://doi.org/10.1186/1745-7580-6-S2-S2.

Emmeche C. Aspects of complexity in life and science. Philosophica. 1997;59:41–68.

Engl HW, Flamm C, Kugler P, Lu J, Muller S, Schuster P. Inverse problems in systems biology. Inverse Prob. 2009;25(12):123014. https://doi.org/10.1088/0266-5611/25/12/123014.

Enshell-Seijffers D, Denisov D, Groisman B, Smelyanski L, Meyuhas R, Gross G, Denisova G, Gershoni JM. The mapping and reconstitution of a conformational discontinuous B-cell epitope of HIV-1. J Mol Biol. 2003;334:87–101.

Ensoli B, Fiorelli V, Ensoli F, Cafaro A, Titti F, et al. Candidate HIV-1 Tat vaccine development: from basic science to clinical trials. AIDS. 2006;20:2245–61.

Ensoli B, Fiorelli V, Ensoli F, Lazzarin A, Visintini R, et al. The therapeutic phase I trial of the recombinant native HIV-1 Tat protein. AIDS. 2008;22:2207–9.

Ensoli B, Fiorelli V, Ensoli F, Lazzarin A, Visintini R, et al. The preventive phase I trial with the HIV-1 Tat-based vaccine. Vaccine. 2009;28:371–8.

Ensoli B, Bellino S, Tripiciano A, Longo O, Francavilla V, et al. Therapeutic immunization with HIV-1 Tat reduces immune activation and loss of regulatory T-cells and improves immune function in subjects on HAART. PLoS One. 2010;5:e13540.

Ensoli F, Cafaro A, Casabianca A, Tripiciano A, Bellino S, et al. HIV-1 Tat immunization restores immune homeostasis and attacks the HAART-resistant blood HIV DNA: Results of a randomized phase II exploratory clinical trial. Retrovirology. 2015;12:33.

Ensoli B, Nchabeleng M, Ensoli F, Tripiciano A, Bellino S, et al. HIV-Tat immunization induces cross-clade neutralizing antibodies and CD4(+) T cell increases in antiretroviral-treated South African volunteers: a randomized phase II clinical trial. Retrovirology. 2016a;13:34.

Ensoli B, Cafaro A, Amicosante M, Andrieu J-M, Boyer JD, Garcia F, Gray G, King MR, Rios A, Sandstorm E, Van Regenmortel MHV. Old and new concepts and strategies in HIV vaccinology: a report from a Workshop held in Rome on 17 June 2016. J AIDS Clin Res. 2016b;7(11):634. https://doi.org/10.4172/2155-6113.1000634.

Ertl HC. Viral vectors as vaccine carriers. Curr Opin Virol. 2016;21:1–8.

Esparza J. The global HIV vaccine enterprise. Int Microbiol. 2005;8:93–101.

Esparza J. A brief history of the global effort to develop an HIV vaccine. Vaccine. 2013a;31:3502–18. https://doi.org/10.1016/j.vaccine.2013.05.018.

Esparza J. What has 30 years of HIV research taught us? Vaccine. 2013b;1:513–26.

Esparza J. A new scientific paradigm may be needed to finally develop an HIV vaccine. Front Immunol. 2015;6:124. https://doi.org/10.3389/fimmu.2015.00124.

Esparza J, Bhamarapravati N. Accelerating the development and future availability of HIV-1 vaccines: why, when, where and how? Lancet. 2000;355:2061–6.

Esparza J, Yamada T. The discovery value of "Big Science". J Exp Med. 2007;204:701–4.

Esper AM, Moss M, Lewis CA, Nisbet R, Mannino DM, Martin GS. The role of infection and comorbidity: factors that influence disparities in sepsis. Crit Care Med. 2006;34:2576–32. https://doi.org/10.1097/01.CCM.0000239114.50519.0Eg3725.

Euler Z, van Gils MJ, Bunnik EM, Phung P, Schweighardt B, Wrin T, et al. Cross-reactive neutralizing humoral immunity does not protect from HIV type 1 disease progression. J Infect Dis. 2010;201:1045–53. https://doi.org/10.1086/651144.

Euler Z, Bunnik EM, Burger JA, Boeser-Nunnink BD, Grijsen ML, Prins JM, et al. Activity of broadly neutralizing antibodies, including PG9, PG16, and VRC01, against recently transmitted subtype B HIV-1 variants from early and late in the epidemic. J Virol. 2011;85:7236–45. https://doi.org/10.1128/JVI.00196-11.

Euler Z, van den Kerkhof TL, van Gils MJ, Burger JA, Edo-Matas D, Phung P, et al. Longitudinal analysis of early HIV-1-specific neutralizing activity in an elite neutralizer and in five patients who developed cross-reactive neutralizing activity. J Virol. 2012;86:2045–55. https://doi.org/10.1128/JVI.06091-11.

Excler JL, Ake J, Robb ML, Kim JH, Plotkin SA. Non neutralizing functional antibodies: a new 'old' paradigm for HIV vaccines. Clin Vaccine Immunol. 2014;21(8):1023–36. https://doi.org/10.1128/CVI.00230-14.

Faburay B, Labeaud AD, McVey DS, Wilson WC, Richt JA. Current status of Rift Valley fever vaccine development. Vaccine. 2017;5:29. https://doi.org/10.3390/vaccines5030029.

Fagerstam LG, Karlsson R. In: Van Oss C, Van Regenmortel MHV, editors. Immunochemistry. New York: Dekker; 1994. p. 949–70.

Fair JM, Rivas AL. Systems biology and ratio-based, real-time disease sur-veillance. Transbound Emerg Dis. 2015;62:437–45. https://doi.org/10.1111/tbed.12162.

Fauci AS. Host factors and the pathogenesis of HIV-induced disease. Nature. 1996;384:529–34.

Fauci AS, Marston HD. Ending AIDS – is an HIV vaccine necessary? N Engl J Med. 2014;370:495–8. https://doi.org/10.1056/NEJMp1313771.

Fauci AS, Marovich MA, Dieffenbach CW, Hunter E, Buchbinder SP. Immune activation with HIV vaccines. Science. 2014;344:49–51. https://doi.org/10.1126/science.1250672.

Feytmans E, Noble D, Peitsch M. Genome size and numbers of biological functions. Translat Comput Syst Biol. 2005;1:44–9.

Finlay BJ, Esteban GF. Can biological complexity be rationalized? Bioscience. 2009;59:333–40.

Fischer N. Sequencing antibody repertoires, the next generation. MAbs. 2011;3:17–20.

Fischer D, Rood D, Barrette RW, et al. Intranasal immunization of guinea pigs with an immunodominant foot-and-mouth disease virus peptide conjugate induces mucosal and humoral antibodies and protection against challenge. J Virol. 2003;77:7486–91.

Folgori A, Tafi R, Meola A, Felici F, Galfré G, Cortese R, Monaci P, Nicosia A. A general strategy to identify mimotopes of pathological antigens using only random peptide libraries and human sera. EMBO J. 1994;13:2236–43.

Foote J, Eisen HN. Kinetic and affinity limits on antibodies produced during immune responses. Proc Natl Acad Sci U S A. 1995;92:1254–6.

Ford N, Shubber Z, Meintjes G, Grinsztejn B, Eholie S, Mills EJ, et al. Causes of hospital admission among people living with HIV worldwide: a systematic review and meta-analysis. Lancet HIV. 2015;2:e438–44. https://doi.org/10.1016/S2352-3018(15)00137-X.

Fox-Keller E. Making sense of life. Explaining biological development with models, metaphors and machines. Cambridge, MA: Harward University Press; 2002.

Frank R. Spot-synthesis: an easy technique for the positionally addressable, parallel chemical synthesis on a membrane support. Tetrahedron. 1992;48:9217–32.

Frank SA. Immunology and evolution of infectious disease. Princeton: Princeton University Press; 2002. p. 42.

Frey G, Peng H, Rits-Volloch S, Morelli M, Cheng Y, Chen B. A fusion-intermediate state of HIV-1 gp41 targeted by broadly neutralizing antibodies. Proc Natl Acad Sci U S A. 2008;105:3739–44.

Friede M, Muller S, Briand JP, Schuber F, Van Regenmortel MHV. Generation of antibodies cross-reactive with proteins by peptide immunization. In: Gosling JP, Reen DJ, editors. Immunotechnology. London: Portland Press; 1993. p. 1–11.

Friede M, Muller S, Briand JP, Plaué S, Fernandes I, Frisch B, Schuber F, Van Regenmortel MHV. Selective induction of protection against influenza virus infection in mice by a lipid-peptide conjugate delivered in liposomes. Vaccine. 1994;12:791–7.

Frison EA, Stace-Smith R. Cross-reacting and heterospecific monoclonal antibodies produced against arabis mosaic nepovirus. J Gen Virol. 1992;73:2525–30.

Gabdoulline RR, Wade RC. On the protein-protein diffusional encounter complex. J Mol Recognit. 1999;12:226–34.

Gallagher RB, Gilder J, Nossal GJV, Salvatore G. Immunity. The making of a modern science. New York: Academic; 1995.

Gandhi RT, O'Neill D, Bosch RJ, Chan ES, Bucy RP, Shopis J, et al. A randomized therapeutic vaccine trial of canarypox-HIV-pulsed dendritic cells vs. canarypox-HIV alone in HIV-1-infected patients on antiretroviral therapy. Vaccine. 2009;27:6088–94. https://doi.org/10.1016/j.vaccine.2009.05.016.

Gannon F. Too complex to comprehend? EMBO Rep. 2007;8:705.

Garces F, Lee JH, de Val N, de la Pena AT, Kong L, Puchades C, et al. Affinity maturation of a potent family of HIV antibodies is primarily focused on accommodating or avoiding glycans. Immunity. 2015;43:1053–63. https://doi.org/10.1016/j.immuni.2015.11.007.

Garcia KC, Teyton L. T cell receptor peptide-MHC interactions: biological lessons from structural studies. Curr Opin Biotechnol. 1998;9:338–43.

Garcia J, Dumy P, Rosen O, Anglister J. Stabilization of the biologically active conformation of the principal neutralizing determinant of HIV-1(IIIB) containing a cis-proline surrogate: 1H NMR and molecular modeling study. Biochemistry. 2006;45:4284–94.

García F, Climent N, Assoumou L, et al. A therapeutic dendritic cell-based vaccine for HIV-1 infection. J Infect Dis. 2011;203:473–8.

García F, Climent N, Guardo AC, Gil C, León A, et al. A dendritic cell-based vaccine elicits T cell responses associated with control of HIV-1 replication. Sci Transl Med. 2013;5:166ra2.

Gaschen B, Taylor J, Yusim K, et al. Diversity considerations in HIV-1 vaccine selection. Science. 2002;296:2354–60.

Gatherer D. So what do we really mean when we say that systems biology is holistic? BMC Syst Biol. 2010;4:22. https://doi.org/10.1186/1752-0509-4-22.

Gauch HG. Scientific method in brief. Cambridge, UK: Cambridge University Press; 2012. p. 1–288.

Gauduin MC, Parren PW, Weir R, Barbas CF, Burton DR, Koup RA. Passive immunization with a human monoclonal antibody protects hu-PBL-SCID mice against challenge by primary isolates of HIV-1. Nat Med. 1997;3:1389–93.

Geoghenan JL, Holmes EC. Predicting virus emergence amid evolutionary noise. Open Biol. 2017;170189. https://doi.org/10.1098/rsob.170189.

Gershell LJ, Atkins JH. A brief history of novel drug discovery technologies. Nat Rev Drug Discov. 2003;2:321–7.

Getzoff ED, Tainer JA, Lerner RA, Geysen HM. The chemistry and mechanism of antibody binding to protein antigens. Adv Immunol. 1988;43:1–98.

Geysen HM, Rodda SJ, Mason TJ. A priori delineation of a peptide which mimics a discontinuous antigenic determinant. Mol Immunol. 1986;23:709–15.

Geysen HM, Rodda SJ, Mason TJ, Tribbick G, Schoofs PG. Strategies for epitope analysis using peptide synthesis. J Immunol Methods. 1987;102:259–74.

Geysen HM, Mason TJ, Rodda SJ. Cognitive features of continuous antigenic determinants. J Mol Recognit. 1988;1:32–41.

Ghilarov AM. The changing place of theory in 20th century ecology: from universal laws to array of methodologies. Oikos. 2001;92:357–62. https://doi.org/10.1034/j.1600-0706.2001.920218.x.

Ghosh G, Cambell AM. Multispecific monoclonal antibodies. Immunol Today. 1986;7:217–22.

Gibbs AJ, Gibbs MJ. A broader definition of the 'virus species'. Arch Virol. 2006;151:1419–22.

Gibbs AJ, Armstrong JS, Gibbs MJ. A type of nucleotide motif that distinguishes tobamovirus species more efficiently than nucleotide signatures. Arch Virol. 2004;149:1941–54.

Giere RN. Explaining science. Chicago: University of Chicago Press; 1988.

Gigerenzer G, Selten R. Bounded rationality. The adaptive toolbox. Cambridge, MA: MIT Press; 2002.

Gill J, Hangartner D. Circular data in political science and how to handle it. Polit Anal. 2010;18:316–36. https://doi.org/10.1093/pan/mpq009.

Glassman RH, Sun AY. Biotechnology: identifying advances from the hype. Nat Rev Drug Discov. 2004;3:177–83.

Glynn LE, Steward MW. Immunochemistry: an advanced textbook. New York: Wiley; 1977.

Go EP, Irungu J, Zhang Y, Dalpathado DS, Liao HX, Sutherland LL, et al. Glycosylation site-specific analysis of HIV envelope proteins (JR-FL and CON-S) reveals major differences in glycosylation site occupancy, glycoform profiles, and antigenic epitopes' accessibility. J Proteome Res. 2008;7:1660–74.

Goh C-S, Milburn D, Gerstein M. Conformational changes associated with protein–protein interactions. Curr Opin Struct Biol. 2004;14:104–9.

Goldbaum FA, Schwarz FP, Eisenstein E, Cauerhff A, Mariuzza RA, Poljak RJ. The effect of water activity on the association constant and the enthalpy of reaction between lysozyme and the specific antibodies D1.3 and D44.1. J Mol Recognit. 1996;9:6–12.

Gomara MJ, Haro I. Synthetic peptides for the immunodiagnosis of human diseases. Curr Med Chem. 2007;14:531–46.

Gonzalez S, González-Rodríguez AP, Suárez-Álvarez B, López-Soto A, Huergo-Zapico L, Lopez-Larrea C. Conceptual aspects of self and nonself discrimination. Self Nonself. 2011;2:19–25.

Gorny MK, Xu J-Y, Karwowska S, Buchbinder A, ZollaPazner S. Repertoire of neutralizing human monoclonal antibodies specific for the V3 domain of HIV-1 gp120. J Immunol. 1993;150:635–43.

Gorny MK, VanCott TC, Hioe C, Israel ZR, Michael NL, Conley AJ, et al. Human monoclonal antibodies to the V3 loop of HIV-1 with intra-and interclade cross-reactivity. J Immunol. 1997;159:5114–22.

Gorny MK, Williams C, Volsky B, Revesz K, Cohen S, Polonis VR, Honnen WJ, Kayman SC, Krachmarov C, Pinter A, et al. Human monoclonal antibodies specific for conformation sensitive epitopes of V3 neutralize human immunodeficiency virus type 1 primary isolates from various clades. J Virol. 2002;76:9035–45.

Gorny MK, Revesz K, Williams C, Volsky B, Louder MK, Anyangwe CA, Krachmarov C, Kayman SC, Pinter A, Nadas A, Nyambi PN, Mascola JR, Zolla-Pazner S. The V3 loop is accessible on the surface of most human immunodeficiency virus type 1 primary isolates and serves as a neutralization epitope. J Virol. 2004;78:2394–404.

Gorny MK, Stamatatos L, Volsky B, Revesz K, Williams C, Wang XH, Cohen S, Staudinger R, Zolla-Pazner S. Identification of a new quaternary neutralizing epitope on human immunodeficiency virus type 1 virus particles. J Virol. 2005;79:5232–7.

Gorny MK, Williams C, Volsky B, Revesz K, Wang XH, Burda S, Kimura T, Konings FA, Nádas A, Anyangwe CA, et al. Cross-clade neutralizing activity of human anti-V3 monoclonal antibodies derived from the cells of individuals infected with non-B clades of human immunodeficiency virus type. J Virol. 2006;80:6865–72.

Graham G. Genes: a philosophical enquiry. London: Routledge; 2002.

Gras-Masse H, Georges B, Estaquier J, et al. Convergent peptide libraries, or mixotopes, to elicit or to identify specific immune responses. Curr Opin Immunol. 1999;11:223–8.

Gray ES, Moore PL, Bibollet-Ruche F, Li H, Decker JM, Meyers T, Shaw GM, Morris L. 4E10-resistant variants in a human immunodeficiency virus type 1 subtype C-infected individual with an anti-membrane-proximal external region-neutralizing antibody response. J Virol. 2008;82:2367–75.

Gray ES, Taylor N, Wycuff D, Moore PL, Tomaras GD, Wibmer CK, et al. Antibody specificities associated with neutralization breadth in plasma from human immunodeficiency virus type 1 subtype C-infected blood donors. J Virol. 2009;83:8925–37.

Gray ES, Madiga MC, Hermanus T, Moore PL, Wibmer CK, Tumba NL, et al. The neutralization breadth of HIV-1 develops incrementally over four years and is associated with CD4+ T cell decline and high viral load during acute infection. J Virol. 2011;85:4828–40. https://doi.org/10.1128/JVI.00198-11.

Gray GE, Moodie Z, Metch B, Gilbert PB, Bekker L-G, Churchyard G, et al. Recombinant Adenovirus type 5 HIV gag/pol/nef vaccine in South Africa: unblinded, long-term follow-up of the phase 2b HVTN503/Phambili study. Lancet Infect Dis. 2014;14:388–96. https://doi.org/10.1016/S1473-3099(14)70020-9.

Greek R. Animal models and the development of an HIV vaccine. J AIDS Clin Res. 2012;1(S8).

Greek R, Shanks N, Rice MJ. The history and implications of testing thalidomide in animals. J Philos Sci Law. 2011;10:1–32.

Green S. Can biological complexity be reverse engineered ? Stud Hist Phil Biol Biomed Sci. 2015;53:73–83.

Greenbaum JA, Andersen PH, Blythe M, Bui H-H, Cachau RE, Crowe J, Davies M, Kolaskar AS, Lund O, Morrison S, et al. Towards a consensus on datasets and evaluation metrics for developing B-cell epitope prediction tools. J Mol Recognit. 2007;20:75–82.

Greenhalgh T, Howick J, Maskrey N. Evidence based medicine: a movement in crisis? BMJ. 2014;348:g3725. https://doi.org/10.1136/bmj.

Greenspan NS. Antigen mimicry with anti-idiotypic antibodies. In: Van Regenmortel MHV, editor. Structure of antigens, vol. 1. Boca Raton, FL: CRC Press; 1992. p. 55–79.

Greenspan NS. Design challenges for HIV-1 vaccines based on humoral immunity. Front Immunol. 2014;5:335. https://doi.org/10.3389/fimmu.2014.00335.

Greenspan NS, Di Cera E. Defining epitopes: it's not as easy as it seems. Nat Biotechnol. 1999;17:936–7. https://doi.org/10.1038/13590.

Gross L. Scientific illiteracy and the partisan takeover of biology. PLoS Biol. 2006;4(5) e167. https://doi.org/10.1371/journal.pbio.0040167.

Grunenberg J. Complexity in molecular recognition. Phys Chem Chem Phys. 2011;13:10136–46.

Gschwend DA, Good AC, Kuntz ID. Molecular docking towards drug discovery. J Mol Recognit. 1996;9:175–86.

Guenaga J, Dosenovic P, Ofek G, Baker D, Schief WR, Kwong PD, Karlsson Hedestam GB, Wyatt RT. Heterologous epitope-scaffold prime: boosting immunofocuses B cell responses to the HIV-1 gp412F5 neutralization determinant. PLoS One. 2011;6:e16074. https://doi.org/10.1371/journal.pone.0016074.

Guenter D, Esparza J, Macklin R. Ethical considerations in International HIV vaccine trials: summary of a consultative process conducted by the joint United Nations programme on HIV/AIDS (UNAIDS). J Med Ethics. 2000;26:37–43.

Gunawardena J. Models in biology: 'accurate descriptions of our pathetic thinking'. BMC Biol. 2014;12:29. https://doi.org/10.1186/1741-7007-12-29.

Haab BB. Antibody arrays in cancer research. Proteomics. 2005;4:377–83.

Haas G, Karaoli G, Ebermayer K, et al. Immunoproteomics of Helicobacter pylori infection and relation to gastric disease. Proteomics. 2002;2:313–24.

Hacking I. Representing and intervening. Princeton, NJ: Cambridge University Press; 1983. p. 1–304.

Hafenstein S, Bowman VD, Sun T, et al. Structural comparison of different antibodies interacting with parvovirus capsids. J Virol. 2009;83:5556–66.

Haimovich J, Tarrab R, Sulica A, Sela M. Antibodies of different specificities in normal rabbit sera. J Immunol. 1970;104:1033–4.

Halperin I, Ma B, Wolfson H, Nussinov R. Principles of docking: an overview of search algorithms and a guide to scoring functions. Proteins. 2002;47:409–43.

Hammer SM, Sobieszczyk ME, Janes H, Karuna ST, Mulligan MJ, Groves D, et al. Efficacy trial of a DNA/rAds HIV-1 preventive vaccine. N Engl J Med. 2013;369:2083–92. https://doi.org/10.1056/NEJMoa1310566.

Hanin V, Déry O, Boquet D, Sagot MA, Créminon C, Couraud JY, Grassi J. Importance of hydropathic complementarity for the binding of the neuropeptide substance P to a monoclonal antibody: equilibrium and kinetic studies. Mol Immunol. 1997;34:829–38.

Hanke D. In: Cornwell J, editor. Explanations. Styles of explanation in science. Oxford: Oxford University Press; 2004. p. 143–55.

Hanke T, Goonetilleke N, McMichael AJ, Dorrell L. Clinical experience with plasmid DNA- and modified vaccinia virus Ankara-vectored human immunodeficiency virus type 1 clade A vaccine focusing on T-cell induction. J Gen Virol. 2007;88:1–12.

Hans D, Young PR, Fairlie DP. Current status of short synthetic peptides as vaccines. Med Chem. 2006;2:627–46.

Hansen SG, Sacha JB, Hughes CM, Ford JC, Burwitz BJ, Scholz I, et al. Cytomegalovirus vectors violate CD8+ T cell epitope recognition paradigms. Science. 2013;340(6135):1237874. https://doi.org/10.1126/science.1237874.

Hansen SG, Wu HL, Burwitz BJ, Hughes CM, Hammond KB, et al. Broadly targeted CD8+ T cell responses restricted by major histocompatibility complex E. Science. 2016;351:714–20.

Harper M, Lema F, Boulot G, Poljak RJ. Antigen specificity and cross-reactivity of monoclonal anti-lysozyme antibodies. Mol Immunol. 1987;24:97–108.

Harris A, Borgnia MJ, Shi D, Bartesaghi A, He H, Pejchal R, Kang YK, Depetris R, Marozsan AJ, Sanders RW, et al. Trimeric HIV-1 glycoprotein gp140 immunogens and native HIV-1 envelope glycoproteins display the same closed and open quaternary molecular architectures. Proc Natl Acad Sci U S A. 2011;108:11440–5.

Harwood NE, Batista FD. Early events in B cell activation. Annu Rev Immunol. 2010;28:185–210.

Haste-Andersen P, Nielsen M, Lund O. Prediction of residues in discontinuous B-cell epitopes using protein 3D structures. Protein Sci. 2006;15:2558–67.

Haurowitz F. lmmunochemistry and the biosynthesis of antibodies. New York: Interscience Publishers, Wiley & Sons; 1968.

Havlasova J, Hernychova L, Halada P, et al. Mapping of immunoreactive antigens of *Francisella tularensis* live vaccine strain. Proteomics. 2002;2:857–67.

Haynes BF, Fleming J, StClair EW, Katinger H, Stiegler G, Kunert R, Robinson J, Scearce RM, Plonk K, Staats HF, Ortel TL, Liao HX, Alam SM. Cardiolipin polyspecific autoreactivity in two broadly neutralizing HIV-1 antibodies. Science. 2005a;308:1906–8.

Haynes BF, Moody MA, Verkoczy L, Kelsoe G, Alam SM. Antibody polyspecificity and neutralization of HIV-1: a hypothesis. Hum Antibodies. 2005b;14:59–67.

Haynes BF, Kelsoe G, Harrison SC, Kepler TB. B-cell-lineage immunogen design in vaccine development with HIV-1 as a case study. Nat Biotechnol. 2012a;30:423–33. https://doi.org/10.1038/nbt.2197.

Haynes BF, Gilbert PB, McElrath MJ, Zolla-Pazner S, Tomaras G, Alam S, et al. Immune-correlates analysis of an HIV-1 vaccine efficacy trial. N Engl J Med. 2012b;366:1275–86. https://doi.org/10.1056/NEJMoa1113425.

Hearty S, Conroy PJ, Ayyar BV, Byrne B, O'Kennedy R. Surface plasmon resonance for vaccine design and efficacy studies: recent applications and future trends. Expert Rev Vaccines. 2010;9:645–64.

Henderson GE. The ethics of HIV "cure" research: what can we learn from consent forms? AIDS Res Hum Retrovir. 2015;31:56–63.

Herzenberg LA, Weir DM, Herzenberg LA, Blackwell C. Experimental immunology. Weir handbook. Malden, MA: Blackwell Science; 1997.

Hilleman MR. Impediments, imponderables and alternatives in the attempt to develop an effective vaccine against AIDS. Vaccine. 1992;10:1053–8.

Hilleman MR. A simplified vaccinologist's vaccinology and the pursuit of a vaccine against AIDS. Vaccine. 1998;16:778–93. https://doi.org/10.1016/S0264-410X(97)00272-7.

Hilleman MR. Vaccines in historic evolution and perspective: a narrative of vaccine discoveries. Vaccine. 2000;18:1436–47.

Hioe CE, Wrin T, Seaman MS, Yu X, Wood B, Self S, et al. Anti-V3 monoclonal antibodies display broad neutralizing activities against multiple HIV-1 subtypes. PLoS One. 2010;5:e10254.

Ho J, Uger RA, Zwick MB, Luscher MA, Barber BH, MacDonald KS. Conformational constraints imposed on a pan-neutralizing HIV-1 antibody epitope result in increased antigenicity but not neutralizing response. Vaccine. 2005;23:1559–73. https://doi.org/10.1016/j.vaccine.2004.09. 037.

Hofmann-Lehmann R, Vlasak J, Rasmussen RA, Smith BA, Baba TW, Liska V, et al. Postnatal passive immunization of neonatal macaques with a triple combination of human monoclonal antibodies against oral simian-human immunodeficiency virus challenge. J Virol. 2001;75:7470–80. https://doi.org/10.1128/JVI.75.16.7470-7480.2001.

Hofnung M, Charbit A. Expression of antigens as recombinant proteins. In: Van Regenmortel MHV, editor. Structure of antigens, vol. 2. BocaRaton, FL: CRC Press; 1993. p. 79–128.

Holland JH. Emergence. Reading, MA: Perseus Books; 1994.

Honnen WJ, Krachmarov C, Kayman SC, Gorny MK, Zolla-Pazner S, Pinter A. Type-specific epitopes targeted by monoclonal antibodies with exceptionally potent neutralizing activities for selected strains of human immunodeficiency virus type 1 map to a common region of the V2 domain of gp120 and differ only at single positions from the clade B consensus sequence. J Virol. 2007;81:1424–32.

Hopp TP. Retrospective: 12 years of antigenic determinant predictions, and more. Pept Res. 1993;6:183–90.

Hopp TP. Different views of protein antigenicity. Pept Res. 1994;7:229–31.

Hopp TP, Woods KR. Prediction of protein antigenic determinants from amino acid sequences. Proc Natl Acad Sci U S A. 1981;78:3824–8.

Horgan J. The end of science. New York: Broadway Books; 1996. p. 41–7.

Horgan J. The end of science. New York: Broadway Books; 2006. p. 32–41.

Horrobin DF. Realism in drug discovery: could Cassandra be right? Nat Biotechnol. 2001;19:1099–100.

Horrobin DF. Modern biomedical research: an internally self-consistent universe with little contact with medical reality? Nat Rev Drug Discov. 2003;2:151–4.

Hoxie JA. Toward an antibody-based HIV-1 vaccine. Annu Rev Med. 2010;61:135–52.

Huang CC, Venturi M, Majeed S, Moore MJ, Phogat S, Zhang MY, Dimitrov DS, Hendrickson WA, Robinson J, Sodroski J, Wyatt R, Choe H, Farzan M, Kwong PD. Structural basis of tyrosine sulfation and VH-gene usage in antibodies that recognize the HIV type 1 coreceptor-binding site on gp120. Proc Natl Acad Sci U S A. 2004;101:2706–11.

Huang CC, Tang M, Zhang MY, Majeed S, Montabana E, Stanfield RL, Dimitrov DS, Korber B, Sodroski J, Wilson IA, Wyatt R, Kwong PD. Structure of a V3-containing HIV-1 gp120 core. Science. 2005;310:1025–8.

Huang J, Kang BH, Ishida E, Zhou T, Griesman T, Sheng Z, et al. Identification of a CD4-binding-site antibody to HIV that evolved nearpan neutralization breadth. Immunity. 2016;45:1108–21. https://doi.org/10.1016/j.immuni.2016.10.027.

Hunziker L, Recher M, Macpherson AJ, Ciurea A, Freigang S, Hengartner H, Zinkernagel RM. Hypergammaglobulinemia and autoantibody induction mechanisms in viral infections. Nat Immunol. 2003;4:343–9.

Iandiorio MJ, Fair JM, Chatzipanagiotou S, Ioannidis A, Trikka-Graphakos E, Charalampaki N, et al. Preventing data ambiguity in infectious diseases with four-dimensional, earlier, personalized, *in vivo* evaluations. PLoS One. 2016;11:e0159001. https://doi.org/10.1371/journal.pone.0159001.

Ideker T, Galitski T, Hood L. A new approach to decoding life: systems biology. Annu Rev Genomics Hum Genet. 2001;2:243–72.

Irving MB, Pan O, Scott JK. Random-peptide libraries and antigen-fragment libraries for epitope mapping and the development of vaccines and diagnostics. Curr Opin Chem Biol. 2001;5:314–24.

Irving MB, Craig L, Menendez A, Gangadhar BP, Montero M, van Houten NE, Scott JK. Exploring peptide mimics for the production of antibodies against discontinuous protein epitopes. Mol Immunol. 2010;47:1137–48.

James LC, Roversi P, Tawfik DS. Antibody multi-specificity mediated by conformational diversity. Science. 2003;299:1362–7.

Janin J. Ångströms and calories. Structure. 1997;5:473–9.

Jardine J, Julien JP, Menis S, Ota T, Kalyuzhniy O, McGuire A, et al. Rational HIV immunogen design to target specific germline B cell receptors. Science. 2013;340:711–6. https://doi.org/10.1126/science.1234150.

Javaherian K, Langlois AJ, McDanal C, Ross KL, Eckler LI, Jellis CL, Profy AT, Rusche JR, Bolognesi DP, Putney SD, et al. Principal neutralizing domain of the human immunodeficiency virus type 1 envelope protein. Proc Natl Acad Sci U S A. 1989;86:6768–72.

Jefferys R, Harrington M. Outstanding questions on HIV vaccine trial. Science. 2004;305:180.

Jemmerson R. Antigenicity and native structure of globular proteins: low frequency of peptide reactive antibodies. Proc Natl Acad Sci U S A. 1987;84:9180–4.

Jerne NK. Immunological speculations. Annu Rev Microbiol. 1960;14:341–58.

Jerne N. Towards a network theory of the immune system. Annu Immunol. 1974;125C:373–89.

Jiang X, Burke V, Totrov M, et al. Conserved structural elements in the V3 crown of HIV-1 gp120. Nat Struct Mol Biol. 2010;17:955–61.

Jimenez R, Salazar G, Baldridge KK, Romesberg FE. Flexibility and molecular recognition in the immune system. Proc Natl Acad Sci U S A. 2003;100:92–7.

Jin L, Wells A. Mutational analysis of antibody binding sites. In: Van Regenmortel MHV, editor. Structure of antigens, vol. 3. Boca Raton: CRC Press; 1996. p. 21–36.

Johnston MI, Fauci AS. An HIV vaccine–evolving concepts. N Engl J Med. 2007;356:2073–81. https://doi.org/10.1056/NEJMra066267.

Johnston MI, Fauci AS. An HIV vaccine development improving on natural immunity. N Engl J Med. 2011;365:873–5. https://doi.org/10.1056/NEJMp1107621.

Johnstone AP, Turner MW. Immunochemistry: a practical approach, vol. I & II. New York: IRL Press, Oxford University Press; 1997.

Jones OW, Nirenberg MW. Degeneracy in the amino acid code. Biochem Biophys Acta. 1966;119:400–6.

Joyce JG, Hurni WM, Bogusky MJ, Garsky VM, Liang X, Citron MP, et al. Enhancement of alpha-helicity in the HIV-1 inhibitory peptide DP178 leads to an increased affinity for human monoclonal antibody 2F5 but does not elicit neutralizing responses in vitro. Implications for vaccine design. J Biol Chem. 2002;277:45811–20. https://doi.org/10.1074/jbc.M205862200.

Julien JP, Bryson S, Nieva JL, Pai EF. Structural details of HIV-1 recognition by the broadly neutralizing monoclonal antibody 2F5: epitope conformation, antigen-recognition loop mobility, and anion-binding site. J Mol Biol. 2008;384:377–92.

Julien JP, Huarte N, Maeso R, Taneva SG, Cunningham A, Nieva JL, Pai EF. Ablation of the complementarity-determining region H3 apex of the anti-HIV-1 broadly neutralizing antibody 2F5 abrogates neutralizing capacity without affecting core epitope binding. J Virol. 2010;84:4136–47.

Julien JP, Cupo A, Sok D, Stanfield RL, Lyumkis D, Deller MC, Klasse PJ, Burton DR, Sanders RW, Moore JP, et al. Crystal structure of a soluble cleaved HIV-1 envelope trimer. Science. 2013;342:1477–83.

Kabat EA. Structural concepts in immunology and immunochemistry. 2nd ed. New York: Holt, Rinehart & Winston; 1976.

Kabat EA, Mayer MM. Experimental immunochemistry. Springfield, IL: Thomas, C.C; 1961.

Kaiser J. Review of vaccine failure prompts to basic science. Science. 2008;320:30–1. https://doi.org/10.1126/science.320.5872.30.

Kang CY, Brunck TK, Kieber-Emmons T, Blalock JE, Kohler H. Inhibition of self-binding antibodies (autobodies) by a VH-derived peptide. Science. 1988;240:1034–6.

Kariuki SM, Selhorst P, Ariën KK, Dorfman JR. The HIV-1 transmission bottleneck. Retrovirology. 2017;14:22. https://doi.org/10.1186/s12977-017-0343-8.

Karlsson R, Fält A. Experimental design for kinetic analysis of protein-protein interactions with surface plasmon resonance biosensors. J Immunol Methods. 1997;200:121–33.

Karlsson R, Roos H. Reaction kinetics. In: Price CP, Newman DJ, editors. Principles and practice of immunoassay. 2nd ed. London: Macmillan; 1997. p. 99–122.

Karlsson-Hedestam GB, Fouchier RA, Phogat S, Burton DR, Sodroski J, Wyatt RT. The challenges of eliciting neutralizing antibodies to HIV-1 and to influenza virus. Nat Rev Microbiol. 2008;6:143–55. https://doi.org/10.1038/nrmicro1819.

Katz JN, King G. A statistical model for multiparty electoral data. Am Polit Sci Rev. 1999;93:15–32. https://doi.org/10.2307/2585758.

Kaumaya PTP, Kobs-Conrad S, DiGoerge AM, Stevens VC. `De novo' engineering of peptide immunogenic and antigenic determinants as potential vaccines. In: Basava G, Anantharamaiah GE, editors. Peptides: design, synthetis and biological activity. Boston, MA: Birkhauser 1994. p. 133–64.

Kauvar LM, Villar HO. Deciphering cryptic similarities in protein binding sites. Curr Opin Biotechnol. 1998;9:390–4.

Kauvar LM, Higgins DL, Villar HO, Sportsman JR, Engqvist GAE, Bukar R, Bauer KE, Diley H, Rocke DM. Predicting ligand binding to proteins by affinity fingerprinting. Chem Biol. 1995;2:107–18.

Kay LE. Who wrote the book of life: a history of the genetic code. Stanford: Stanford University Press; 2000.

Kayman SC, Wu Z, Revesz K, Chen H, Kopelman R, Pinter A. Presentation of native epitopes in the V1/V2 and V3 regions of human immunodeficiency virus type 1 gp120 by fusion glycoproteins containing isolated gp120 domains. J Virol. 1994;68:400–10.

Keele BF, Tazi L, Gartner S, Liu Y, Burgon TB, Estes JD, et al. Characterization of the follicular dendritic cell reservoir of human immunodeficiency virus type 1. J Virol. 2008;82:5548–61. https://doi.org/10.1128/JVI.00124-08.

Kell DB, Oliver SG. Here is the evidence, now what is the hypothesis? The complementary roles of inductive and hypothesis-driven science in the post-genomic era. BioEssays. 2004;26:99–105.

Keller PM, Arnold BA, Shaw AR, et al. Identification of HIV vaccine candidate peptides by screening random phage epitope libraries. Virology. 1993;193:709–16.

Kelley RF. Thermodynamics of antigen binding and evaluation of humanized antibodies. In: Van Regenmortel MHV, editor. Structure of antigens, vol. 3. Boca Raton: CRC Press; 1996. p. 1–20.

Kelley L, Scott M. The evolution of biology. A shift towards the engineering and prediction-generating tools and away from traditional research practice. EMBO Rep. 2008;9:1163–7.

Keskin O, Gursoy A, Ma B, Nussinov R. Principles of protein-protein interactions: what are the preferred ways for proteins to interact? Chem Rev. 2008;108:1225–44.

Khan T, Salunke DM. Structural elucidation of the mechanistic basis of degeneracy in the primary humoral response. J Immunol. 2012;188:1819–27.

Kim J. Making sense of emergence. Philos Stud. 1999;95:3–36.

Kim YB, Han DP, Cao C, Cho MW. Immunogenicity and ability of variable loop-deleted human immunodeficiency virus type 1 envelope glycoproteins to elicit neutralizing antibodies. Virology. 2003;305:124–37.

Kim M, Sun ZY, Rand KD, Shi X, Song L, Cheng Y, Fahmy AF, Majumdar S, Ofek G, Yang Y, Kwong PD, Wang JH, Engen JR, Wagner G, Reinherz EL. Antibody mechanics on a membrane-bound HIV segment essential for GP41-targeted viral neutralization. Nat Struct Mol Biol. 2011;18:1235–43.

Kimura T, Wang XH, Williams C, Zolla-Pazner S, Gorny MK. Human monoclonal antibody 2909 binds to pseudovirions expressing trimers but not monomeric HIV-1 envelope proteins. Hum Antibodies. 2009;18:35–40.

King A. Comments to proposed modification to code rule 3.21 (defining virus species). ICTV Discussions; 2012. https://talk.ictvonline.org/ictv1/f/general_ictv_discussions-20/3930/comments-toproposed-modification-to-code-rule-3-21-defining-virus-species. Accessed 23 July 2018.

King MR. Commentary: basic research in HIV vaccinology is hampered by reductionist thinking. Front Immunol. 2016;7:42.

Kingsmore SF. Multiplexed protein measurement: technologies and applications of protein and antibody arrays. Nat Rev Drug Discov. 2006;5:310–20.

Kirschner MW, Gerhart JC. The plausibility of life: resolving Darwin's dilemma. New Haven, CT: Yale University Press; 2005.

Kistler M. 2003. Available at: http://www.institutnicod.org/reduction.htm

Kitano H. Systems biology: a brief overview. Science. 2002;295:1662–4.

Kitcher P. Abusing science: the case against creationism. Cambridge, MA: MIT Press; 1982.

Kitcher P. Function and design. In: Hull DL, Ruse ME, editors. The philosophy of biology. New York: Oxford University Press; 1998. p. 258–79.

Klade CS. Proteomics approaches towards antigen discovery and vaccine development. Curr Opin Mol Ther. 2002;4:216–23.

Klasse PJ. The molecular basis of HIV entry. Cell Microbiol. 2012;14:1183–92.

Klasse PJ. How to assess the binding strength of antibodies elicited by vaccination against HIV and other viruses. Expert Rev Vaccines. 2016;15:295–311.

Klasse PJ, Sanders RW, Cerutti A, et al. How can HIV-type-1-Env immunogenicity be improved to facilitate antibody-based vaccine development? AIDS Res Hum Retrovir. 2012;28:1–15.

Klatt NR, Chahroudi A, Silvestri G. Human immunodeficiency viruses: pathogenesis. In: Mahy BWJ, Van Regenmortel MHV, editors. Desk encyclopedia of human and medical virology. Oxford: Academic Press, Elsevier; 2010. p. 345–53.

Klee R. Introduction to the philosophy of science. Cutting nature at its seams. New York: Oxford University Press; 1997. p. 41–61.

Klein JS, Bjorkman PJ. Few and far between: how HIV may be evading antibody avidity. PLoS Pathog. 2010;6:e1000908.

Klein JS, Gnanapragasam PN, Galimidi RP, Foglesong CP, West AP Jr, Bjorkman PJ. Examination of the contributions of size and avidity to the neutralization mechanisms of the anti-HIV antibodies b12 and 4E10. Proc Natl Acad Sci U S A. 2009;106:7385–90.

Klein F, Halper-Stromberg A, Horwitz JA, Gruell H, Scheid JF, Bournazos S, et al. HIV therapy by a combination of broadly neutralizing antibodies in humanized mice. Nature. 2012;492:118–22. https://doi.org/10.1038/nature11604.

Klein F, Diskin R, Scheid JF, Gaebler SC, Mouquet H, Georgiev IS, Pancera M, Zhou T, Incesu R-B, Fu BZ, et al. Somatic mutations of the immunoglobulin framework are generally required for broad and potent HIV-1 neutralization. Cell. 2013;153:126–38.

Klinke DJ. Validating a dimensionless number for glucose homeostasis in humans. Ann Biomed Eng. 2009;37:1886–96. https://doi.org/10.1007/s10439-009-9733-y.

Knowlton ER, Lepone LM, Li J, Rappocciolo G, Jenkins FJ, Rinaldo CR. Professional antigen presenting cells in human herpesvirus 8 infection. Front Immunol. 2013;3:427. https://doi.org/10.3389/fimmu.2012.00427.

Kong L, Sattentau QJ. Antigenicity and immunogenicity in HIV-1 antibody-based vaccine design. J AIDS Clin Res. 2012. https://doi.org/10.4172/2155-6113.

Korber B, Gnanakaran S. The implications of patterns in HIV diversity for neutralizing antibody induction and susceptibility. Curr Opin HIV AIDS. 2009;4:408–17.

Korkut A, Hendrickson WA. Structural plasticity and conformational transitions of HIV envelope glycoprotein gp120. PLoS One. 2012;7:e52170.

Koup RA, Graham BS, Douek DC. The quest for a T cell-based immune correlate of protection against HIV: a story of trials and errors. Nat Rev Immunol. 2011;11:65–70.

Kowalczewska M, Fenoller F, Lafitte D, et al. Identification of candidate antigen in Whipple's disease using a serological proteomic approach. Proteomics. 2006;6:3294–305.

Krachmarov C, Pinter A, Honnen WJ, Gorny MK, Nyambi PN, Zolla-Pazner S, Kayman SC. Antibodies that are cross-reactive for human immunodeficiency virus type 1 clade A and clade B V3 domains are common in patient sera from Cameroon, but their neutralization activity is usually restricted by epitope masking. J Virol. 2005;79:780–90.

Kringelum JV, Nielsen M, Padkjær SB, Lund O. Structural analysis of B-cell epitopes in antibody: protein complexes. Mol Immunol. 2013;53:24–34.

Kubinyi H. Drug research: myths, hype and reality. Nat Rev Drug Discov. 2003;2:665–8.

Kuhn TS. The structure of scientific revolutions. Chicago, IL: University of Chicago Press; 1962. p. 1–264.

Kulkarni-Kale U, Bhosle S, Kolaskar AS. CEP: a conformational epitope prediction server. Nucleic Acids Res. 2005;33(Web server issue):W168–71.

Kumar R, Visciano ML, Li H, Hioe C. Targeting a neutralizing epitope of HIV envelope gp 20 by immune complex vaccine. J AIDS Clin Res. 2012;S8(2):5512.

Kunik V, Ofran Y. The indistinguishability of epitopes from protein surface is explained by the distinct binding preferences of each of the six antigen-binding loops. Protein Eng Des Sel. 2013;26:599–609.

Kunik V, Ashkenazi S, Ofran Y. Paratome: an online tool for systematic identification of antigen-binding regions in antibodies based on sequence or structure. Nucleic Acids Res. 2012a;40: W521–4.

Kunik V, Peters B, Ofran Y. Structural consensus among antibodies defines the antigen binding site. PLoS Comput Biol. 2012b;8:e1002388.

Kuntz ID. Structure-based strategies for drug design and discovery. Science. 1992;257:1078–82.

Kurakin A. Self-organization versus Watchmaker: ambiguity of molecular recognition and design charts of cellular circuitry. J Mol Recognit. 2007;20:205–114.

Kurakin A. Scale-free flow of life: on the biology, economics, and physics of the cell. Theor Biol Med Model. 2009;6:6–34.

Kurakin A. Order without design. Theor Biol Med Model. 2010;7:12–22.

Kusnezow W, Hoheisel JD. Solid supports for microarray immunoassays. J Mol Recog. 2003;16:165–76.

Kusters I, Almond JW. Vaccine strategies. In: Mahy BWJ, Van Regenmortel MHV, editors. Desk encyclopedia of general virology. Oxford: Elsevier Academic Press; 2010. p. 381–9.

Kwong PD, Mascola JR. Human antibodies that neutralize HIV-1: identification, structures, and B cell ontogenies. Immunity. 2012;37:412–25. https://doi.org/10.1016/j.immuni.2012.08.012.

Kwong PD, Wyatt R, Robinson J, Sweet RW, et al. Structure of an HIV gp120 envelope glycoprotein in complex with the CD4 receptor and a neutralizing human antibody. Nature. 1998;393:648–59.

Kwong PD, Wyatt R, Majeed S, Robinson J, Sweet RW, Sodroski J, Hendrickson WA. Structures of HIV-1 gp 120 envelope glycoproteins from laboratory-adapted and primary isolates. Struct Fold Des. 2000;8:1329–39.

Kwong PD, Doyle ML, Casper DJ, Cicala C, Leavitt SA, Majeed S, Steenbeke TD, Venturi M, Chaiken I, Fung M, et al. HIV-1 evades antibody-mediated neutralization through conformational masking of receptor-binding sites. Nature. 2002;420:678–82.

Laal S, Burda S, Gorny MK, Karwowska S, Buchbinder A, Zolla-Pazner S. Synergistic neutralization of human immunodeficiency virus type 1 by combinations of human monoclonal antibodies. J Virol. 1994;68:4001–8.

Labrijn AF, Poignard P, Raja A, Zwick MB, Delgado K, Franti M, Binley J, Vivona V, Grundner C, Huang CC, Venturi M, Petropoulos CJ, Wrin T, Dimitrov DS, Robinson J, Kwong PD, Wyatt RT, Sodroski J, Burton DR. Access of antibody molecules to the conserved coreceptor binding site on glycoprotein gp120 is sterically restricted on primary human immunodeficiency virus type 1. J Virol. 2003;77:10557–65.

Ladbury JE. Just add water! The effect of water on the specificity of protein-ligand binding sites and its potential application to drug design. Chem Biol. 1996;3:973–80.

LaFemina RL. Antiviral research strategies in antiviral drug discovery. Washington, DC: ASM Press; 2009.

Lam KS, Lake D, Salmon SE, Smith J, Chen ML, Wade S, Abdul-Latif F, Knapp RJ, Leblova Z, Ferguson RD, Krchnak VV, Sepetov NF, Lebl M. A one-bead one-peptide combinatorial library method for B-cell epitope mapping. Methods. 1996;9:482–93.

Lambert DM, Hughes AJ. Keywords and concepts in structuralist and functionalist biology. J Theor Biol. 1988;133:133–45.

Landsteiner K. The specificity of serological reactions. New York: Dover Publications; 1962. p. 330.

Langeveld JPM, Casal JI, Osterhaus ADME, Cortès E, De Swart R, Vela C, Dalsgaard K, Puijk WC, Schaaper WMM, Meloen RH. First peptide vaccine providing protection against viral infection in the target animal: studies of canine parvovirus in dogs. J Virol. 1994;68:4506–13.

Langman RE. The specificity of immunological reactions. Mol Immunol. 2000;37:555–61.

Largent E. For love and money: the need to rethink benefits in HIV cure studies. J Med Ethics. 2017;43:96–9.

Larralde OG, Martinez R, Camacho F, Amin N, Aguilar A, Talavera A, Stott DL, Perez EM. Identification of hepatitis A virus mimotopes by phage display, antigenicity and immunogenicity. J Virol Methods. 2007;140:49–58.

Larsericsdotter H, Jansson O, Zhukov A, et al. Optimizing the surface plasmon resonance/mass spectrometry interface for functional proteomics applications: how to avoid and utilize nonspecific adsorption. Proteomics. 2006;6:2355–64.

Laubichler M. Tinkering: a conceptual and historical evaluation. Tinkering: the microevolution of development: Novartis Foundation Symposium, vol. 284. Chichester: Wiley; 2007. p. 20–9.

Laune D, Molina F, Ferrieres G, Mani JC, Cohen P, Simon D, Bernardi T, Piechaczyk M, Pau B, Granier C. Systematic exploration of the antigen binding activity of synthetic peptides isolated from the variable regions of immunoglobulins. J Biol Chem. 1997;272:30937–44.

Laver WG, Air GM, Webster RG, Smith-Gill SJ. Epitopes on protein antigens: misconceptions and realities. Cell. 1990;61:553–6.

Lavoie TB, Drohan WN, Smith-Gill SJ. Experimental analysis by site-directed mutagenesis of somatic mutation effects on affinity and fine specificity in antibodies specific for lysozyme. J lmmunol. 1992;148:503–13.

Law M, Cardoso RM, Wilson IA, Burton DR. Antigenic and immunogenic study of membrane-proximal external region-grafted gp120 antigens by a DNA prime-protein boost immunization strategy. J Virol. 2007;81:4272–85.

Lebrun SJ, Petchput WN, Hui A, et al. Development of a sensitive, colorimetric microarray assay for allergen-responsive human IgE. J Immunol Methods. 2005;300:24–31.

Lecca P, Mura I, Re A, Barker GC, Ihekwaba AEC. Time series analy sis of the *Bacillus subtilis* sporulation network reveals low dimensional chaotic dynamics. Front Microbiol. 2016;7:1760. https://doi.org/10.3389/fmicb.2016.01760.

Leder L, Berger C, Bornhauser S, Wendt H, Ackermann F, Jelesarov I, et al. Spectroscopic, calorimetric, and kinetic demonstration of conformational adaptation in peptide-antibody recognition. Biochemistry. 1995;34:16509–18.

Lefranc MP. IMGT, the international ImMunoGeneTics database. Nucleic Acid Res. 2003;31:307–10.

Lefranc MP, Giudicelli V, Kaas Q, Duprat E, Jabado-Michaloud J, Scaviner D, Ginestoux C, Clément O, Chaume D, Lefranc G. IMGT, the international ImMunoGeneTics information system. Nucleic Acids Res. 2005;33:D593–7.

Leinikki P, Lehtinen M, Hyöty H, Parkkonen P, Kantanen ML, Hakulinen J. Synthetic peptides as diagnostic tools in virology. Adv Virus Res. 1993;42:149–86.

Leitner G, Blum S, Rivas AL. Visualizing the indefinable: three-dimensional complexity of 'infectious diseases'. PLoS One. 2015;10:e0123674. https://doi.org/10.1371/journal.pone.01236742015.

Leng Q, Bentwich Z. Beyond self and nonself: Fuzzy recognition of the immune system. Scand J Immunol. 2002;56:224–32.

Leonelli S. Understanding in biology: the impure nature of biological knowledge. In: De Regt HW, Leonelli S, Eigner K, editors. Scientific understanding. Pittsburgh: University of Pittsburgh Press; 2009. p. 189–209.

Lerner RA. Antibodies of predetermined specificity in biology and medicine. Adv Immunol. 1984;36:1–44.

Lescar J, Pellegrini M, Souchon H, Tello D, Poljak RJ, Peterson N, Greene M, Alzari PM. Crystal structure of a cross-reaction complex between Fab F9.13.7 and guinea fowl lysozyme. J Biol Chem. 1995;270:18067–76.

Letvin NL. Moving forward in HIV vaccine development. Science. 2009;326:1196–8. https://doi.org/10.1126/science.1183278.

Levi M, Sällberg M, Rudén U, Herlyn D, Maruyama H, Wigzell H, Marks J, Wahren B. A complementarity-determining region synthetic peptide acts as a miniantibody and neutralizes human immunodeficiency virus type 1 in vitro. Proc Natl Acad Sci U S A. 1993;90:4374–8.

Lewis GK. Challenges of antibody-mediated protection against HIV-1. Expert Rev Vaccines. 2010;9:683–7.

Lewontin R. The triple helix. Gene, organism and environment. Cambridge, MA: Harvard University Press; 2000.

Li A, Baba TW, Sodroski J, et al. Synergistic neutralization of a chimeric SIV/HIV type 1 virus with combinations of human anti-HIV type 1 envelope monoclonal antibodies or hyperimmune globulins. AIDS Res Hum Retrovir. 1997a;13:647–56.

Li H, Dunn JJ, Luft BJ, Lawson CL. Crystal structure of Lyme disease antigen outer surface protein A complexed with an Fab. Proc Natl Acad Sci U S A. 1997b;94:3584–9.

Li A, Katinger H, Posner MR, et al. Synergistic neutralization of simian-human immunodeficiency virus SHIV-vpu+ by triple and quadruple combinations of human monoclonal antibodies and high-titer anti-human immunodeficiency virus type 1 immunoglobulins. J Virol. 1998;72:3235–40.

Li Y, Migueles SA, Welcher B, Svehla K, Phogat A, Louder MK, Wu X, Shaw GM, Connors M, Wyatt RT, Mascola JR. Broad HIV-1 neutralization mediated by CD4-binding site antibodies. Nat Med. 2007;13:1032–4.

Li Y, Svehla K, Louder MK, Wycuff D, Phogat S, Tang M, et al. Analysis of neutralization specificities in polyclonal sera derived from human immunodeficiency virus type 1-infected individuals. J Virol. 2009;83:1045–59.

Li JZ, Brumme ZL, Brumme CJ, et al. Factors associated with viral rebound in HIV-1-infected individuals enrolled in a therapeutic HIV-1 gag vaccine trial. J Infect Dis. 2011;203:976–83.

Li X, Song B, Chen X, Wang Z, Zeng M, Yu D, Hu D, Chen Z, Jin L, Yang S, Yang C, Chen B. Crystal structure of a four-layer aggregate of engineered TMV CP implies the importance of terminal residues for oligomer assembly. PLoS One. 2013;8:e77717.

Liang S, Zheng D, Zhang C, Zacharias M. Prediction of antigenic epitopes on protein surfaces by consensus scoring. BMC Bioinf. 2009;10:302.

Liao HX, Chen X, Munshaw S, Zhang R, Marshall DJ, Vandergrift N, Whitesides JF, Lu X, Yu JS, Hwang KK, Gao F, Markowitz M, Heath SL, Bar KJ, Goepfert PA, Montefiori DC, Shaw GC, Alam SM, Margolis DM, Denny TN, Boyd SD, Marshal E, Egholm M, Simen BB, Hanczaruk B, Fire AZ, Voss G, Kelsoe G, Tomaras GD, Moody MA, Kepler TB, Haynes BF. Initial antibodies binding to HIV-1 gp41 in acutely infected subjects are polyreactive and highly mutated. J Exp Med. 2011;208:2237–49.

Liao HX, Lynch R, Zhou T, Gao F, Alam SM, Boyd SD, et al. Co-evolution of a broadly neutralizing HIV-1 antibody and founder virus. Nature. 2013a;496:469–76. https://doi.org/10.1038/nature12053.

Liao HX, Tsao CY, Alam SM, Muldoon M, Vandergrift N, Ma BJ, et al. Antigenicity and immunogenicity of transmitted/founder, consensus, and chronic envelope glycoproteins of human immunodeficiency virus type 1. J Virol. 2013b;87:4185–201. https://doi.org/10.1128/JVI.02297-12.

Lifson JD, Rossio JL, Piatak M Jr, Bess J Jr, Chertova E, Schneider DK, et al. Evaluation of the safety, immunogenicity, and protective efficacy of whole inactivated simian immunodeficiency virus (SIV) vaccines with conformationally and functionally intact envelope glycoproteins. AIDS Res Hum Retrovir. 2004;20:772–87.

Lipschultz CA, Yee A, Mohan S, Yili Li S, Smith-Gill J. Temperature differentially affects encounter and docking thermodynamics of antibody-antigen association. J Mol Recognit. 2002;15:44–52.

Lipton P. What good is an explanation? In: Cornwell J, editor. Explanations. Styles of explanation in science. Oxford: Oxford University Press; 2004. p. 1–21.

Liu J, Bartesaghi A, Borgnia MJ, Sapiro G, Subramaniam S. Molecular architecture of native HIV-1 gp120 trimers. Nature. 2008;455:109–13.

Longo O, Tripiciano A, Fiorelli V, Bellino S, Scoglio A, et al. Phase I therapeutic trial of the HIV-1 Tat protein and long term follow-up. Vaccine. 2009;27:3306–12.

Loor F. On the existence of heterospecific antibodies in sera from rabbits immunized against tobacco mosaic virus determinants. Immunology. 1971;21:557–64.

Loscalzo J, Barabasi AL. Systems biology and the future of medicine. Wiley Interdiscip Rev Syst Biol Med. 2011;3:619–27. https://doi.org/10.1002/wsbm.144.

Love JC, Ronan JL, Grotenberg GM, et al. A microengraving method for rapid selection of single cells producing antigen-specific antibodies. Nat Biotechnol. 2006;24:703–7.

Lu S, Wyatt R, Richmond JF, et al. Immunogenicity of DNA vaccines expressing human immunodeficiency virus type 1 envelope glycoprotein with and without deletions in the V1/2 and V3 regions. AIDS Res Hum Retrovir. 1998;14:151–5.

Lu W, Wu X, Lu Y, Guo W, Andrieu JM. Therapeutic dendritic-cell vaccine for simian AIDS. Nat Med. 2003;9:27–32.

Lu W, Arraes LC, Ferreira WT, Andrieu JM. Therapeutic dendritic-cell vaccine for chronic HIV-1 infection. Nat Med. 2004;10:1359–65.

Lu W, Chen S, Lai C, Guo W, Andrieu J-M. Induction of CD8+ regulatory T cells protects macaques against SIV challenge. Cell Rep. 2012;2:1736–46. https://doi.org/10.1016/j.celrep.2012.11.016.

Lucas K, Roosen P. Emergence, analysis and evolution of structures. Berlin: Springer Verlag; 2010.

Ludewig B, Krebs P, Metters H, Tatzel J, Türeci O, Sahin U. Molecular characterization of virus-induced autoantibody responses. J Exp Med. 2004;200:637–46.

Lynch RM, Rong R, Boliar S, Sethi A, Li B, Mulenga J, Allen S, Robinson JE, Gnanakaran S, Derdeyn CA. The B cell response is redundant and highly focused on V1V2 during early subtype C infection in a Zambian seroconverter. J Virol. 2011;85:905–15.

Ma B, Shatsky M, Wolfson HJ, Nussinov R. Multiple diverse ligands binding at a single protein site: a matter of pre-existing populations. Protein Sci. 2002;11:184–97.

Ma BJ, Alam SM, Go EP, Lu X, Desaire H, Tomaras GD, Bowman C, Sutherland LL, Scearce RM, Santra S, Letvin NL, Kepler TB, Liao HX, Haynes BF. Envelope deglycosylation enhances antigenicity of HIV-1 gp41 epitopes for both broad neutralizing antibodies and their unmutated ancestor antibodies. PLoS Pathog. 2011;7:e1002200. https://doi.org/10.1371/journal.ppat.1002200.

MacCallum RM, Martin AC, Thornton JM. Antibody-antigen interactions: contact analysis and binding site topography. J Mol Biol. 1996;262:732–45.

Macdonald GT. Reduction and evolutionary biology. In: Charles D, Lennon K, editors. Reduction, explanation and realism. Oxford: Clarendon Press; 1992. p. 69–96.

Macklem PT, Seely A. Towards a definition of life. Perspect Biol Med. 2010;53:330–40. https://doi.org/10.1353/pbm.0.0167.

Madhumathi J, Prince P, Rao D, Karande A, Reddy M, Kaliraj P. Epitope mapping of Brugia malayi ALT-2 and the development of a multi-epitope vaccine for lymphatic filariasis. J Helminthol. 2017;91:43–54. https://doi.org/10.1017/S0022149X16000055.

Maggiorella MT, Baroncelli S, Michelini Z, Fanales-Belasio E, Moretti S, et al. Long-term protection against SHIV89.6P replication in HIV-1 Tat vaccinated cynomolgus monkeys. Vaccine. 2004;22:3258–69.

Mahner M. The philosophy of mind needs a better metaphysics. In: Bunge M, editor. Science in the light of philosophy. Singapore: World Scientific Publishers; 2017. p. 177–97.

Mahner M, Bunge M. Foundations of biophilosophy. Springer: Berlin; 1997.

Mäkelä O. Single lymph node cells producing heteroclitic bacteriophage antibody. J Immunol. 1965;95:378–86.

Malby RL, Tulip WR, Harley VR, McKimm-Breschkin JL, Graeme Laver W, Webster RG, Colman PM. The structure of a complex between the NC10 antibody and influenza virus neuraminidase and comparison with the overlapping binding site of the NC41 antibody. Structure. 1994;2:733–46.

Manivel V, Sahoo NC, Salunke DM, Rao KV. Maturation of an antibody response is governed by modulations inflexibility of the antigen-combining site. Immunity. 2000;13:611–20.

Manivel V, Bayiroglu F, Siddiqui Z, Salunke DM, Rao KV. The primary antibody repertoire represents a linked network of degenerate antigen specificities. J Immunol. 2002;169:888–97.

Marchalonis JJ, Kaveri S, Lacroix-Desmazes S, Kazatchkine MD. Natural recognition repertoire and the evolutionary emergence of the combinatorial immune system. FASEB J. 2002;16:842–8.

Margineanu DG. Neuropharmacology beyond reductionism – a likely prospect. Biosystems. 2016;141:1–9. https://doi.org/10.1016/j.biosystems.2015.11.010.

Mariuzza RA. Multiple paths to multispecificity. Immunity. 2006;24:359–61.

Mariuzza RA, Phillips SE, Poljak RJ. The structural basis of antigen-antibody recognition. Annu Rev Biophys Chem. 1987;16:139–59.

Markgren PO, Hamalainen M, Danielson UH. Kinetic analysis of the interaction between HIV-1 protease and inhibitors using optical biosensor technology. Anal Biochem. 2000;279:71–8.

Marks JD. Maturation of the immune response. In: Van Oss CJ, Van Regenmortel MHV, editors. Immunochemistry. New York: Marcel Dekker; 1994. p. 1039–51.

Markus G, Tritsch GL, Parthasarathy R. A model for hydropathybased peptide interactions. Arch Biochem Biophys. 1989;272:433–9.

Martin ACR, Orengo CA, Hutchinson EG, Jones S, Karmirantzou M, Laskowski RA, Mitchell JBO, Taroni C, Thornton JM. Protein folds and functions. Structure. 1998;6:875–84.

Mascola JR. Herpes simplex virus vaccines–why don't antibodies protect? J Am Med Assoc. 1999;281:379–80.

Mascola JR. HIV/AIDS: allied responses. Natuzwickre. 2007;449:29–30. https://doi.org/10.1038/449029a.

Mascola JR, Haynes BF. HIV-1 neutralizing antibodies: understanding nature's pathways. Immunol Rev. 2013;254:225–44.

Mascola JR, Montefiori DC. The role of antibodies in HIV vaccines. Annu Rev Immunol. 2010;28:413–44.

Mascola JR, Louder MK, VanCott TC, Sapan CV, Lambert JS, Muenz LR, Bunow B, Birx DL, Robb ML. Potent and synergistic neutralization of human immunodeficiency virus (HIV) type 1 primary isolates by hyperimmune anti-HIV immunoglobulin combined with monoclonal antibodies 2F5 and 2G12. J Virol. 1997;71:7198–206.

Mascola JR, Lewis MG, Stiegler G, Harris D, VanCott TC, Hayes D, et al. Protection of macaques against pathogenic simian/human immunodeficiency virus 89.6PD by passive transfer of neutralizing antibodies. J Virol. 1999;73:4009–18.

Mascola JR, Stiegler G, Van Cott TC, Katinger H, Carpenter CB, Hanson CE, Beary H, Hayes D, Frankel SS, Birx DL, Lewis MG. Protection of macaques against vaginal transmission of a pathogenic HIV-1/SIV chimeric virus by passive infusion of neutralizing antibodies. Nat Med. 2000;6:207–10.

Maynard-Smith J. The concept of information in biology. Philos Sci. 2000;67:177–94.

Mayr E. Cause and effect in biology. Science. 1961;134:1501–6.

Mayr E. The growth of biological thought. Diversity, evolution, and inheritance. Cambridge, MA: Harvard University Press; 1982. p. 974.

Mazumdar PH. Species and specificity. Cambridge: Cambridge University Press; 1995.

Mazzocchi F. Complexity in biology. EMBO Rep. 2008;9:10–4.

Mazzocchi F. Complexity and the reductionism–holism debate in systems biology. Wiley Interdiscip Rev Syst Biol Med. 2012;4:413–27. https://doi.org/10.1002/wsbm.1181.

Mc Farland BJ, Strong RK. Thermodynamic analysis of degenerate recognition by the NKG2D immunoreceptor: not induced fit but rigid adaptation. Immunity. 2003;19:803–12.

Mc Lennan IC. Germinal centers. Annu Rev Immunol. 1994;12:117–39.

McBurney SP, Ross TM. Human immunodeficiency virus-like particles with consensus envelopes elicited broader cell-mediated peripheral and mucosal immune responses than polyvalent and monovalent Env vaccines. Vaccine. 2009;27:4337–49.

McElrath MJ, Haynes BF. Induction of immunity to human immunodeficiency virus type-1 by vaccination. Immunity. 2010;33:542–54.

McGaughey GB, Citron M, Danzeisen RC, Freidinger RM, Garsky VM, Hurni WM, et al. HIV-1 vaccine development: constrained peptide immunogens show improved binding to the anti-HIV-1 gp41 Mab. Biochemistry. 2003;42:3214–23.

McLellan JS, Pancera M, Carrico C, Gorman J, Julien JP, Khayat R, Louder R, Pejchal R, Sastry M, Dai K, O'Dell S, Patel N, Shahzadul-Hussan S, Yang Y, Zhang B, Zhou T, Zhu J, Boyington JC, Chuang GY, Diwanji D, Georgiev I, Kwon YD, Lee D, Louder MK, Moquin S, Schmidt SD, Yang ZY, Bonsignori M, Crump JA, Kapiga SH, Sam NE, Haynes BF, Burton DR, Koff WC, Walker LM, Phogat S, Wyatt R, Orwenyo J, Wang LX, Arthos J, Bewley CA, Mascola JR, Nabel GJ, Schief WR, Ward AB, Wilson IA, Kwong PD. Structure of HIV-1 gp120 V1/V2 domain with broadly neutralizing antibody PG9. Nature. 2011;480:336–43.

McLellan JS, Chen M, Joyce MG, Sastry M, Stewart-Jones GB, Yang Y, et al. Structure-based design of a fusion glycoprotein vaccine for respiratory syncytial virus. Science. 2013;342:592–8. https://doi.org/10.1126/science.1243283.

McMahon MJ, O'Kennedy R. Polyreactivity as an acquired artefact, rather than a physiologic property, of antibodies: evidence that monoreactive antibodies may gain the ability to bind to multiple antigens after exposure to low pH. J Immunol Methods. 2000;241:1–10.

McMichael AJ, Borrow P, Tomaras GD, Goonetilleke N, Haynes BF. The immune response during acute HIV-1 infection: clues for vaccine development. Nat Rev Immunol. 2010;10:11–23. https://doi.org/10.1038/nri2674.

McNeil JG, Johnston MI, Birx DL, Tramont EC. HIV vaccine trial justified. Science. 2004;303:961.

McNeill D, Freiberger P. Fuzzy logic. New York: Simon and Schuster; 1993. p. 319.

Medawar PB, Medawar JS. The life science. London: Granada Publishing; 1978.

Melendez-Hevia E, Waddell TG, Cascante M. The puzzle of the Krebs citric acid cycle: assembling the pieces of chemically feasible reactions, and opportunism in the design of metabolic pathways during evolution. J Mol Evol. 1996;43:293–303.

Meloen RH, Casal JI, Dalsgaard K, Langeveld JPM. Synthetic peptide vaccines: success at last. Vaccine. 1995;13:885–6.

Meloen RH, Puyk WC, Sloostra JW. Mimotopes: realization of an unlikely concept. J Mol Recognit. 2000;13:352–9.

Mengistu M, Tang A-H, Foulke JS, Blanpied TA, et al. Patterns of conserved gp120 epitope presentation on attached HIV-1 virions. Proc Natl Acad Sci U S A. 2017;114(46):E9893–902.

Menzies P, Price H. Causation as a secondary quality. Br J Philos Sci. 1993;44:187–203.

Mester B, Manor R, Mor A, Arshava B, Rosen O, Ding FX, Naider F, Anglister J. HIV-1 peptide vaccine candidates: selecting constrained V3 peptides with highest affinity to antibody 447–52D. Biochemistry. 2009;48:7867–77.

Mian IS, Bradwell AR, Olson AJ. Structure, function and properties of antibody binding sites. J Mol Biol. 1991;217:133–51.

Michaud GA, Salcius M, Zhou F, et al. Analyzing antibody specificity with whole proteome microarrays. Nat Biotechnol. 2003;21:1509–12.

Miller KR. In: Dembski WA, Ruse M, editors. Debating design: from Darwin to DNA. Cambridge: Cambridge University Press; 2004. p. 81–97.

Mills E, Cooper C, Guyatt G, Gilchrist A, Rachlis B, et al. Barriers to participating in an HIV vaccine trial: a systematic review. AIDS. 2004;18:2235–42.

Miska D. Biotech's twentieth birthday blues. Nat Rev Drug Discov. 2003;2:231–3.

Misumi S, Nakayama D, Kusaba M, Iiboshi T, Mukai R, Tachibana K, Nakasone T, Umeda M, Shibata H, Endo M, Takamune N, Shoji S. Effects of immunization with CCR5-based cycloimmunogen on Simian/HIVSF162P3 challenge. J Immunol. 2006;176:463–71.

Mokyr J. The gifts of Athena. Princeton, NJ: Princeton University Press; 2002a. 359 p.

Mokyr J. Technology and the problem of human knowledge. In: The gifts of Athena. Historical origins of the knowledge economy. Princeton: Princeton University Press; 2002b. p. 1–27.

Molina F, Laune D, Gougat C, Pau B, Granier C. Improved performances of spot multiple peptide synthesis. Pept Res. 1996;9:151–5.

Monaco-Malbet S, Berthet-Colominas C, Novelli A, Battai' N, Piga N, Cheynet V, Mallet F, Cusack S. Mutual conformational adaptations in antigen and antibody upon complex formation between an Fab and HIV-1 capsid protein p24. Struct Fold Des. 2000;8:1069–77.

Monini P, Cafaro A, Srivastava IK, Moretti S, Sharma VA, et al. HIV-1 Tat promotes integrin-mediated HIV transmission to dendritic cells by binding Env spikes and competes with neutralization by anti-HIV antibodies. PLoS One. 2012;7:e48781.

Montefiori DC, Mascola JR. Neutralizing antibodies against HIV-1: can we elicit them with vaccines and how much do we need? Curr Opin HIV AIDS. 2009;4:347–51.

Montero M, van Houten NE, Wang X, Scott JK. The membrane-proximal external region of the human immunodeficiency virus type 1 envelope: dominant site of antibody neutralization and target for vaccine design. Microbiol Mol Biol Rev. 2008;72:54–84. https://doi.org/10.1128/MMBR.00020-07.

Moodie SL, Mitchell JBO, Thornton JM. Protein recognition of adenylate: an example of a fuzzy recognition template. J Mol Biol. 1996;263:486–500.

Moore JP, Ho DD. Antibodies to discontinuous or conformationally sensitive epitopes on the gp120 glycoprotein of human immunodeficiency virus type 1 are highly prevalent in sera of infected humans. J Virol. 1993;67:863–75.

Moore PL, Gray ES, Sheward D, Madiga M, Ranchobe N, Lai Z, Honnen, WJ, Nonyane M, Tumba N, Hermanus T, Sibeko S, Mlisana K, AbdoolKarim SS, Williamson C, Pinter A, Morris L, CAPRISA 002 Study. Potent and broad neutralization of HIV-1subtype C by plasma antibodies targeting a quaternary epitope including residues in the V2 loop. J Virol. 2011;85:3128–41.

Moore PL, Crooks ET, Porter L, Zhu P, Cayanan CS, Grise H, et al. Nature of nonfunctional envelope proteins on the surface of human immunodeficiency virus type 1. J Virol. 2006;80:2515–28.

Moore PL, Gray ES, Wibmer CK, et al. Evolution of an HIV glycan-dependent broadly neutralizing antibody epitope through immune escape. Nat Med. 2012;18:1688–92.

Mor A, Segal E, Mester B, Arshava B, Rosen O, Ding FX, Russo J, Dafni A, Schvartzman F, Scherf T, et al. Mimicking the structure of the V3 epitope bound to HIV-1 neutralizing antibodies. Biochemistry. 2009;48:3288–303.

Moran MJ, Andris JS, Matsumato Y, Capra JD, Hersh EM. Variable region genes of anti-HIV human monoclonal antibodies: non-restricted use of the V gene repertoire and extensive somatic mutation. Mol Immunol. 1993;30:1543–51.

Morange M. A history of molecular biology. Cambridge, MA: Harvard University Press; 1998.

Morange M. A successful form of reductionism. Biochemist. 2001a;23:37–9.

Morange M. The misunderstood gene. Cambridge, MA: Harvard University Press; 2001b.

Morcock DR, Thomas JA, Gagliardi TD, Gorelick RJ, Roser JD, Chertova EN, et al. Elimination of retroviral infectivity by N-ethylmaleimide with preservation of functional envelope glycoproteins. J Virol. 2005;9:1533–42.

Moreau V, Granier C, Villard S, Laune D, Molina F. Discontinuous epitope prediction based on mimotope analysis. Bioinformatics. 2006;22:1088–95.

Moreau V, Fleury C, Piquer D, Nguyen C, Novali N, Villard S, Laune D, Granier C, Molina F. PEPOP: computational design of immunogenic peptides. BMC Bioinf. 2008;9:71.

Morel-Montero A, Delaage M. Immunochemistry of pharmacological substances. In: Van Oss CJ, Van Regenmortel MHV, editors. Immunochemistry. New York: Marcel Dekker 1994. p. 357–72.

Morgan C, Marthas M, Miller C, Duerr A, Cheng-Mayer C, Desrosiers R, Flores J, Haigwood N, Hu SL, Johnson RP, Lifson J, Montefiori D, Moore J, Robert-Guroff M, Robinson H, Self S, Corey L. The use of non-human primate models in HIV vaccine development. PLoS Med. 2008;5:e173. https://doi.org/10.1371/journal.pmed.0050173.

Morowitz HJ. The emergence of everything. How the world became complex. Oxford: Oxford University Press; 2002.

Morris L, Chen X, Alam M, Tomaras G, Zhang R, Marshall DJ, Chen B, Parks R, Foulger A, Jaeger F, Donathan M, Bilska M, Gray ES, AbdoolKarim SS, Kepler TB, Whitesides J, Montefiori D, Moody MA, Liao HX, Haynes BF. Isolation of a human anti-HIV gp41 membrane proximal region neutralizing antibody by antigen-specific single B cell sorting. PLoS One. 2011;6:e23532. https://doi.org/10.1371/journal.pone.0023532.

Moseri A, Tantry S, Sagi Y, Arshava B, Naider F, Anglister J. An optimally constrained V3 peptide is a better immunogen than its linear homolog or HIV-1 gp120. Virology. 2010;401:293–304.

Moss L. What genes can't do. Cambridge, MA: MIT Press; 2003.

Moticka EJ. The non-specific stimulation of immunoglobulin secretion following specific stimulation of the immune system. Immunology. 1974;27:401–12.

Moudgil KD, Sercarz EE, Grewal IS. Modulation of the immunogenicity of antigenic determinants by their flanking residues. Immunol Today. 1998;19:217–20. https://doi.org/10.1016/S0167-5699(97)01233-4.

Mouquet H, Nussenzweig MC. Polyreactive antibodies in adaptive immune responses to viruses. Cell Mol Life Sci. 2012;69:1435–45.

Mouquet H, Nussenzweig MC. HIV: roadmaps to a vaccine. Nature. 2013;496:441–2. https://doi.org/10.1038/nature12091.

Mouquet H, Scheid JF, Zoller MJ, Krogsgaard M, Ott RG, Shukair S, Artyomov MN, Pietzsch J, Connors M, Pereyra F, Walker BD, Ho DD, Wilson PC, Seaman MS, Eisen HN, Chakraborty AK, Hope TJ, Ravetch JV, Wardemann H, Nussenzweig MC. Polyreactivity increases the apparent affinity of anti-HIV antibodies by heteroligation. Nature. 2010;467:591–5.

Mouquet H, Klein F, Scheid JF, Warncke M, Pietzsch J, Oliveira TY, Velinzon K, Seaman MS, Nussenzweig MC. Memory B cell antibodies to HIV-1 gp140 cloned from individuals infected with clade A and B viruses. PLoS One. 2011;6:e24078. https://doi.org/10.1371/journal.pone.0024078.

Mullen LM, Nair SP, Ward JM, Rycroft AN, Henderson B. Phage display in the study of infectious diseases. Trends Microbiol. 2006;14:141–7.

Muller S. Immunization with peptides. In: Van Regenmortel MHV, Muller S, editors. Synthetic peptides as antigens. Amsterdam: Elsevier; 1999a. p. 113–77.

Muller S. Use of antipeptide antibodies in molecular and cellular biology. In: Van Regenmortel MHV, Muller S, editors. Synthetic peptides as antigens. Amsterdam: Elsevier; 1999b. p. 215–35.

Muller S, Plaue S, Couppez M, Van Regenmortel MHV. Comparison of different methods for localizing antigenic regions in histone H2A. Mol Immunol. 1986;23:593–601.

Muller S, Plaue S, Samara JP, Valette M, Briand JP, Van Regenmortel MHV. Antigenic properties and protective capacity of a cyclic peptide corresponding to site A of influenza virus haemagglutinin. Vaccine. 1990;8:307–14.

Munro JB, Mothes W. Structure and dynamics of the native HIV-1 Env trimer. J Virol. 2015;89:5752–5.

Munro JB, Gorman J, Ma X, Zhou Z, Arthos J, Burton DR, Koff WC, Courter JR, Smith AB III, Kwong PD, et al. Conformational dynamics of single HIV-1 envelope trimers on the surface of native virions. Science. 2014;346:759–63.

Münz C, Lünemann JD, Getts MT, Miller SD. Antiviral immune responses: triggers of or triggered by autoimmunity? Nat Rev Immunol. 2009;9:246–58.

Murzin AG, Patthy L. Sequences and topology: from sequence to structure to function. Curr Opin Struct Biol. 1999;9:359–62.

Muster T, Steindl F, Purtscher M, Trkola A, Klima A, Himmler G, Rüker F, Katinger H. A conserved neutralizing epitope on gp41 of human immunodeficiency virus type1. J Virol. 1993;67:6642–7.

Muster T, Guinea R, Trkola A, Purtscher M, Klima A, Steindl F, et al. Cross-neutralizing activity against divergent human immunodeficiency virus type 1 isolates induced by the gp41 sequence ELDKWAS. J Virol. 1994;68:4031–4.

Myers SR, Leigh IM, Navsaria H. Epidermal repair results from activation of follicular and epidermal progenitor keratinocytes mediated by a growth factor cascade. Wound Repair Regen. 2007;15:693–701. https://doi.org/10.1111/j.1524-475X.2007.00297.x.

Myszka DG. Survey of the 1998 optical biosensor literature. J Mol Recognit. 1999;12:390–408.

Nabel GJ, Kwong PD, Mascola JR. Progress in the rational design of an AIDS vaccine. Philos Trans R Soc Lond B Biol Sci. 2011;366:2759–65.

Nagel E. The structure of science. London: Routledge, Keegan Paul; 1961.

Nagel E. Teleology revisited and other essays in the philosophy and history of science. New York: Columbia University Press; 1979. p. 352.

Nagel T. Reductionism and antireductionism. In: The limits of reductionism in biology. Novartis Foundation Symposium 213. Chichester: Wiley; 1998. p. 3–14.

Nagorsen D, Marinola FM, Panelli MC. Cytokine and chemokine expression profiles of maturing dendritic cells using multiprotein platform arrays. Cytokine. 2004;25:31–5.

Nakaya HI, Pulendran B. Systems vaccinology: its promise and challenge for HIV vaccine development. Curr Opin HIV AIDS. 2012;7:24–31.

Nara PL, Garrity R. Deceptive imprinting: a cosmopolitan strategy for complicating vaccination. Vaccine. 1998;16:1780–7.

Nargi F, Kramer E, Mezencio J, et al. Protection of swine from foot-and-mouth disease with one dose of an all-D retro peptide. Vaccine. 1999;17:2888–93.

Nelson JD, Brunel FM, Jensen R, Crooks ET, Cardoso RMF, Wang M, Hessel A, Wilson IA Zwick MB. An affinity-enhanced neutralizing antibody against the membrane-proximal external region of human immunodeficiency virus type 1 gp41 recognizes an epitope between those of 2F5 and 4E10. J Virol. 2007;81:4033–43.

Ness JE, Welch M, Giver L, Bueno M, Cherry JR, Borchert TV, et al. DNA shuffling of subgenomic sequences of subtilisin. Nat Biotechnol. 1999;17:893–6.

Neurath AR. Antibody-mediated immunity to viruses. In: Mahy BWJ, Van Regenmortel MHV, editors. Desk encyclopedia of general virology. Amsterdam: Elsevier; 2008. p. 361–75.

Neurath AR, Rubin BA. Viral structural components as immunogens of prophylactic value. In: Monographs in virology. Basel: Karger; 1971. p. 88.

Neurath AR, Strick N, Lin K, Jiang S. Multifaceted consequences of anti-gp41 monoclonal antibody 2F5 binding to HIV type 1 virions. AIDS Res Hum Retrovir. 1995;11:687–95.

Newton-Smith W. The rationality of science. Boston, MA: Routledge, KeganPaul; 1981.

Nezlin R. Immunoglobulin structure and function. In: van Oss CJ, Van Regenmortel MHV, editors. Immunochemistry. New York: Marcel Dekker; 1994. p. 3–45.

Nicholson BH. Synthetic vaccines. Oxford: Blackwell Scientific Publishers; 1994.

Nieva JL, Apellaniz B, Huarte N, Lorizate M. A new paradigm in molecular recognition? Specific antibody binding to membrane-inserted HIV-1 epitopes. J Mol Recognit. 2011;24:642–6.

Noble D. The music of life: biology beyond the genes. New York: Oxford University Press; 2006.

Noble D. Genes and causation. Philos Transact A Math Phys Eng Sci. 2008a;366:3001–15.

Noble D. Claude Bernard, the first systems biologist, and the future of physiology. Exp Physiol. 2008b;93:16–26. https://doi.org/10.1113/expphysiol.2007.038695.

Noble D. Biophysics and systems biology. Phil Trans R Soc A. 2010;368:1125–39. https://doi.org/10.1098/rsta.2009.0245.

Noble D. A theory of biological relativity: no privileged level of causation. Interface Focus. 2012;2:55–64.

Nola R, Sankey H. Theories of scientific method. Stocksfield: Acumen Publishers; 2007. p. 1–381.

Norrby E, editor. Immunochemistry of AIDS. Switzerland: Karger; 1993.

North B, Lehmann A, Dunbrack RL Jr. A new clustering of antibody CDR loop conformations. J Mol Biol. 2011;406:228–56.

Notkins AL. Polyreactivity of antibody molecules. Trends Immunol. 2004;25:174–9.

Novotny J, Bruccoleri RE, Carlson WD, Handschumacher M, Haber E. Antigenicity of myohemerythrin. Science. 1987;238:1584–6.

Novotny J, Bruccoleri E, Saul FA. On the attribution of binding energy in antigen-antibody complexes McPC 603, D1.3, and HyHEL-5. Biochemistry. 1989;28:4735–49.

Nowak MA, McMichael AJ. How HIV defeats the immune system. Sci Am. 1995;273:42–9.

O'Malley MA, Dupré J. Fundamental issues in systems biology. BioEssays. 2005;27:1270–6.

Obeid OE, Steward MW. The potential of immunization with synthetic peptides to overcome the immunosuppressive effect of maternal anti-measles virus antibodies in young mice. Immunology. 1994;82:16–21.

Obeid OE, Partidos CD, Howard CR, Steward MW. Protection against morbillivirus-induced encephalitis by immunization with a rationally designed synthetic peptide vaccine containing B-and T-cell epitopes from the fusion protein of measles virus. J Virol. 1995;69:1420–8.

Obeid OE, Stanley CM, Steward MW. Immunological analysis of the protective responses to the chimeric synthetic peptide representing T- and B-cell epitopes from the fusion protein of measles virus. Virus Res. 1996;42:173–80.

Oberg AL, Kennedy RB, Li P, Ovsyannikova IG, Poland GA. Systems biology approaches to new vaccine development. Curr Opin Immunol. 2011;23:436–43.

Oberle CS, Joos B, Rusert P, et al. Tracing HIV-1 transmission: envelope traits of HIV-1 transmitter and recipient pairs. Retrovirology. 2016;13:62.

Ochsenbein AF, Fehr T, Lutz C, Suter M, Brombacher F, Hengartner H, Zinkernagel RM. Control of early viral and bacterial distribution and disease by natural antibodies. Science. 1999;286:2156–9.

Odorico M, Pellequer JL. BEPITOPE: predicting the location of continuous epitopes and patterns in proteins. J Mol Recognit. 2003;16:20–2.

Ofek G, Tang M, Sambor A, Katinger H, Mascola JR, Wyatt R, Kwong PD. Structure and mechanistic analysis of the anti-human immunodeficiency virus type 1 antibody 2F5 in complex with its gp41 epitope. J Virol. 2004;78:10724–37.

Ofek G, Guenaga FJ, Schief WR, Skinner J, Baker D, Wyatt R, Kwong PD. Elicitation of structure-specific antibodies by epitope scaffolds. Proc Natl Acad Sci U S A. 2010a;107:17880–9887.

Ofek G, McKee K, Yang Y, Yang ZY, Skinner J, Guenaga FJ, Wyatt R, Zwick MB, Nabel GJ, Mascola JR, Kwong PD. Relationship between antibody 2F5 neutralization of HIV-1 and hydrophobicity of its heavy chain third complementarity-determining region. J Virol. 2010b;84:2955–62.

Ofran Y, Schlessinger A, Rost B. Automated identification of complementarity determining regions (CDRs) reveals peculiar characteristics of CDRs and B cell epitopes. J Immunol. 2008;181:6230–5.

Ohara R, Knappik A, Shimada K, et al. Antibodies for proteomic research: comparison of traditional immunization with recombinant antibody technology. Proteomics. 2006;6:2638–46.

Okulicz JF. Elite controllers and long-term non progressors: models for HIV vaccine development? J AIDS Clin Res. 2012;3:139.

Okulicz JF, Marconi VC, Landrum ML, et al. Clinical outcomes of elite controllers, viremic controllers, and long-term nonprogressors in the US Department of Defense HIV natural history study. J Infect Dis. 2009;200:1714–23.

Olshansky SJ, Carnes BA, Butler RN. If humans were built to last. Sci Am. 2003;284:50–5.

Oomen CJ, Hoogerhout P, Bonvin AMJJ, Kuipers B, Brugghe H, Timmermans H, Haseley SR, Van Alphen L, Gros P. Immunogenicity of peptide-vaccine candidates predicted by molecular dynamics simulations. J Mol Biol. 2003;328:1083–9.

Oppezzo P, Dumas G, Bouvet JP, Robello C, Cayota A, Pizarro JC, Dighiero G, Pritsch O. Somatic mutations can lead to a loss of superantigenic and polyreactive binding. Eur J Immunol. 2004;34:1423–32.

Osinga HM, Sherman A, Tsaneva-Atanasova K. Cross-currents between biology and mathematics: the codimension of pseudo-plateau bursting. Discrete Contin Dyn Syst Ser A. 2012;32:2853–77. https://doi.org/10.3934/dcds.2012.32.2853.

Oue S, Okamoto A, Yano T, Kagamiyama H. Redesigning the substrate specificity of an enzyme by cumulative effects of the mutations of non-active site residues. J Biol Chem. 1999;274:2344–9.

Overbaugh J, Morris L. The antibody response against HIV-1. Cold Spring Harb Perspect Med. 2012;2:a007039.

Oyama S. The ontogeny of information. 2nd ed. Durham, NC: Duke University Press; 2009.

Oyama S, Griffiths PE, Gray RD, editors. Cycles of contingency: developmental systems and evolution. Cambridge, MA: The MIT Press; 2001.

Padlan EA. On the nature of antibody combining sites: unusual structural features that may confer on these sites an enhanced capacity for binding ligands. Proteins Struct Funct Genet. 1990;7:112–24.

Padlan EA, Abergel C, Tipper JP. Identification of specificity determining residues in antibodies. FASEB J. 1995;9:133–9.

Pajon R, Beernink PT, Granoff DM. Design of meningococcal factor H binding protein mutant vaccines that do not bind human complement factor H. Infect Immun. 2012;80:2667–77.

Palma P, Gudmundsdotter L, Finocchi A, Eriksson LE, Mora N, Santilli V, Aquilani A, Manno EC, Zangari P, Romiti ML, Montesano C, Grifoni A, Brave A, Ljungberg K, Blomberg P, Bernardi S, Sandström E, Hejdeman B, Rossi P, Wahren B. Immunotherapy with an HIV-DNA vaccine in children and adults. Vaccine. 2014;2:563–80.

Panagi M, Georgila K, Eliopoulos AG, Apidianakis Y. Constructing personalized longitudinal holo'omes of colon cancer-prone humans and their modeling in flies and mice. Oncotarget. 2015;10(41):4224. https://doi.org/10.18632/oncotarget.6463.

Pancera M, McLellan JS, Wu X, Zhu J, Changela A, Schmidt SD, Yang Y, Zhou T, Phogat S, Mascola JR, Kwong PD. Crystal structure of PG16 and chimeric dissection with somatically related PG9: structure-function analysis of two quaternary-specific antibodies that effectively neutralize HIV-1. J Virol. 2010;84:8098–110.

Pancera M, Zhou T, Druz A, Georgiev IS, Soto C, Gorman J, Huang J, Acharya P, Chuang GY, Ofek G, et al. Structure and immune recognition of trimeric pre-fusion HIV-1 Env. Nature. 2014;514:455–61.

Pandey A, Mann M. Proteomics to study genes and genomes. Nature. 2000;405:837–46.

Pantophlet R. Antibody epitope exposure and neutralization of HIV-1. Curr Pharm Des. 2010;16:3729–43.

Pantophlet R, Burton DR. Immunofocusing: antigen engineering to promote the induction of HIV-neutralizing antibodies. Trends Mol Med. 2003;9:468–73.

Pantophlet R, Wilson IA, Burton DR. Hyperglycosylated mutants of human immunodeficiency virus (HIV) type1 monomeric gp120 as novel antigens for HIV vaccine design. J Virol. 2003a;77:5889–901.

Pantophlet R, Ollmann Saphire E, Poignard P, Parren PW, Wilson IA, Burton DR. Fine mapping of the interaction of neutralizing and non neutralizing monoclonal antibodies with the CD4 binding site of human immunodeficiency virus type 1 gp120. J Virol. 2003b;77:642–58.

Pantophlet R, Aguilar-Sino RO, Wrin T, Cavacini LA, Burton DR. Analysis of the neutralization breadth of the anti-V3 antibody F425-B4e8 and re-assessment of its epitope fine specificity by scanning mutagenesis. Virology. 2007;364:441–53.

Pantophlet R, Wrin T, Cavacini LA, Robinson JE, Burton DR. Neutralizing activity of antibodies to the V3 loop region of HIV-1 gp120 relative to their epitope fine specificity. Virology. 2008;381:251–60.

Parnes O. From interception to incorporation: degeneracy and promiscuous recognition as precursors of a paradigm shift in immunology. Mol Immunol. 2004;40:985–91.

Parren PW, Burton DR. The antiviral activity of antibodies in vitro and in vivo. Adv Immunol. 2001;77:195–262.

Parrish NF, Gao F, Li H, Giorgi EE, Barbian HJ, Parrish EH, et al. Phenotypic properties of transmitted founder HIV-1. Proc Natl Acad Sci U S A. 2013;110:6626–33. https://doi.org/10.1073/pnas.1304288110.

Parry NR, Barnett PV, Ouldridge EJ, Rowlands DJ, Brown F. Neutralizing epitopes of type O foot-and-mouth disease virus. II. Mapping three conformational sites with synthetic peptide reagents. J Gen Virol. 1989;70:1493–503.

Parry N, Fox G, Rowlands D, Brown F, Fry E, Acharya R, et al. Structural and serological evidence for a novel mechanism of antigenic variation in foot-and-mouth disease virus. Nature. 1990;347:569–72.

Partidos CD, Steward MW. The effects of a flanking sequence on the immune response to a B-cell and a T-cell epitope from the fusion protein of measles virus. J Gen Virol. 1992;73:1987–94.

Paul WE. Immunity. Baltimore: Johns Hopkins University Press; 2015.

Pearson H. Surviving a knockout blow. Nature. 2002;415:8–9.

Pejchal R, Wilson IA. Structure-based vaccine design in HIV: blind men and the elephant? Curr Pharm Des. 2010;16:3744–53.

Pejchal R, Gach JS, Brunel FM, Cardoso RM, Stanfield RL, Dawson PE, et al. A conformational switch in human immunodeficiency virus gp41 revealed by the structures of overlapping epitopes recognized by neutralizing antibodies. J Virol. 2009;83:8451–62.

Pellequer JL, Westhof E, Van Regenmortel MHV. Predicting the location of continuous epitopes in proteins from their primary structures. Methods Enzymol. 1991;203:176–201.

Pellequer JL, Westhof E, Van Regenmortel MHV. Correlation between the location of antigenic sites and the prediction of turns in proteins. Immunol Lett. 1993;36:83–99.

Pellequer JL, Westhof E, Van Regenmortel MHV. Epitope predictions from the primary structure of proteins. In: Wisdom GB, editor. Peptide antigens: a practical approach. Oxford: JRL; 1994. p. 7–25.

Pennock RT. In: Dembski WA, Ruse M, editors. Debating design: from Darwin to DNA. Cambridge: Cambridge University Press; 2004. p. 130–48.

Phogat S, Svehla K, Tang M, Spadaccini A, Muller J, Mascola J, et al. Analysis of the human immunodeficiency virus type 1 gp41 membrane proximal external region arrayed on hepatitis B surface antigen particles. Virology. 2008;373:72–84.

Pilgrim D. The biopsychosocial model in health research: its strengths and limitations for critical realists. J Crit Realism. 2015;14:164–80. https://doi.org/10.1179/1572513814Y.0000000007.

Pinilla C, Appel JR, Houghten RA. Functional importance of amino acid residues making up peptide antigenic determinants. Mol Immunol. 1993;30(6):577–85.

Pinter A, Honnen WJ, He Y, Gorny MK, Zolla-Pazner S, Kayman SC. The V1/V2 domain of gp120 is a global regulator of the sensitivity of primary human immunodeficiency virus type 1 isolates to neutralization by antibodies commonly induced upon infection. J Virol. 2004;78:5205–15.

Pirofski L, Casadevall A. What is infectiveness and how is it involved in infection and immunity? BMC Immunol. 2015;16:13. https://doi.org/10.1186/s12865-015-0076-1.

Poignard P, Klasse PJ, Sattentau QJ. Antibody neutralization of HIV-1. Immunol Today. 1996;17:239–46.

Poignard P, Saphire EO, Parren PW, Burton DR. gp120: biologic aspects of structural features. Annu Rev Immunol. 2001;19:253–74.

Poignard P, Moulard M, Golez E, Vivona V, Franti M, Venturini S, et al. Heterogeneity of envelope molecules expressed on primary human immunodeficiency virus type 1 particles as probed by the binding of neutralizing and non neutralizing antibodies. J Virol. 2003;77:353–65.

Pomorska-Mól M, Pejsak Z. Effects of antibiotics on acquired immunity in vivo – current state of knowledge. Pol J Vet Sci. 2012;15:583–9. https://doi.org/10.2478/v10181-012-0089-0.

Ponomarenko JV, Van Regenmortel MHV. B cell epitope prediction. In: Gu J, Bourne PE, editors. Structural bioinformatics. 2nd ed. Hoboken, NJ: John Wiley; 2009. p. 849–79.

Poon B, Safrit JT, McClure H, Kitchen C, Hsu JF, Gudeman V, Petropoulos C, Wrin T, Chen ISY, Grovit-Ferbas K. Induction of humoral immune responses following vaccination with envelope-containing, formaldehyde-treated, thermally inactivated human immunodeficiency virus type 1. J Virol. 2005a;79:4927–35.

Poon B, Hsu JF, Gudeman V, Chen ISY, Grovit-Ferbas K. Formaldehyde-treated, heat-inactivated virions with increased human immunodeficiency virus type 1 env can be used to induce high-titer neutralizing antibody responses. J Virol. 2005b;79:10210–7.

Popper K. The logic of scientific discovery. London: Hutchinson; 1959. p. 1–513.

Popper KR. Conjectures and refutations. London: Routledge and Kegan Paul; 1969. p. 1–412.

Poropatich K, Sullivan DJ Jr. Human immunodeficiency virus type 1 long-term non-progressors: the viral, genetic and immunological basis for disease non-progression. J Gen Virol. 2011;92:247–68.

Prabakaran P, Chen W, Dimitrov DS. The antibody germline/maturation hypothesis, elicitation of broadly neutralizing antibodies against HIV-1 and cord blood IgM repertoires. Front Immunol. 2014;5:398.

Pressman D, Grossberg AL. The structural basis of antibody specificity. New York: Benjamin W.A; 1968.

Presta L. Antibody engineering for therapeutics. Curr Opin Struct Biol. 2003;13:519–25.

Prezzi C, Nuzzo M, Meola A, Delmastro P, Galfrè G, Cortese R, Nicosia A, Monaci P. Selection of antigenic and immunogenic mimics of hepatitis C virus using sera from patients. J Immunol. 1996;156:4504–13.

Psillos S. Causation and explanation. Chesham/Montreal: Acumen/McGill-Queen's University Press; 2002.

Pulendran B. Systems vaccinology: probing humanity's diverse immune systems with vaccines. Proc Natl Acad Sci U S A. 2014;111:12300–6. https://doi.org/10.1073/pnas.1400476111.

Pulendran B, Li S, Nakaya HI. Systems vaccinology. Immunity. 2010;33:516–29.

Puntoriero G, Meola A, Lahm A, et al. Towards a solution for hepatitis C virus hypervariability: mimotopes of the hyper-variable region 1 can induce antibodies cross-reacting with a large number of viral variants. EMBO J. 1998;17:3521–33.

Purcell AW, Gorman J. Immunoproteomics: mass spectrometry-based methods to study the targets of the immune response. J Mol Cell Proteomics. 2004;3:193–208.

Qu Z, Garfinkel A, Weiss JN, Nivala M. Multi-scale modeling in biology: how to bridge the gaps between scales? Prog Biophys Mol Biol. 2011;107:21–31. https://doi.org/10.1016/j.pbiomolbio.2011.06.004.

Querec TD, Akondy RS, Lee EK, Cao W, Nakaya HI, Teuwen D, Pirani A, Gernert K, Deng J, Marzolf B, Kennedy K, Wu H, Bennouna S, Oluoch H, Miller J, Vencio RZ, Mulligan M, Aderem A, Ahmed R, Pulendran B. Systems biology approach predicts immunogenicity of the yellow fever vaccine in humans. Nat Immunol. 2009;10:116–25.

Quesniaux V, Briand JP, Van Regenmortel MHV. Immuno-chemical studies of turnip yellow mosaic virus II: localization of a viral epitope in the N-terminal residues of the coat protein. Mol Immunol. 1983;20:179–85.

Quesniaux VFJ, Schmitter D, Schreier M, Van Regenmortel MHV. Monoclonal antibodies to Cyclosporine are representative of the major antibody populations present in antisera of immunized mice. Mol Immunol. 1990;27:227–36.

Race E, Stein CA, Wigg MD, Baksh A, Addawe M, Frezza P, et al. A multistep procedure for the chemical inactivation of human immunodeficiency virus for use as an experimental vaccine. Vaccine. 1995;13:1567–75.

Raghunathan G, Smart J, Williams J, Almagro JC. Antigen-binding site anatomy and somatic mutations in antibodies that recognize different types of antigens. J Mol Recognit. 2012;25:103–13.

Rajewsky K. Clonal selection and learning in the antibody system. Nature. 1996;381:751–8.

Ramsland PA, Guddat LW, Edmundson AB, Raison RL. Diverse binding site structures revealed in homology models of polyreactive immunoglobulins. J Comput Aided Mol Des. 1997;11:453–61.

Rappuoli R. Reverse vaccinology, a genome-based approach to vaccine development. Vaccine. 2001;19:2688–91.

Rappuoli R, Bagnoli F. Vaccine design: innovative approaches and novel strategies. Norfolk: Caister Academic Press; 2011.

Rauffer-Bruyère N, Chatellier J, Weiss E, Van Regenmortel MHV, Altschuh D. Cooperative effects of mutations in a recombinant Fab on the kinetics of antigen binding. Mol Immunol. 1997;34:165–73.

Raviv Y, Viard M, Bess JW Jr, Chertova E, Blumenthal R. Inactivation of retroviruses with preservation of structural integrity by targeting the hydrophobic domain of the viral envelope. J Virol. 2005;79:12394–400.

Reading SA, Dimmock NJ. Neutralization of animal virus infectivity by antibody. Arch Virol. 2007;152:1047–59.

Rees AR, Pedersen JT, Searle SMJ, Henry AH, Webster DM. In: Van Oss CJ, Van Regenmortel MHV, editors. Immunochemistry. New York: Dekker; 1994. p. 615.

Rerks-Ngarm S, Pitisuttithum P, Nitayaphan S, Kaewkungwal J, Chiu J, Paris R, Premsri N, Namwat C, de Souza M, Adams E, et al. Vaccination with ALVAC and AIDSVAX to prevent HIV-1 infection in Thailand. N Engl J Med. 2009;361:2209–20.

Rezacova P, Lescar J, Brynda J, Fabry M, Horejsi M, Sedlacek J, Bentley GA. Structural basis of HIV-1 and HIV-2 protease inhibition by a monoclonal antibody. Structure. 2001;9:887–95.

Rezza G, Fiorelli V, Dorrucci M, Ciccozzi M, Tripiciano A, et al. The presence of anti-Tat antibodies is predictive of long-term nonprogression to AIDS or severe immunodeficiency: findings in a cohort of HIV-1 seroconverters. J Infect Dis. 2005;191:1321–4.

Rich RL, Myszka DG. Survey of the 1999 surface plasmon resonance biosensor literature. J Mol Recognit. 2000;13:388–407.

Rich RL, Myszka DG. Survey of the year 2000 commercial optical biosensor literature. J Mol Recognit. 2001;14:273–94.

Rich RL, Myszka DG. Spying on HIV with SPR. Trends Microbiol. 2003;11:124–33.

Rich R, Myszka D. Survey of the year 2003 commercial optical biosensor literature. J Mol Recognit. 2005a;18:1–39.

Rich R, Myszka D. Survey of the year 2004 commercial optical biosensor literature. J Mol Recognit. 2005b;18:431–78.

Richalet-Secordel P, Zeder-Lutz G, Plaue S, Sommermeyer-Leroux G, Van Regenmortel MHV. Cross-reactivity of monoclonal antibodies to a chimeric V3 peptide of HIV-1 with peptide analogues studied by biosensor technology and ELISA. J Immunol Methods. 1994;176:221–34.

Richalet-Sécordel PM, Rauffer-Bruyère N, Christensen LL, Ofenloch-Haehnle B, Seidel C, Van Regenmortel MHV. Concentration measurement of unpurified proteins using biosensor technology under conditions of partial mass transport limitation. Anal Biochem. 1997;249:165–73.

Richards FF, Konigsberg WH. Speculations. How specific are antibodies? Immunochemistry. 1973;10:545–53. https://doi.org/10.1016/0019-2791(73)90227-9.

Richards FF, Konigsberg WH, Rosenstein RW, Varga JM. On the specificity of antibodies. Science. 1975;187:130–7.

Rini JM, Schulze-Gahmen U, Wilson IA. Structural evidence for induced fit as a mechanism for antibody-antigen recognition. Science. 1992;255:959–65.

Rios A. Fundamental challenges to the development of a preventive HIV vaccine. Curr Opin Virol. 2018;29:26–32.

Rios A, Poteet EC, Siwak EB, Anderson DW, Yao QC. HIV inactivation: time for a second look. AIDS. 2015;29:129–31.

Rios A, Pottet EC, Siwak EB, Anderson DW, Yao QC. The human immune response to HIV and its impact in the potential development of an inactivated HIV vaccine. AIDS Rev. 2016;18:151–7.

Rivas AL, Jankowski MD, Piccinini R, Leitner G, Schwarz D, Anderson KL, et al. Feedback-based, system-level properties of vertebrate-microbial interac-tions. PLoS One. 2013;8:e53984. https://doi.org/10.1371/journal.pone.0053984.

Rivas AL, Hoogesteijn AL, Piccinini R. Beyond numbers: the informative patterns of staphylococcal dynamics. Curr Pharm Des. 2015;21:2122–30. https://doi.org/10.2174/1381612821666150310104053.

Rivas AL, Leitner G, Jankowski MD, Hoogesteijn AL, Iandiorio MJ, Chatzipanagiotou S, et al. Nature and consequences of biological reductionism for the immunological study of infectious diseases. Front Immunol. 2017;8:612. https://doi.org/10.3389/fimmu.2017.00612.

Roberts VA, Getzoff ED, Tainer JA. In: Van Regenmortel MHV, editor. Structure of antigens, vol. 3. Boca Raton, FL: CRC; 1993. p. 31–53.

Roberts AEL, Kragh KN, Bjarnsholt T, Diggle SP. The limitations of *in vitro* experimentation in understanding biofilms and chronic infection. J Mol Biol. 2015;427:3646–61. https://doi.org/10.1016/j.jmb.2015.09.002.

Robinson WH, Steinman L, Utz PJ. Protein arrays for autoantibody profiling and fine-specificity mapping. Proteomics. 2003;3:2077–84.

Robinson JE, Franco K, Elliott DH, Maher MJ, Reyna A, Montefiori DC, Zolla-Pazner S, Gorny MK, Kraft Z, Stamatatos L. Quaternary epitope specificities of anti-HIV-1 neutralizing antibodies generated in rhesus macaques infected by the simian/human immunodeficiency virus SHIVSF162P4. J Virol. 2010;84:3443–53.

Robson B. The dragon on the gold: myths and realities for data mining in biomedicine and biotechnology using digital and molecular libraries. J Proteome Res. 2005;3:1113–9. https://doi.org/10.1021/pr0499242.

Rodda SJ, Tribbick G. Antibody-defined epitope mapping using the multipin method of peptide synthesis. Methods. 1996;9:473–81.

Roivanen M, Piirainen L, Rysa T, Narvanen A, Hovi T. An immunodominant N-terminal region of VP1 protein of poliovirus that is buried in crystal structure can be exposed in solution. Virology. 1993;195:762–5.

Rolland M, Gilbert P. Evaluating immune correlates in HIV type1 vaccine efficacy trials: what RV144 may provide. AIDS Res Hum Retrovir. 2012;28:400–4.

Ronacher K, Joosten SA, van Crevel R, Dockrell HM, Walzl G, Ottenhoff THM. Acquired immunodeficiencies and tuberculosis: focus on HIV/ AIDS and diabetes mellitus. Immunol Rev. 2015;264:121–37. https://doi.org/10.1111/imr.12257.

Roos H, Karlsson R, Nilshans H, Persson A. Thermodynamic analysis of protein interactions with biosensor technology. J Mol Recognit. 1998;11:204–10.

Root-Bernstein RS. Discovering. Inventing and solving problems at the frontiers of scientific knowledge. Cambridge, MA: Harvard University Press; 1991. p. 501.

Rose S. Lifelines. London: Penguin; 1997.

Rose S. What is wrong with reductionist explanations of behavior? In: Bock G, Goode J, editors. The limits of reductionism in biology. Novartis Foundation Symposium No 213. Chichester: Wiley; 1998. p. 176.

Rose H, Rose S. Alas poor Darwin. Arguments against evolutionary psychology. London: Vintage; 2000.

Rosen R. Life itself. New York: Columbia University Press; 1991.

Rosen O, Chill J, Sharon M, Kessler N, Mester B, Zolla-Pazner S, Anglister J. Induced fit in HIV-neutralizing antibody complexes: evidence for alternative conformations of the gp120 V3 loop and the molecular basis for broad neutralization. Biochemistry. 2005;44:7250–77258.

Rosenberg A. The structure of biological science. Cambridge: Cambridge University Press; 1985. p. 281.

Rosenberg A. Instrumental biology or the disunity of science. Chicago, IL: University of Chicago Press; 1994.

Rosenberg A. Philosophy of science. London: Routledge; 2000.

Rossio JL, Esser MT, Suryanarayana K, Schneider DK, Bess JW Jr, Vasquez GM, et al. Inactivation of human immunodeficiency virus type 1 infectivity with preservation of conformational and functional integrity of virion surface proteins. J Virol. 1998;72:7992–8001.

Rowlands DJ, Clarke BE, Carroll AR, et al. Chemical basis of antigenic variation in foot-and-mouth disease virus. Nature. 1983;306:694–7.

Rubinstein ND, Mayrose I, Halperin D, Yekutieli D, Gershoni JM, Pupko T. Computational characterization of B-cell epitopes. Mol Immunol. 2008;45:3477–89.

Ruse M. All my love if for individuals. Evolution. 1998;52:283–8.

Ruse M. Darwin and design: does evolution have a purpose? Cambridge, MA: Harvard University Press; 2002.

Ruse M. The evolution-creation struggle. Cambridge, MA: Harvard University Press; 2005.

Rybicki EP. Plant-produced vaccines: promise and reality. Drug Discov Today. 2009;14:16–24.

Sadegh-Zadeh K. The fuzzy revolution: goodbye to the Aristotelian Weltanschauung. Artif Intell Med. 2001;21:1–25.

Sällberg M, Sherefa K, Zhang ZX. The antigen/antibody specificity exchanger: a new peptide based tool for re-directing antibodies of other specificities to recognize the V3 domain of HIV-1 gp120. Biochem Biophys Res Commun. 1994;205:1386–90.

Salmon WC. Causality and explanation. Oxford: Oxford University Press; 1998.

Sandberg JK, Franksson L, Sundbäck J, Michaelsson J, Petersson M, Achour A, Wallin RPA, Sherman NE, Bergman T, Jörnvall H, et al. T cell tolerance based on avidity thresholds rather than complete deletion allows maintenance of maximal repertoire diversity. J Immunol. 2000;165:25–33.

Sandstrom E, Nilsson C, Hejdeman B, Brave A, Bratt G, et al. Broad immunogenicity of a multigene, multiclade HIV-1 DNA vaccine boosted with heterologous HIV-1 recombinant modified vaccinia virus Ankara. J Infect Dis. 2008;198:1482–90.

Saphire EO, Parren PWHI, Pantophlet R, Zwick MB, et al. Crystal structure of a neutralizing human IGG against HIV-1: a template for vaccine design. Science. 2001;293:1155–9.

Saphire EO, Montero M, Menendez A, van Houten NE, Irving MB, Pantophlet R, Zwick MB, Parren PWHI, Burton DR, Scott JK, Wilson IA. Crystal structure of a broadly neutralizing anti-HIV-1 antibody in complex with a peptide: mechanism of gp120 cross-reactivity. J Mol Biol. 2007;369:696–709.

Sattentau QJ, McMichael AJ. New templates for HIV-1 antibody-based vaccine design. F1000 Biol Rep. 2010;2:60. https://doi.org/10.3410/b2-60.

Sattentau QJ, Moore JP. Human immunodeficiency virus type 1 neutralization is determined by epitope exposure on the gp120 oligomer. J Exp Med. 1995;182:185–96.

Sattentau QJ, Moore JP, Vignaux F, Traincard F, Poignard P. Conformational changes induced in the envelope glycoproteins of the human and simian immunodeficiency viruses by soluble receptor binding. J Virol. 1993;67:7383–93.

Sattler R. Bio-philosophy. Analytic and holistic perspectives. Berlin: Springer Verlag; 1986. p. 284.

Saunders CJ, McCaffrey RA, Zharkikh I, Kraft Z, Malenbaum SE, Burke B, Cheng-Mayer C, Stamatatos L. The V1, V2, and V3 regions of the human immunodeficiency virus type 1 envelope differentially affect the viral phenotype in an isolate-dependent manner. J Virol. 2005;79:9069–80.

Scala G, Chen X, Liu W, Telles J-N, Cohen OJ, Vaccarezza M, Igarashi T, Fauci AS. Selection of HIV-specific immunogenic epitopes by screening random peptide libraries with HIV-1-positive sera. J Immunol. 1999;162:6155–61.

Scanlan CN, Pantophlet R, Wormald MR, Ollmann Saphire E, Stanfield R, Wilson IA, Katinger H, Dwek RA, Rudd PM, Burton DR. The broadly neutralizing anti-human immunodeficiency virus type1 antibody 2G12 recognizes a cluster of alpha 1→2 mannose residues on the outer face of gp120. J Virol. 2002;76:7306–21.

Scarselli M, Aricò B, Brunelli B, Savino S, Di Marcello F, Palumbo E, Veggi D, Ciucchi L, Cartocci E, Bottomley MJ, et al. Rational design of a meningococcal antigen inducing broad protective immunity. Sci Transl Med. 2011;3:91ra62.

Schaffner K. Discovery and explanation in biology and medicine. Chicago, IL: Chicago University Press; 1993.

Scheid JF, Mouquet H, Feldhahn N, Seaman MS, Velinzon K, Pietzsch J, Ott RG, Anthony RM, Zebroski H, Hurley A, et al. Broad diversity of neutralizing antibodies isolated from memory B cells in HIV-infected individuals. Nature. 2009;458:636–40.

Scheid JF, Mouquet H, Ueberheide B, Diskin R, Klein F, Oliveira TY, Pietzsch J, Fenyo D, Abadir A, Velinzon K, Hurley A, Myung S, Boulad F, Poignard P, Burton DR, Pereyra F, Ho DD, Walker BD, Seaman MS, Bjorkman PJ, Chait BT, Nussenzweig MC. Sequence and structural convergence of broad and potent HIV antibodies that mimic CD4 binding. Science. 2011;333:1633–7.

Schellekens GA, Lasonder E, Feijlbrief M, et al. Identification of the core residues of the epitope of a monoclonal antibody raised against glycoprotein D of herpes simplex virus type 1 by screening of a random peptide library. Eur J Immunol. 1994;24:3188–93.

Scherer EM, Leaman DP, Zwick MB, McMichael AJ, Burton DR. Aromatic residues at the edge of the antibody combining site facilitate viral glycoprotein recognition through membrane interactions. Proc Natl Acad Sci U S A. 2010;107:1529–34.

Schief WR, Ban YE, Stamatatos L. Challenges for structure-based HIV vaccine design. Curr Opin HIV AIDS. 2009;4:431–40.

Schiffner T, Sattentau QJ, Dorrell L. Development of prophylactic vaccines against HIV-1. Retrovirology. 2013;10:72. https://doi.org/10.1186/1742-4690-10-72.

Schildbach JF, Near RI, Bruccoleri RE, Haber E, Jeffrey PD, Novotny J, et al. Modulation of antibody affinity by a non-contact residue. Protein Sci. 1993;2:206–14. https://doi.org/10.1002/pro.5560020209.

Schiller J, Chackerian B. Why HIV virions have low numbers of envelope spikes: implications for vaccine development. PLoS Pathog. 2014;10:e1004254.

Schreiber G. Kinetic studies of protein-protein interactions. Curr Opin Struct Biol. 2002;12:41–7.

Schroer JA, Bender T, Feldmann T, Kim KJ. Mapping epitopes on the insulin molecule using monoclonal antibodies. Eur J Immunol. 1983;13:693–700.

Schubert W. Systematic, spatial imaging of large multimolecular assemblies and the emerging principles of supramolecular order in biological systems. J Mol Recognit. 2014;27:3–13.

Schubert W, Bonnekoh B, Pommer AJ, Philipsen L, Böckelmann R, Malykh Y, Gollnick H, Friedenberger M, Bode M, Dress AWM. Analyzing proteome topology and function by automated multi-dimensional fluorescence microscopy. Nat Biotechnol. 2006;24:1270–8.

Schuck P. Use of surface plasmon resonance to probe the equilibrium and dynamic aspects of interactions between biological macromolecules. Annu Rev Biophys Biomol Struct. 1997;26:541–66.

Schwartz FP, Tello D, Goldbaum FA, Mariuzza RA, Poljak RJ. Thermodynamics of antigen-antibody binding using specific anti-lysozyme antibodies. Eur J Biochem. 1995;228:388–94.

Scudamore JM, Harris DM. Control of foot-and-mouth disease: lessons from the experience of the outbreak in Great Britain in 2001. Rev Sci Tech. 2002;21:699–710.

Sela-Culang I, Kunik V, Ofran Y. The structural basis of antibody-antigen recognition. Front Immunol. 2013;4:302. https://doi.org/10.3389/fimmu.2013.00302.

Sela-Culang I, Benhnia MR, Matho MH, Kaever T, Maybeno M, Schlossman A, Nimrod G, Li S, Xiang Y, Zajonc D, Crotty S, Ofran Y, Peters B. Using a combined computational-experimental approach to predict antibody-specific B cell epitopes. Structure. 2014;22(4):646–57. https://doi.org/10.1016/j.str.2014.02.003.

Sela-Culang I, Ofran Y, Peters B. Antibody specific epitope prediction-emergence of a new paradigm. Curr Opin Virol. 2015;11:98–102.

Selvarajah S, Puffer B, Pantophlet R, Law M, Doms RW, Burton DR. Comparing antigenicity and immunogenicity of engineered gp120. J Virol. 2005;79:12148–63.

Seo HS. Application of radiation technology in vaccines development. Clin Exp Vaccine Res. 2015;4:145–58.

Sethi DK, Agarwal A, Manivel V, Rao KV, Salunke DM. Differential epitope positioning within the germline antibody enhances promiscuity in the primary immune response. Immunity. 2006;24:429–438.

Sercarz EE, Berzofsky JA. Immunogenicity of protein antigens: repertoire and regulation, vol. 1 & 2. Boca Raton, FL: CRC Press; 1987.

Sercarz EE, Maverakis E. Recognition and function in a degenerate immune system. Mol Immunol. 2004;40:1003–108.

Serruto D, Serino L, Masignani V, Pizza M. Genome-based approaches to develop vaccines against bacterial pathogens. Vaccine. 2009;27:3245–50.

Sette A, Rappuoli R. Reverse vaccinology: vaccines in the era of genomics. Immunity. 2010;33:530–41.

Shanks N, Greek R, Greek J. Are animal models predictive for humans? Philos Ethics Humanit Med. 2009;4:2.

Shapiro JA. Revisiting the central dogma in the 21st century. Ann N Y Acad Sci. 2009;1178:6–28.

Sharon M, Kessler N, Levy R, Zolla-Pazner S, Goërlach M, Anglister J. Alternative conformations of HIV-1 V3 loops mimic β hairpins in chemokines, suggesting a mechanism for coreceptor selectivity. Structure. 2003;11:225–36.

Shedlock DJ, Silvestri G, Weiner DB. Monkeying around with HIV vaccines: using rhesus macaques to define 'gatekeepers' for clinical trials. Nat Rev Immunol. 2009;9:717–28.

Sheinerman FB, Norel R, Honig B. Electrostatic aspects of protein-protein interactions. Curr Opin Struct Biol. 2000;10:153–9.

Shepard JF, Secor GA, Purcifull DE. Immunochemical cross-reactivity between the dissociated capsid proteins of PVY group plant viruses. Virology. 1974;58:464–75.

Shepherd NE, Hoang HN, Abbenante G, Fairlie DP. Single turn peptide alpha helices with exceptional stability in water. J Am Chem Soc. 2004;127:2974–83.

Sheppard H. Inactivated-or killed-virus HIV/AIDS vaccines. Curr Drug Targets Infect Disord. 2005;5:131–41.

Sheppard HW, Dorman BP. Time for a systematic look at inactivated HIV vaccines. AIDS. 2015;29:125–7.

Shoshan SH, Admon A. Proteomics in cancer vaccine development. Expert Rev Proteomics. 2005;2:229–41.

Shrager J. The fiction of function. Bioinformatics. 2003;19:1934–6.

Silverstein AM. History of immunology: development of the concept of immunologic specificity: II. Cell Immunol. 1982;71:183–95.

Silverstein AM, Rose NR. On the implications of polyclonal B cell activation. Nat Immunol. 2003;4:931–2.

Simek MD, Rida W, Priddy FH, Pung P, Carrow E, Laufer DS, Lehrman JK, Boaz M, Tarragona-Fiol T, Miiro G, Birungi J, Pozniak A, McPhee DA, Manigart O, Karita E, Inwoley A, Jaoko W, Dehovitz J, Bekker LG, Pitisuttithum P, Paris R, Walker LM, Poignard P, Wrin T, Fast PE, Burton DR, Koff WC. Human immunodeficiency virus type1 elite neutralizers: individuals with broad and potent neutralizing activity identified by using a high-throughput neutralization assay together with an analytical selection algorithm. J Virol. 2009;83:7337–48.

Simmonds P. A clash of ideas - the varying uses of the 'species' term in virology and their utility for classifying viruses in metagenomic datasets. J Gen Virol. 2018;99:277–87.

Simmonds P, Adams MJ, Benkő M, Breitbart M, Brister JR, et al. Consensus statement: virus taxonomy in the age of metagenomics. Nat Rev Microbiol. 2017;15:161–8.

Simon H. The sciences of the artificial. 3rd ed. Cambridge, MA: MIT Press; 1996.

Simonton DKAE. scientific genius is extinct. Nature. 2013;493:602. https://doi.org/10.1038/493602a.

Sivalingam GN, Sheperd AJ. An analysis of B-cell epitope discontinuity. Mol Immunol. 2012;51:304–9.

Skinner JE. Low-dimensional chaos in biological systems. Biotechnology. 1994;12:596–600. https://doi.org/10.1038/nbt0694-596.

Soga S, Kuroda D, Shirai H, Kobori M, Hirayama N. Use of amino acid composition to predict epitope residues of individual antibodies. Protein Eng Des Sel. 2010;23:441–8.

Sok D, Doores KJ, Briney B, Le KM, Saye-Francisco KL, Ramos A, et al. Promiscuous glycan site recognition by antibodies to the high-mannose patch of gp120 broadens neutralization of HIV. Sci Transl Med. 2014;6:236ra63. https://doi.org/10.1126/scitranslmed.3008104.

Sok D, Pauthner M, Briney B, Lee JH, Saye-Francisco KL, Hsueh J, et al. A prominent site of antibody vulnerability on HIV envelope incorporates a motif associated with CCR5 binding and its camouflaging glycans. Immunity. 2016;45:31–45. https://doi.org/10.1016/j.immuni.2016.06.026.

Sok D, Le KM, Vadnais M, Saye-Francisco KL, Jardine JG, Torres JL, et al. Rapid elicitation of broadly neutralizing antibodies to HIV by immunization in cows. Nature. 2017;548:108–11. https://doi.org/10.1038/nature23301.

Sonnenschein C, Soto AM. The society of cells. Cancer and control of cell proliferation. New York: Springer; 1999.

Sonnenschein C, Soto AM. The somatic mutation theory of carcinogenesis: why it should be dropped and replaced. Mol Carcinog. 2000;29:1–7.

Soria I, Quattrochi V, Langellotti C, Gammelia M, et al. Dendrimeric peptides can confer protection against foot-and-mouth disease virus in cattle. PLoS One. 2017;12:e0185184. https://doi.org/10.1371/journal.pone.0185184.

Spangler BD. Binding to native proteins by antipeptide monoclonal antibodies. J Immunol. 1991;146:1591–5.

Spear GR, Takefman DM, Sharpe S, Ghassemi M, Zolla-Pazner S. Antibodies to the HIV-1 V3 loop in serum from infected persons contribute a major proportion of immune effect or functions including complement activation, antibody binding, and neutralization. Virology. 1994;204:609–15.

Sperling R, Francus T, Siskind GW. Degeneracy of antibody specificity. J Immunol. 1983;131:882–5.

Stamatatos L, Morris L, Burton DR, Mascola JR. Neutralizing antibodies generated during natural HIV-1 infection: good news for an HIV-1 vaccine? Nat Med. 2009;15:866–70. https://doi.org/10.1038/nm.1949.

Stamatatos L, Pancera M, McGuire AT. Germline-targeting immunogens. Immunol Rev. 2017;275:203–16.

Stanfield RL, Wilson IA. Antigen-induced conformational changes in antibodies: a problem for structural prediction and design. Trends Biotechnol. 1994;12:275–9.

Stanfield R, Cabezas E, Satterthwait A, Stura E, Profy A, Wilson I. Dual conformations for the HIV-1 gp120 V3 loop in complexes with different neutralizing fabs. Structure. 1999;7 131–42.

Stanfield RL, Gorny MK, Williams C, Zolla-Pazner S, Wilson IA. Structural rationale for the broad neutralization of HIV-1 by human monoclonal antibody 447-52D. Structure. 2004;12:193–204.

Stanfield RL, Gorny MK, Zolla-Pazner S, Wilson IA. Crystal structures of human immunodeficiency virus type 1 (HIV-1) neutralizing antibody 2219 in complex with three different V3 peptides reveal a new binding mode for HIV-1 cross-reactivity. J Virol. 2006;80:6093–105.

Steward MW, Stanley CM, Obeid OE. A mimotope from a solid-phase peptide library induces a measles virus-neutralizing and protective antibody response. J Virol. 1995;69:7668–73.

Stewart J. Immunoglobulins did not arise in evolution to fight infection. Immunol Today. 1992;13:396–9.

Stiegler G, Kunert R, Purtscher M, Wolbank S, Voglauer R, Steindl F, et al. A potent cross-clade neutralizing human monoclonal antibody against a novel epitope on gp41 of human immunodeficiency virus type 1. AIDS Res Hum Retrovir. 2001;17:1757–65.

Stokes DE. Pasteur's quadrant. Basic science and technological innovation. Washington, DC: Brooking Institution Press; 1997. p. 1–196.

Sturtevant JM. The thermodynamic effects of protein mutations. Curr Opin Struct Biol. 1994;4:69–78.

Suan D, Sundling C, Brink R. Plasma cell and memory B cell differentiation from the germinal center. Curr Opin Immunol. 2017;45:97–102. https://doi.org/10.1016/j.coi.2017.03.006.

Sun ZY, Oh KJ, Kim M, Yu J, Brusic V, Song L, et al. HIV-1 broadly neutralizing antibody extracts its epitope from a kinked gp41 ectodomain region on the viral membrane. Immunity. 2008;28:52–63.

Sundaram R, Dakappagari NK, Kaumaya P. Synthetic peptides as cancer vaccines. Biopolymers. 2002;66:200–16.

Sundaram R, Lynch MP, Rawale SV, Sun Y, Kazanji M, Kaumaya PT. De novo design of peptide immunogens that mimic the coiled coil region of human T-cell leukemia virus type-1 glycoprotein 21 transmembrane subunit for induction of native protein reactive neutralizing antibodies. J Biol Chem. 2004;279:24141–51.

Sundberg EJ. Structural basis of antibody-antigen interactions. In: Reineke U, Schutkowski M, editors. Methods in molecular biology, epitope mapping protocols, vol. 524. New York: Humana Press; 2009. p. 23–36.

Sundberg EJ, Mariuzza RA. Molecular recognition in antibody-antigen complexes. Adv Protein Chem. 2002;61:119–60.

Sweredoski MJ, Baldi P. PEPITO: improved discontinuous Bcell epitope prediction using multiple distance thresholds and half sphere exposure. Bioinformatics. 2008;24:1459–60.

Tainer JA, Getzoff ED, Paterson Y, Olson AJ, Lerner RA. The atomic mobility component of protein antigenicity. Annu Rev Immunol. 1985;3:501–35.

Talmage DW. Immunological specificity, unique combinations of selected natural globulins provide an alternative to the classical concept. Science. 1959;129:1643–8.

Tarlinton DM, Smith KG. Dissecting affinity maturation: a model explaining selection of antibody-forming cells and memory B cells in the germinal centre. Immunol Today. 2000;21:436–41.

Taub R, Gould RJ, Garsky VM, Ciccarone TM, Hoxie J, Friedman PA, Shattil SJ. A monoclonal antibody against the platelet fibrinogen receptor contains a sequence that mimics a receptor recognition domain in fibrinogen. J Biol Chem. 1989;264:259–65.

Tauber AI. The immune self: theory or metaphor. Cambridge: Cambridge University Press; 1997. p. 1–354.

Thali M, Olshevsky U, Furman C, Gabuzda D, Posner M, Sodroski J. Characterization of a discontinuous human immunodeficiency virus type 1 gp120 epitope recognized by a broadly reactive neutralizing human monoclonal antibody. J Virol. 1991;65:6188–93.

Thomas C. Road blocks in HIV research: five questions. Nat Med. 2009;15:855–9.

Thornhill RH, Ussery DW. A classification of possible routes of Darwinian evolution. J Theor Biol. 2000;203:111–6.

Thornton JM, Sibanda BL. Amino and carboxy-terminal regions in globular proteins. J Mol Biol. 1983;167:443–60.

Thornton JM, Edwards MS, Taylor WR, Barlow DJ. Location of "continuous" antigenic determinants in the protruding regions of proteins. EMBO J. 1986;5:409–13.

Thorpe IF, Brooks CL III. Molecular evolution of affinity and flexibility in the immune system. Proc Natl Acad Sci U S A. 2007;104:8821–6.

Tian Y, Ramesh CV, Ma X, Naqvi S, Patel T, Cenizal T, Tiscione M, Diaz K, Crea T, Arnold E, Arnold GF, Taylor JW, et al. Structure-affinity relationships in the gp41 ELDKWA epitope for the HIV-1 neutralizing monoclonal antibody 2F5: effects of side-chain and backbone modifications and conformational constraints. J Pept Res. 2002;59:264–76.

Tieri P, Grignolio A, Zaikin A, Mishto M, Remondini D, Castellani GC, et al. Network, degeneracy and bow tie. Integrating paradigms and architectures to grasp the complexity of the immune system. Theor Biol Med Model. 2010;7:32. https://doi.org/10.1186/1742-4682-7-32.

Tilley SA, Honnen WJ, Racho ME, Chou TC, Pinter A. Synergistic neutralization of HIV-1 by human monoclonal antibodies against the V3 loop and the CD4 binding site of gp120. AIDS Res Hum Retrovir. 1992;8:461–7.

Timmerman P, Beld J, Puijk WC, Meloen RH. Rapid and quantitative cyclization of multiple peptide loops onto synthetic scaffolds for structural mimicry of protein surfaces. Chembiochem. 2005;6:821–4.

Timmerman P, Puijk WC, Meloen RH. Functional reconstruction and synthetic mimicry of a conformational epitope using CLIPS technology. J Mol Recognit. 2007;20:283–99.

Tobin MB, Gustafson C, Huisman GW. Directed evolution: the 'rational' basis for 'irrational' design. Curr Opin Struct Biol. 2000;10:421–7.

Tobin GJ, Trujillo JD, Bushnell RV, et al. Deceptive imprinting and immune refocusing in vaccine design. Vaccine. 2008;26:6189–99.

Tomaras GD, Plotkin SA. Complex immune correlates of protection in HIV-1 vaccine efficacy trials. Immunol Rev. 2017;275:245–61.

Tomaras GD, Binley JM, Gray ES, Crooks ET, Osawa K, Moore PL, Tumba N, Tong T, Shen X, Yates NL, Decker J, Wibmer CK, Gao F, Alam SM, Easterbrook P, AbdoolKarim S, Kamanga G, Crump JA, Cohen M, Shaw GM, Mascola JR, Haynes BF, Montefiori DC, Morris L. Polyclonal B cell responses to conserved neutralization epitopes in a subset of HIV-1-infected individuals. J Virol. 2011;85:11502–19.

Tompa P, Szász C, Buday L. Structural disorder throws new light on moonlighting. Trends Biochem Sci. 2005;30:484–9.

Tonini G, Koch C. The neural correlates of consciousness. An update. Ann N Y Acad Sci. 2008;1124:239–51. https://doi.org/10.1196/annals.1440.004.

Trifilieff E, Dubs MC, Van Regenmortel MHV. Antigenic cross-reactivity potential of synthetic peptides immobilized on polyethylene rods. Mol Immunol. 1991;28:889–96.

Trkola A, Purtscher M, Muster T, Ballaun C, Buchacher A, Sullivan N, et al. Human monoclonal antibody 2G12 defines a distinctive neutralization epitope on the gp120 glycoprotein of human immunodeficiency virus type 1. J Virol. 1996;70:1100–8.

Tropsha A, Kizer JS, Chaiken IM. Making sense from antisense: a review of experimental data and developing ideas on sense-antisense peptide recognition. J Mol Recognit. 1992;5:43–54.

Tsai CJ, Ma B, Nussinov R. Protein-protein interaction networks: how can a hub protein bind so many different partners? Trends Biochem Sci. 2009;34:594–600.

Tsimring LS. Noise in biology. Rep Prog Phys. 2014;77:29. https://doi.org/10.1088/0034-4885/77/2/026601.

Tugarinov V, Zvi A, Levy R, Hayek Y, Matsushita S, Anglister J. NMR structure of an anti-gp 120 antibody complex with a V3 peptide reveals a surface important for co-receptor binding. Struct Fold Des. 2000;8:385–95.

Uddowla S, Hollister J, Pacheco JM, Rodriguez LL, Rieder E. A safe foot-and-mouth disease vaccine platform with two negative markers for differentiating infected from vaccinated animals. J Virol. 2012;86:11675–85.

Umotoy J, Bagaya BS, Joyce C, Schiffner T, Menis S, et al. Rapid and focused maturation of a VRC01-class HIV broadly neutralizing antibody lineage involves both binding and accommodation of the N276-glycan. Immunity. 2019;51:141–54.

UNAIDS. Joint United Nations Programme on HIV/AIDS (UNAIDS), AIDS by the Numbers. Geneva, Switzerland; 2015. http://www.unaids.org/sites/default/files/media_asset/AIDS_by_the_numbers_2015_en.pdf

Underwood PA. Theoretical considerations of the ability of monoclonal antibodies to detect antigenic differences between closely related variants, with particular reference to heterospecific reactions. J Immunol Methods. 1985;85:295–307.

Upadhyay C, Mayr LM, Zhang J, Kumar R, Gorny MK, Nádas A, Zolla-Pazner S, Hioe CE. Distinct mechanisms regulate exposure of neutralizing epitopes in the V2 and V3 loops of HIV-1 envelope. J Virol. 2014;88:12853–65.

Urbain-Vansanten G. Concomitant synthesis, in separate cells, of non-reactive immunoglobulins and specific antibodies after immunization with tobacco mosaic virus. Immunology. 1970;19:783–97.

Uversky VN, Oldfield CJ, Dunker AK. Showing your ID: intrinsic disorder as an ID for recognition, regulation and cell signaling. J Mol Recognit. 2005;18:343–84.

Uversky VN, Davé V, Iakoucheva LM, Malaney P, Metallo SJ, Pathak RR, et al. Pathological unfoldomics of uncontrolled chaos: intrinsically disordered proteins and human diseases. Chem Rev. 2014;114:6844–79. https://doi.org/10.1021/cr400713r.

Valenzuela P, Medina A, Rutter WJ, Ammerer G, Hall BD. Synthesis and assembly of hepatitis B virus surface antigen particles in yeast. Nature. 1982;298:347–50. https://doi.org/10.1038/298347a0.

Van Bubnoff A. Vaccines to antibodies: grow up. IAVI Report; 2010. July–August.

Van Cott TC, Bethke FR, Polonis VR, Gorny MK, Zolla-Pazner S, Redfield RR, Birx DL. Dissociation rate of antibody-gp120 binding interactions is predictive of V3-mediated neutralization of HIV-1. J Immunol. 1994;153:449–59.

Van Cott TC, Polonis VR, Loomis LD, Michael NL, Nara PL, Birx DL. Differential role of V3-specific antibodies in neutralization assays involving primary and laboratory-adapted isolates of HIV type 1. AIDS Res Hum Retrovir. 1995a;11:1379–91.

Van Cott TC, Bethke FR, Burke DS, Redfield RR, Birx DL. Lack of induction of antibodies specific for conserved, discontinuous epitopes of HIV-1 envelope glycoprotein by candidate AIDS vaccines. J Immunol. 1995b;155:4100–10.

Van der Werf S, Briand JP, Plaue S, Burckard J, Girard M, Van Regenmortel MHV. Ability of linear and cyclic peptides of neutralization antigenic site 1 of poliovirus type 1 to induce virus cross-reactive and neutralizing antibodies. Res Virol. 1994;145:349–59.

Van Fraassen BC. Experience. In: The empirical stance. New Haven, CT: Yale University Press; 2002. p. 111–52.

van Gils MJ, Bunnik EM, Burger JA, Jacob Y, Schweighardt B, Wrin T, et al. Rapid escape from preserved cross-reactive neutralizing humoral immunity without loss of viral fitness in HIV-1-infected progressors and long-term non progressors. J Virol. 2010;84:3576–85. https://doi.org/10.1128/JVI.02622-09.

Van Oss CJ. Antigen-antibody interaction. In: Van Regenmortel MHV, editor. Structure of antigens, vol. 1. Boca Raton, FL: CRC Press; 1992. p. 99–125.

Van Oss CJ. Hydrophobic, hydrophilic and other interactions in epitope-paratope binding. Mol Immunol. 1995;32:199–211.

Van Oss CJ, Van Regenmortel MHV. Immunochemistry. New York: Marcel Dekker; 1994.

Van Regenmortel MHV. Plant virus serology MHV. Adv Virus Res. 1966;12:207–71.

Van Regenmortel MHV. Serology and immunochemistry of plant viruses. New-York: Academic; 1982. p. 268.

Van Regenmortel MHV. Structural and functional approaches to the study of protein antigenicity. Immunol Today. 1989a;10:266–72.

Van Regenmortel MHV. The concept and operational definition of protein epitopes. Philos Trans R Soc Lond B. 1989b;323:451–66.

Van Regenmortel MHV. The conformational specificity of viral epitopes. FEMS Microbiol Lett. 1992a;100:483–7.

Van Regenmortel MHV. Molecular dissection of protein antigens. In: Van Regenmortel MHV, editor. Structure of antigens, vol. 1. Boca Raton, FL: CRC; 1992b. p. 1–27.

Van Regenmortel MHV. Transcending the structuralist paradigm in immunology-affinity and biological activity rather than purely structural considerations should guide the design of synthetic peptide epitopes. Biomed Pept Proteins Nucleic Acids. 1995;1:109–16.

Van Regenmortel MHV. Mapping epitope structure and activity: from one-dimensional prediction to four-dimensional description of antigenic specificity. Methods. 1996;9:465–72.

Van Regenmortel MHV. From absolute to exquisite specificity. Reflections on the fuzzy nature of species, specificity and antigenic sites. J Immunol Methods. 1998;216:37–48.

Van Regenmortel MHV. Molecular dissection of protein antigens and the prediction of epitopes. In: Van Regenmortel MHV, Muller S, editors. Synthetic peptides as antigens. Amsterdam: Elsevier; 1999a. p. 1–78.

Van Regenmortel MHV. Molecular design versus empirical discovery in peptide-based vaccines. Coming to terms with fuzzy recognition sites and ill-defined structure-function relationships in immunology. Vaccine. 1999b;18:216–21. https://doi.org/10.1016/S0264-410X(99)00192-9.

Van Regenmortel MHV. Molecular dissection of protein antigens and the prediction of epitopes. In: Van Regenmortel MHV, Muller S, editors. Synthetic peptides as antigens. Amsterdam: Elsevier; 1999c. p. 281–317.

Van Regenmortel MHV. Are there two distinct research strategies for developing biologically active molecules: rational design and empirical selection. J Mol Recognit. 2000;13:1–4.

Van Regenmortel MHV. Antigenicity and immunogenicity of synthetic peptides. Biologicals. 2001a;29:209–13.

Van Regenmortel MHV. Pitfalls of reductionism in the design of peptide-based vaccines. Vaccine. 2001b;19:2369–74.

Van Regenmortel MHV. Proteomics versus genomics. What type of structure-function relationship are we looking for? J Mol Recognit. 2001c;14:321–2.

Van Regenmortel MHV. Analysing structure-function relationships with biosensors. Cell Mol Life Sci. 2001d;58:794–800.

Van Regenmortel MHV. Reductionism and the search for structure-function relationships in antibody molecules. J Mol Recognit. 2002a;15:240–7.

Van Regenmortel MHV. A paradigm shift is needed in proteomics: 'structure determines function' should be replaced by 'binding determines function. J Mol Recognit. 2002b;15:349–51.

Van Regenmortel MHV. Pitfalls of reductionism in immunology. In: Van Regenmortel MHV, Hull DL, editors. Promises and limits of reductionism in the biomedical sciences. Chichester: John Wiley; 2002c.

Van Regenmortel MHV. Biological complexity emerges from the ashes of genetic reductionism. J Mol Recognit. 2004a;17:145–8.

Van Regenmortel MHV. Reductionism and complexity in molecular biology. EMBO J. 2004b;5:1016–20.

Van Regenmortel MHV. The contribution of optical biosensors to the analysis of structure-function relationships in proteins. In: Kamp RM, Calvette JJ, Choli-Papadopoulou T, editors. Methods in proteome and protein analysis. Berlin: Springer; 2004c. p. 93–101.

Van Regenmortel MHV. Immunoinformatics may lead to a reappraisal of the nature of B cell epitopes and of the feasibility of synthetic peptide vaccines. J Mol Recognit. 2006;19:183–7.

Van Regenmortel MHV. The rational design of biological complexity: a deceptive metaphor. Proteomics. 2007;7:965–75.

Van Regenmortel MHV. Synthetic peptide vaccines and the search for neutralization of B cell epitopes. Open Vaccine J. 2009a;2:33–44. https://doi.org/10.2174/1875035401002010033.

Van Regenmortel MHV. What is a B cell epitope? In: Reineke U, Schutkowski M, editors. Methods in molecular biology, epitope mapping protocols, vol. 524. New York: Humana Press; 2009b. p. 3–20.

Van Regenmortel MHV. Antigenicity and immunogenicity of viral proteins. In: Mahy BWJ, Van Regenmortel MHV, editors. Desk encyclopedia of general virology. Oxford: Academic press, Elsevier; 2010. p. 343–9.

Van Regenmortel MHV. Two meanings of reverse vaccinology and the empirical nature of vaccine science. Vaccine. 2011a;29:7875.

Van Regenmortel MHV. Limitations to the structure-based design of HIV-1 vaccine immunogens. J Mol Recognit. 2011b;24:741–53.

Van Regenmortel MHV. Requirements for empirical immunogenicity trials, rather than structure-based design, for developing an effective HIV vaccine. Arch Virol. 2012a;157:1–20.

Van Regenmortel MHV. Basic research in HIV vaccinology is hampered by reductionist thinking. Front Immunol. 2012b;3:194. https://doi.org/10.3389/fimmu.2012.00194.

Van Regenmortel MHV. An oral tolerogenic vaccine protects macaques from SIV infection without eliciting SIV-specific antibodies nor CTLs. J AIDS Clin Res. 2013;4:e112. https://doi.org/10.4172/2155-6113.1000e112.

Van Regenmortel MHV. Specificity, polyspecificity, and heterospecificity of antibody-antigen recognition. J Mol Recognit. 2014a;27:627–39.

Van Regenmortel MHV. An outdated notion of antibody specificity is one of the major detrimental assumptions of the structure-based reverse vaccinology paradigm, which prevented it from developing an effective HIV-1 vaccine. Front Immunol. 2014b;5:593. https://doi.org/10.3389/fimmu.1014.00593.

Van Regenmortel MHV. Why does the molecular structure of broadly neutralizing monoclonal antibodies isolated from individuals infected with HIV-1 not inform the rational design of an HIV-1 vaccine? AIMS Public Health. 2015a;2:183–93.

Van Regenmortel MHV. Paradigm changes are required in HIV vaccine research. Front Immunol. 2015b;6:326. https://doi.org/10.3389/fimmu.2015.00326.

Van Regenmortel MHV. The metaphor that viruses are living is alive and well, but it is no more than a metaphor. Stud Hist Philos Sci Part C Stud Hist Philos Biol Biomed Sci. 2016a;59:117–24. https://doi.org/10.1016/j.shpsc.2016.02.017.

Van Regenmortel MHV. Structure-based reverse vaccinology failed in the case of HIV because it disregarded accepted immunological theory. Int J Mol Sci. 2016b;17:1591–625. https://doi.org/10.3390/ijms17091591.

Van Regenmortel MHV. Immune systems rather than antigenic epitopes elicit and produce protective antibodies against HIV. Vaccine. 2017;35:1985–6. https://doi.org/10.1016/j.vaccine.2017.03.017.

Van Regenmortel MHV. Development of a preventive HIV vaccine requires solving inverse problems which is unattainable by rational vaccine design. Front Immunol. 2018a;8:2009.

Van Regenmortel MHV. Viral species, viral genomes and HIV vaccine design: is the rational design of biological complexity a utopia? Arch Virol. 2018b;137:2047–54.

Van Regenmortel MHV. The species problem in virology. Adv Virus Res. 2018c;100:1–18.

Van Regenmortel MHV, Azimzadeh A. Determination of antibody affinity. In: Van Oss CJ, Van Regenmortel MHV, editors. Immunochemistry. New York: Marcel Dekker; 1994. p. 805–28.

Van Regenmortel MHV, Hardie G. Immunochemical studies of tobacco mosaic virus--II. Univalent and monogamous bivalent binding of IgG antibody. Immunochemistry. 1976;13:503–7.

Van Regenmortel MHV, Hull D. Promises and limits of reductionism in the biomedical sciences. Chichester: Wiley; 2002. p. 1–377.

Van Regenmortel MHV, Muller S. Synthetic peptides as antigens. Amsterdam: Elsevier; 1999. p. 1–381.

Van Regenmortel MHV, Neurath AR. Immunochemistry of viruses. Amsterdam: Elsevier; 1985.

Van Regenmortel MHV, Pellequer J-L. Predicting antigenic determinants in proteins: looking for unidimensional solutions to a three-dimensional problem? Pept Res. 1994;7:224–8.

Van Regenmortel, M.H.V. In: Immunochemistry, Van Oss, C.J. and Van Regenmortel, M.H.V., eds, Dekker, New York, 1994, p. 277.

Van Regenmortel MHV, Andrieu JM, Dimitrov DS, Ensoli B, Hioe CE, Moog C, Ruprecht RM. Paradigm changes and the future of HIV vaccine research: a summary of a workshop held in Baltimore on 20 November 2013. J AIDS Clin Res. 2014;5:281. https://doi.org/10.4172/2155-6113.1000281.

Verkoczy L, Diaz M. Autoreactivity in HIV-1 broadly neutralizing antibodies: implications for their function and induction by vaccination. Curr Opin HIV AIDS. 2014;9:224–34.

Verkoczy L, Kelsoe G, Moody MA, Haynes BF. Role of immune mechanisms in induction of HIV-1 broadly neutralizing antibodies. Curr Opin Immunol. 2011;23:383–90.

Verkoczy L, Kelsoe G, Haynes BF, et al. HIV-1 envelope gp41 broadly neutralizing antibodies: hurdles for vaccine development. PLoS Pathog. 2014;10:e1004073.

Verkoczy L, Alt FW, Tian M. Human Ig knocking mice to study the development and regulation of broadly neutralizing antibodies. Immunol Rev. 2017;275:89–107.

Vijh-Warrier S, Pinter A, Honnen WJ, et al. Synergistic neutralization of human immunodeficiency virus type 1 by a chimpanzee monoclonal antibody against the V2 domain of gp120 in combination with monoclonal antibodies against the V3 loop and the CD4-binding site. J Virol. 1996;70:4466–75.

Villen J, de Oliviera E, Nunez JI, Molina N, Sobrino F, Andreu D. Towards a multi-site synthetic vaccine to foot-and-mouth disease: addition of a discontinuous site peptide mimic increases the neutralization response in immunized animals. Vaccine. 2004;22:3523–9.

Villen J, Rodriguez-Mias RA, Nunez JI, et al. Rational dissection of binding surfaces for mimicking of discontinuous antigenic sites. Chem Biol. 2006;13:815–23.

Virgin HW, Walker BD. Immunology and the elusive AIDS vaccine. Nature. 2010;464:224–31.

von Bertalanffy L. The theory of open systems in physics and biology. Science. 1950;111:23–9. https://doi.org/10.1126/science.111.2872.23.

Von Sengbusch P, Wittman HG. Serological and physicochemical properties of the wild strain and two mutants of tobacco mosaic virus with the same amino acid exchange in different positions of the protein chain. Biochem Biophys Res Commun. 1965;18:780–7.

Vuilleumier S, Mutter M. In: Van Regenmortel MHV, editor. Structure of antigens. Boca Raton, FL: CRC Press; 1992. p. 43.

Vyas JM, Van der Veen AG, Ploegh HL. The known unknowns of antigen processing and presentation. Nat Rev Immunol. 2008;8:607–18.

Wabl M, Cascalho M, Steinberg C. Hypermutation in antibody affinity maturation. Curr Opin Immunol. 1999;11:186–9.

Wagner A. Causality in complex systems. Biol Philos. 1999;14:83–101.

Wagner A. Robustness and evolvability in living systems. Princeton: Princeton University Press; 2005.

Wainwright M, Maisch T, Nonell S, Plaetzer K, Almeida A, Tegos GP, et al. Photoantimicrobials – are we afraid of the light? Lancet Infect Dis. 2017;17:e49–55. https://doi.org/10.1016/S1473-3099(16)30268-7.

Walker LM, Burton DR. Rational antibody-based HIV-1 vaccine design: current approaches and future directions. Curr Opin Immunol. 2010;22:358–66.

Walker LM, Phogat SK, Chan-Hui PY, Wagner D, Phung P, Goss JL, et al. Broad and potent neutralizing antibodies from an African donor reveal a new HIV-1 vaccine target. Science. 2009;326:285–9. https://doi.org/10.1126/science.1178746.

Wallis R. The glory of gravity – Halley comet 1759. Ann Sci. 1964;41:279–86. https://doi.org/10. 1080/00033798400200271.

Walter G. Production and use of antibodies against synthetic peptides. J Immunol Methods. 1986;88:149–61.

Wang IM, Bett AJ, Cristescu R, Loboda A, Meulen J. Transcriptional profiling of vaccine-induced immune responses in humans and non-human primates. Microbiol Biotechnol. 2012;5:177–87. https://doi.org/10.1111/j.1751-7915.2011.00317.x.

Ward AB, Wilson IA. The HIV-1 envelope glycoprotein structure: nailing down a moving target. Immunol Rev. 2017;275:21–32. https://doi.org/10.1111/imr.12507.

Wardemann H, Yurasov S, Schaefer A, Young JW, Meffre E, Nussenzweig MC. Predominant autoantibody production by early human B cell precursors. Science. 2003;301:1374–7.

Watkins DI, Burton DR, Kallas EG, Moore JP, Koff WC. Nonhuman primate models and the failure of the Merck HIV-1 vaccine in humans. Nat Med. 2008;14:617–21.

Watkins JD, Sholukh AM, Mukhtar MM, Siddappa NB, Lakhashe SK, et al. Anti-HIV IgA isotypes: differential virion capture and inhibition of transcytosis are linked to prevention of mucosal R5 SHIV transmission. AIDS. 2013;27:F13–20.

Weber BH, Depew DJ. In: Dembski WA, Ruse M, editors. Debating design: from Darwin to DNA. Cambridge: Cambridge University Press; 2004. p. 173–90.

Wedemayer GJ, Patten PA, Wang LH, Schultz PG, Stevens R. Structural insights into the evolution of an antibody combining site. Science. 1997;276:1665–9.

Weidenhaupt M, Ben Khalifa M, Hugo N, Choulier L, Altschuh D, Vernet T. Functional mapping of conserved, surface-exposed charges of antibody variable domains. J Mol Recognit. 2002;15:94–103.

Weinstein IB. Addiction to oncogenes. The Achilles heel of cancer. Science. 2002;297:63–4.

Weliky DP, Bennett AE, Zvi A, Anglister J, Steinbach PJ, Tycko R. Solid-state NMR evidence for an antibody-dependent conformation of the V3 loop of HIV-1 gp120. Nat Struct Biol. 1999;6:141–5.

Welling GW, van Gorkum J, Damhof RA, Drijfhout JW, Bloemhoff W, Welling-Wester S. A ten-residue fragment of an antibody (mini-antibody) directed against lysozyme as ligand in immunoaffinity chromatography. J Chromatogr. 1991;548:235–42.

Weng G, Bhalla US, Iyengar R. Complexity in biological signalling systems. Science. 1999;284:92–6.

Wertheimer A. Rethinking the ethics of clinical research: widening the lens. New York: Oxford University Press; 2011.

West AP Jr, Scharf L, Scheid JF, et al. Structural insights on the role of antibodies in HIV-1 vaccine and therapy. Cell. 2014;156:633–48.

Westhof E, Altschuh D, Moras D, Bloomer AC, Mondragon A, Klug A, Van Regenmortel MHV. Correlation between segmental mobility and the location of antigenic determinants in proteins. Nature. 1984;311:123–6.

Wilkins AS. Why the philosophy of science actually does matter. BioEssays. 2001;23:561–2.

Willey S, Aasa-Chapman MM. Humoral immunity to HIV-1: neutralisation and antibody effector functions. Trends Microbiol. 2008;16:596–604.

Williams DA, Baum C. Gene therapy: new challenges ahead. Science. 2003;302:400–1.

Williams WV, Guy HR, Rubin DH, Robey F, Myers JN, Kieber-Emmons T, Weiner DB, Greene MI. Sequences of the cell-attachment sites of reovirus type 3 and its anti-idiotypic/antireceptor antibody: modeling of their three-dimensional structures. Proc Natl Acad Sci U S A. 1988;85:6488–92.

Williams WV, Kieber-Emmons T, VonFeldt J, Greene MI, Weiner DB. Design of bioactive peptides based on antibody hypervariable region structures. Development of conformationally constrained and dimeric peptides with enhanced affinity. J Biol Chem. 1991;266:5182–90.

Williamson NA, Purcell AW. Use of proteomics to define targets of T-cell immunity. Expert Rev Proteomics. 2005;2:367–80.

Wilson IA, Stanfield RL. Antibody-antigen interactions: new structures and new conformational changes. Curr Opin Struct Biol. 1994;4:857–67.

Wimsatt WC. Re-engineering philosophy for limited beings. piecewise approximations to reality. Cambridge, MA: Harvard University Press; 2007.

Wlodawer A. Rational approach to AIDS drug design through structural biology. Annu Rev Med. 2002;53:595–14.

Wolkenhauer O, Green S. The search for organizing principles as a cure against reductionism in systems medicine. FEBS J. 2013;280:5938–48. https://doi.org/10.1111/febs.12311.

Wong-Staal F, Gallo RC, editors. Aids vaccine research. Marcel Dekker: New York; 2002.

Woodger JH. Biological principles. New York: Humanities Press; 1967.

Woodward J. Making things happen: a theory of causal explanation. Oxford: Oxford University Press; 2003. p. 410.

Wu TT, Kabat EA. An analysis of the sequences of the variable regions of Bence Jones proteins and myeloma light chains and their implications for antibody complementarity. J Exp Med. 1970;132:211–50.

Wu L, Zhou T, Yang ZY, Svehla K, O'Dell S, Louder MK, et al. Enhanced exposure of the CD4-binding site to neutralizing antibodies by structural design of a membrane-anchored human immunodeficiency virus type 1 gp120 domain. J Virol. 2009;83:5077–86.

Wu X, Yang ZY, Li Y, Hogerkorp CM, Schief WR, Seaman MS, Zhou T, Schmidt SD, Wu L, Xu L, et al. Rational design of envelope identifies broadly neutralizing human monoclonal antibodies to HIV-1. Science. 2010;329:856–61.

Wu X, Changela A, O'Dell S, Schmidt SD, Pancera M, Yang Y, Zhang B, Gorny MK, Phogat S, Robinson JE, Stamatatos L, Zolla-Pazner S, Kwong PD, Mascola JR. Immunotypes of a quaternary site of HIV-1 vulnerability and their recognition by antibodies. J Virol. 2011a;85:4578–85.

Wu X, Zhou T, Zhu J, Zhang B, Georgiev I, Wang C, et al. Focused evolution of HIV-1 neutralizing antibodies revealed by structures and deep sequencing. Science. 2011b;333:1593–602. https://doi.org/10.1126/science.1207532.

Wucherpfennig KW, Allen PM, Celada F, Cohen IR, De Boer R, Garcia KC, Goldstein B, Greenspan R, Hafler D, Hodgkin P, et al. Polyspecificity of T cell and B cell receptor recognition. Semin Immunol. 2007;19:216–24.

Wyatt R, Sodroski J. The HIV-1 envelope glycoproteins: fusogens, antigens, and immunogens. Science. 1998;280:1885–8.

Wyatt R, Kwong PD, Desjardins E, Sweet RW, Robinson J, Hendrickson WA, Sodroski JG. The antigenic structure of the HIV gp120 envelope glycoprotein. Nature. 1998;393:705–11.

Xavier KA, McDonald SM, McCammon JA, Willson RC. Association and dissociation kinetics of bobwhite quail lysozyme with monoclonal antibody HyHEL-5. Protein Eng. 1999;12:79–83.

Xiao X, Chen W, Feng Y, Zhu Z, Prabakaran P, Wang Y, Zhang MY, Longo NS, Dimitrov DS. Germline-like predecessors of broadly neutralizing antibodies lack measurable binding to HIV-1 envelope glycoproteins: implications for evasion of immune responses and design of vaccine immunogens. Biochem Biophys Res Commun. 2009;390:404–9.

Xu H, Song L, Kim M, Holmes MA, Kraft Z, Sellhorn G, et al. Interactions between lipids and human anti-HIV antibody 4E10 can be reduced without ablating neutralizing activity. J Virol. 2010;84:1076–84.

Yao B, Zhang L, Liang S, Zhang C. SVMTriP: a method to predict antigenic epitopes using support vector machine to integrate tri-peptide similarity and propensity. PLoS One. 2012;7:e45152.

Yin J, Beuscher AE IV, Andryski SE, Stevens RC, Schultz PG. Structural plasticity and the evolution of antibody affinity and specificity. J Mol Biol. 2003;330:651–6.

Ylikoski P. The illusion of depth of understanding in science. In: De Regt HW, Leonelli S, Eigner K, editors. Scientific understanding. Pittsburgh: University of Pittsburgh Press; 2009. p. 100–19.

Yordanov Y, Dechartres A, Porcher R, Boutron I, Altman DG, Ravaud P. Avoidable waste of research related to inadequate methods in clinical trials. BMJ. 2015;35:h809. https://doi.org/10.1136/bmj.h809.

Young M, Edis T, editors. Why intelligent design fails. Piscataway, NJ: Rutgers University Press; 2004.

Young KR, McBurney SP, Karkhanis LU, Ross TM. Virus-like particles: designing an effective AIDS vaccine. Methods. 2006;40:98–117.

Yu L, Guan Y. Immunologic basis for long HCDR3s in broadly neutralizing antibodies against HIV-1. Front Immunol. 2014;5:250.

Yu MW, Scott JK, Fournier A, Talbot PJ. Characterization of murine coronavirus neutralization epitopes with phage-displayed peptides. Virology. 2000;271:182–96.

Yuan W, Bazick J, Sodroski J. Characterization of the multiple conformational states of free monomeric and trimeric human immunodeficiency virus envelope glycoproteins after fixation by cross-linker. J Virol. 2006;80:6725–37.

Yuan W, Li X, Kasterka M, Gorny MK, Zolla-Pazner S, Sodroski J. Oligomer-specific conformations of the human immunodeficiency virus (HIV-1) gp41 envelope glycoprotein ectodomain recognized by human monoclonal antibodies. AIDS Res Hum Retrovir. 2009;25:319–28.

Zajac AJ, Harrington LE. Cell-mediated immunity to viruses. In: Mahy BWJ, Van Regenmortel MHV, editors. Desk encyclopedia of general virology. Amsterdam: Elsevier; 2008. p. 354–61.

Zeder-Lutz G, Wenger R, Van Regenmortel MHV, Altschuh D. Interaction of cyclosporin A with an Fab fragment or cyclophilin: affinity measurements and time-dependent changes in binding. FEBS Lett. 1993;326:153–7.

Zeder-Lutz G, Benito A, Van Regenmortel MHV. Active concentration measurements of recombinant biomolecules using biosensor technology. J Mol Recognit. 1999;12:300–9.

Zeder-Lutz G, Hoebeke J, Van Regenmortel MHV. Differential recognition of epitopes present on monomeric and oligomeric forms of gp160 glycoprotein of human immunodeficiency virus type 1 by human monoclonal antibodies. Eur J Biochem. 2001;268:2856–66.

Zhang X, Piatier-Tonneau D, Auffray C, Murali R, Mahapatra A, Zhang F, Maier CC, Saragovi H, Greene MI. Synthetic CD4 exocyclic peptides antagonize CD4 holoreceptor binding and T cell activation. Nat Biotechnol. 1996;14:472–5.

Zhang Q, Wang P, Kim Y, Haste-Andersen P, Beaver J, Bourne PE, Bui HH, Buus S, Frankild S, Greenbaum J, et al. Immune epitope database analysis resource (IEDB-AR). Nucleic Acids Res. 2008;36:W513–8.

Zhang W, Xiong Y, Zhao M, Zou H, Ye X, Liu J. Prediction of conformational B-cell epitopes from 3D structure by random forests with a distant-based feature. BMC Bioinf. 2011;12:341.

Zhou ZH, Tzioufas AG, Notkins AL. Properties and function of polyreactive antibodies and polyreactive antigen-binding B cells. J Autoimmun. 2007a;29:219–28.

Zhou ZH, Zhang Y, Hu YF, Wahl LM, Cisar JO, Notkins AL. The broad antibacterial activity of the natural antibody repertoire is due to polyreactive antibodies. Cell Host Microbe. 2007b;1:51–61.

Zhou T, Xu L, Dey B, Hessell AJ, VanRyk D, Xiang SH, Yang X, Zhang MY, Zwick MB, Arthos J, Burton DR, Dimitrov DS, Sodroski J, Wyatt R, Nabel GJ, Kwong PD. Structural definition of a conserved neutralization epitope on HIV-1 gp120. Nature. 2007c;445:732–7.

Zhou T, Georgiev I, Wu X, Yang ZY, Dai K, Finzi A, Kwon YD, Scheid JF, Shi W, Xu L, Yang Y, Zhu J, Nussenzweig MC, Sodroski J, Shapiro L, Nabel GJ, Mascola JR, Kwong PD. Structural basis for broad and potent neutralization of HIV-1 by antibody VRC01. Science. 2010;329:811–7.

Zhou T, Doria-Rose NA, Cheng C, Stewart-Jones GBE, Chuang GY, Chambers M, et al. Quantification of the impact of the HIV-1-glycan shield on antibody elicitation. Cell Rep. 2017;19:719–32. https://doi.org/10.1016/j.celrep.2017.04.013.

Zhu P, Liu J, Bess J Jr, Chertova E, Lifson JD, Grisé H, Ofek GA, Taylor KA, Roux KH. Distribution and three-dimensional structure of AIDS virus envelope spikes. Nature. 2006;441:847–52.

Zhu Z, Bossart KN, Bishop KA, Crameri G, Dimitrov AS, McEachern JA, et al. Exceptionally potent cross-reactive neutralization of Nipah and Hendra viruses by a human monoclonal antibody. J Infect Dis. 2008a;197:846–53. https://doi.org/10.1086/528801.

Zhu P, Winkler H, Chertova E, Taylor KA, Roux KH. Cryoelectron tomography of HIV-1 envelope spikes: further evidence for tripod-like legs. PLoS Pathog. 2008b;4:e1000203.

Zhu Z, Qin HR, Chen W, Zhao Q, Shen X, Schutte R, et al. Cross-reactive HIV-1-neutralizing human monoclonal antibodies identified from a patient with 2F5-like antibodies. J Virol. 2011;85:11401–8. https://doi.org/10.1128/JVI.05312-11.

Zimmermann D, Van Regenmortel MHV. Spurious cross-reactions between plant viruses and monoclonal antibodies can be overcome by saturating ELISA plates with milk proteins. Arch Virol. 1989;106:15–22.

Zolla-Pazner S. Identifying epitopes of HIV-1 that induce protective antibodies. Nat Rev Immunol. 2004;4:199–210.

Zolla-Pazner S, Cardozo T. Structure-function relationships of HIV-1 envelope sequence-variable regions refocus vaccine design. Nat Rev Immunol. 2010;10:527–35.

Zolla-Pazner S, Kong XP, Jiang X, Cardozo T, Nádas A, Cohen S, Totrov M, Seaman MS, Wang S, Lu S. Cross-clade HIV-1 neutralizing antibodies induced with V3-scaffold protein immunogens following priming with gp120 DNA. J Virol. 2011;85:9887–98.

Zolla-Pazner S, Cohen SS, Boyd D, Kong XP, Seaman M, Nussenzweig M, Klein F, Overbaugh J, Totrov M. Structure/function studies involving the V3 region of the HIV-1 envelope delineate multiple factors that affect neutralization sensitivity. J Virol. 2016;90:636–49.

Zotos D, Tarlington DM. Determining germinal centre B cell fate. Trends Immunol. 2012;33:281–8.

Zwick MB. The membrane-proximal external region of HIV-1 gp41: a vaccine target worth exploring. AIDS. 2005;19:1725–37.

Zwick MB, Burton DR. HIV-1 neutralization: mechanisms and relevance to vaccine design. Curr HIV Res. 2007;5:608–24.

Zwick MB, Bonnycastle LL, Menendez A, Irving MB, Barbas CF III, Parren PW, Burton DR, Scott JK. Identification and characterization of a peptide that specifically binds the human, broadly neutralizing anti-human immunodeficiency virus type 1 antibody B12. J Virol. 2001a;75:6692–9.

Zwick MB, Wang M, Poignard P, Stiegler G, Katinger H, Burton DR, et al. Neutralization synergy of human immunodeficiency virus type 1 primary isolates by cocktails of broadly neutralizing antibodies. J Virol. 2001b;75:12198–208. https://doi.org/10.1128/JVI.75.24.12198-12208. 2001.

Zwick MB, Labrijn AF, Wang M, Spenlehauer C, Saphire EO, Binley JM, et al. Broadly neutralizing antibodies targeted to the membrane-proximal external region of human immunodeficiency virus type 1 glycoprotein gp41. J Virol. 2001c;75:10892–905.

Zwick MB, Parren PW, Saphire EO, Church S, Wang M, Scott JK, Dawson PE, Wilson IA, Burton DR. Molecular features of the broadly neutralizing immunoglobulin G1 b12 required for recognition of human immunodeficiency virus type 1 gp120. J Virol. 2003a;77:5863–76.

Zwick MB, Kelleher R, Jensen R, et al. A novel human antibody against human immunodeficiency virus type 1 gp120 is V1, V2, and V3 loop-dependent and helps delimit the epitope of the broadly neutralizing antibody immunoglobulin G1 b12. J Virol. 2003b;77:6965–78.

Zwick MB, Komori HK, Stanfield RL, Church S, Wang M, Parren PW, et al. The long third complementarity-determining region of the heavy chain is important in the activity of the broadly neutralizing anti-human immunodeficiency virus type 1 antibody 2F5. J Virol. 2004;78:3155–61.

Zwick MB, Jensen R, Church S, Wang M, Stiegler G, Kunert R, Katinger H, Burton DR. Anti-human immunodeficiency virus type 1 (HIV-1) antibodies 2F5 and 4E10 require surprisingly few crucial residues in the membrane-proximal external region of glycoprotein gp41 to neutralize HIV-1. J Virol. 2005;79:1252–61.